IS THERE A MORAL OBLIGATION TO SAVE THE FAMILY FARM?

IS THERE A MORAL OBLIGATION TO SAVE THE FAMILY FARM?

EDITED BY **GARY COMSTOCK**

IS **IOWA STATE UNIVERSITY PRESS, AMES**

GARY COMSTOCK is assistant professor of philosophy and religious studies, Iowa State University. He is the founder and coordinator of the Religious Ethics and Technological Change conference series at Iowa State.

© 1987 Iowa State University Press, Ames, Iowa 50010

Composed by Iowa State University Press
Printed in the United States of America

First edition, 1987

Library of Congress Cataloging-in-Publication Data

Is there a moral obligation to save the family farm?

    Bibliography: p.
    Includes index.
    1. Family farms—Government policy—United States.   2. United States—Rural conditions.   3. Agriculture—Economic aspects—United States.   4. Agriculture and state—United States.   I. Comstock, Gary, 1954–

HD1476.U518   1987   338.1'0973   87-13399
ISBN 0–8138–0999–1
ISBN 0–8138–1000–0 (pbk.)

J. H. and Elizabeth Brown Pippert,
Charles Williams and Emma Pelzer Pippert,
Harry Vernette and Magda Halsor Pippert,
Harold Halsor and Sandra Ballhagen Pippert,
and Annette, Heather, Jason, and Jennifer. . . .

five generations on the Pippert family farm,
Mason City, Iowa

CONTENTS

FOREWORD

LAUREN SOTH

A FOREWORD OR PREFACE to a book is supposed to tell an incipient reader what to expect in the pages that follow and why she or he should read the book. Mark Twain, in prefacing the uniform edition of *The Innocents Abroad,* said wryly, "Prefaces wear many disguises . . . but I think that upon examination, we are quite sure to find that their errand is always the same: they are there to apologize for the book; in other words, to furnish reasons for its publication. This often insures brevity." As the designated prefacer for this book, I have no qualms about performing the conventional duty, and I'll be brief. I believe this compilation to be worth reading, not because it is a coherent, consistent discussion of the family farm—which society is or is not morally obligated to "save"—but because the book is *not* that.

It is a collection of a wide variety of uneven, contradictory views on and around the subject, ranging from ideological passion to scientific detachment, from conservative to liberal (whatever meaning you choose for these labels), and from academic stolidity to political imaginativeness.

The book is unique in this. It combines the thinking of farm people with that of economists, sociologists, philosophers, politicians, and theologians—under one cover. The family farm stereotype so common in our political life—politicians never talk about farms, only about "family" farms—is here disassembled and chewed over, occasionally including reference to facts about the changing structure of the agricultural industry, along with discordant biblical interpretations, nostalgia, and ideological prejudice.

"Saving" the family farm has not been an issue in America until the last few decades. Prior to World War II, while everyone talked about the virtues of the independent, family-sized farm, no one regarded this hallowed institution as in any kind of danger. The leap in farming technology after about 1940 began a process of reorganization or restructuring of the farming industry. The new technology of machines and chemicals replaced farm labor and made economic larger business units – larger in land and capital employed and larger in volume of marketings per farm. New household technology made a difference, too. The self-sufficient "family" farm, with its diversified crop and livestock production, home canning, home clothes-making, and so forth, began to decline while large, specialized, wholly commercialized grain or livestock or vegetable or fruit or cotton farms became more common. The essays in this book describe the changes and the reasons for the waning of what used to be the typical family farm. The landmark study of the structure of the farm industry carried on by former agriculture secretary Bob Bergland and his staff in the 1970s is still the basic textbook of the process.

As many of the pieces in this volume explain, the changes are not so much the *elimination* of the family farm as the evolution of new *kinds* of family farms. Some have referred to the "bimodal" character of the new farming industry: concentration of large-business farming organizations at one end and a sector of numerous small, usually part-time farms at the other end. The midsize farm is a shrinking proportion of the declining total number of farms.

I must not stray at length into reasons for this, for that would be to go beyond prefacing. Nor will I expose the unwary reader to my own (very sound) views about the moral values of the family farm, since the authors herein will explore such values in depth.

As the writer of a preface, I should not try to become one of the essayists, but I cannot resist a brief general comment. I think the idea of the part-time farm, with family members engaged in nonfarm work, has not been given the attention it deserves. Part-time farming is nothing new. Nineteenth and twentieth century family-farming communities were rife with farmers going into and out of various other occupations: drayers, blacksmiths, carpenters, miners, seamstresses, barbers, restaurateurs, cobblers, and so on. After a long period of specialization, we are now returning to a greater mixture of rural jobs.

There is nothing more familial than a part-time farm family. The concept of part-time farming has been gathering force with the increase in labor-saving farm methods, which, paradoxically, also increase specialized, large-scale farming. It is feasible for a weekend farmer to handle quite a large grain operation and, with family help, some livestock raising or feeding may become possible. Is this new kind of family farm the shape of the rural future? Should rural communities put more attention on supplementary jobs for people living on farms who need extra work and income? The economists in this book point out that it takes a large farm to fully occupy the time of just one person, let alone a family of four or five.

The modernization of farm technology and the urbanization of farm living have tended to undercut some of the arguments in behalf of the historic family farm. The arguments that small farming results in more careful soil management and protection of natural resources do not stand up under examination. Farm people, like city people, are not fixed in their occupations or their places to live. The economic impulses to exploit the soil are much the same for big and small. Family farming alone cannot do what society at large must do about maintaining the productivity of the nation's resources.

On the other hand, equity in land distribution and equity in sharing the national farm income are goals which family-sized farm units help to advance. Even if it were true that big farms achieve great economies of scale (most studies of production efficiency indicate this is not true), society might well benefit more in general stability and social calm from programs to favor the smaller farming units.

One more point. The American society of the late twentieth century is so mobile, so urbanized, that it is really misleading to talk about the traditional family farm passing through the generations as in the past. Young farm people go into other occupations; they don't feel obliged to farm the old home place. The diversity of the farming business and of the rural community makes the family farm something different from what we idealize from the past.

I hope this will not be taken to mean that I believe the family farm must give way altogether to industrial, large-scale farming. Not by a long shot! Family-sized farming should be encouraged and not discriminated against by the tax system, government programs, agricultural research, education, and agribusiness. Family farming will be mostly part-time farming in the future. It is now. That is as noble as full-time family farming ever was and probably much better for sustaining and enhancing the small-town, rural society everyone seems to be anxious to "save."

INTRODUCTION

THE HAROLD PIPPERT FAMILY (to whom this book is dedicated) received a right to cure letter, the first step in foreclosure proceedings, on part of their farm in 1987. Like thousands of other farmers, my Uncle Harold and Aunt Sandy are now struggling to keep their land. They believe that the equity they have built up in the farm, along with the support they have received from their family, will help them to weather the current crisis. Whether the farm can support a sixth generation of Pipperts, however, is a darker question, especially when thousands of farm families have already been displaced from the farmsteads of their great-great-grandparents.

Suffering is nothing new to farmers. But it should be pointed out, neither is continuing to love one's land in the midst of suffering. In one of the oldest and greatest works of the western literary tradition, the author of Job tells of Job's happy past and the development of his present misfortune. At the height of his manhood, "in the days of fruit gathering" as the narrator puts it, Job has been stricken. Meeting with the affliction that will make his name synonymous with suffering, he sits penniless, wifeless, childless, almost speechless. But he is not defeated. Before Yahweh climactically answers him out of the whirlwind, Job asserts his innocence one last time. In a speech filled with pride and tragedy, he challenges his earthly accusers. Find me guilty, he taunts them, in the face of the unfailing care I have continued to lavish on my soil.

> If my land has cried out against me,
>   and its furrows have wept together;
> if I have eaten its yield without payment,
>   and caused the death of its owners;
> let thorns grow instead of wheat,
>   and foul weeds instead of barley.
>   (Job 31:38–40)

In the face of unspeakable misery, Job refuses to accept the insinuations of his friends. Throughout his suffering, he boldly asserts, he has persisted in growing wheat and barley in such a way that the land has received its due. He has not taken from the land without giving it proper payment. He has continued in his attempts not to "cause the death of its owners"; he has remained faithful to his creditors. Job will not hang his head, even if his closest friends lay the blame, mistakenly, at his feet.

It is often said that American farmers, like Job, are proud and righteous people. I doubt that all of them are as righteous as he was. American farmers are no different from anyone else; some of them love their land, some of them do not. Some of them are as loyal to their creditors as was their ancient forerunner. Some of them are not. Some of them are as victimized and blameless as he was; some of them are not. But for all their differences, American farmers would all agree with Job on one point. To be a good *farmer,* one must care for one's land.

Suffering among those who work the land did not end with the agrarian society of the ancient Hebrews, nor with the rapid growth of medieval towns. It did not disappear with the development of labor-saving devices in the nineteenth century or with the New Deal of the 1940s. Indeed, the present crisis in rural America is badly named; it is not a crisis, for a crisis is a problem of high intensity and short duration. But the problems facing my aunt and uncle have a long history; they are chronic problems, now reaching a point of unusual intensity. As the Dan Rathers, Jessica Langes, and *New York Times* of the world have made us painfully aware, suffering in America's rural areas may be at an all-time high. Where people live on family farms — farms owned and worked by families — there appears to be more misery in the mid-1980s than there has been since the Great Depression.

America has lost over four million farms since 1935, the vast majority of them family farms. While the transition has been marked by alternating periods of calm and crisis, the changes lately have been dramatic. Newspaper headlines announce "Farmer Slays Ex-Fiancee, Then Turns Shotgun on Self" and "Officers Say Farmer Left Notes Planning Suicide." Television brings us stories from Georgia, Alabama, Texas, Illinois, and Minnesota: foreclosures on mortgages, the boarding up of main-street businesses, bank closings, high interest rates, the lowest real commodity prices since the depression, plummeting land values, shrinking export markets, piles of unwanted corn and beans, psychological depression, alcoholism, family breakdown, and even suicide.

Most farmers show great dignity as they are forced, in the disquieting language of our agricultural economists, "to exit" agriculture. But should they, and should we, be content to have them forced out? An

Office of Technology Assessment report predicts that by the year 2000, the number of farmers will be cut in half: from 2.1 to 1.1 million. And yet an ABC News poll in February of 1986 concluded that over half of all Americans would prefer to live in a rural community and would even like to farm for a living if it were possible to do so. If the poll is even vaguely accurate, there are millions of us who would like to be farmers. Why is it, then, that barely more than two percent of us can now afford to live on family farms? Is this a healthy society in which the apparent desires of so many are not being met? Should we not be encouraging *more* people to be farmers, and instituting federal policies to make that possible, rather than allowing another thirty thousand farms to go out of business this year? (Collum 1986, 14)

The chronic problems of the family farm are real, and the suffering palpable. Many articles and books have been devoted to analyzing the economic features of the collapse of small town America. Some have tried to suggest plans to counteract it. But not one has dealt explicitly with the moral and religious values that ought to guide us as we make difficult choices in these areas; that is the goal of this book. The authors are concerned to lay out the facts of the current crisis. They know better than to try to make moral judgments without first having all of the relevant information at their fingertips. But they are not content with reporting historical, sociological, and economic statistics. They go on to take up the ethical issues raised by the facts.

Americans, both rural and urban, are having to make tough decisions regarding agricultural policy, taxes, food prices, corporate monopolies, and federal budget deficits. What ethical principles should guide us in these areas? Is it fair to encourage increased productivity of food if that means moving people off of small- and medium-sized farms? Is it just to allow economic power to concentrate in the hands of a few when it could be dispersed to the hands of many? Should universities and social scientists be held responsible for the advice they give to farmers and workers? If so, how? If not, why not? Does it show compassion to allow an economic trend to run its "natural course" when that course inevitably displaces people from homes and jobs? Would it not be more compassionate to intervene with specific public policies and financial measures designed to keep people employed?

On the other hand, what makes farmers so special? Why didn't we write books on behalf of automobile dealers in 1981, when 758 of them went bankrupt? Why single out farmers for attention when unemployment problems seem so much worse in, say, the manufacturing industries? The moral questions haunt us. So do other spiritual, religious concerns. Do we have duties to generations to come? To what extent are we responsible for the pollution of air and water? Does the land or atmosphere itself have any rights? If there is a God, does that God regard abuse of land as sin? If there is not a God, should we be good stewards anyway? Is there anything intrinsically wrong with trying to make agriculture a profitable business? Why shouldn't someone try to maximize profits and productivity? If technological achievements lead "naturally" to fewer farmers with more efficient operations, what is objectionable about that?

These are difficult and complex questions. Easy slogans will not do for answers. Experts in economics, sociology, ethics, and theology — as we will see — disagree about the answers. For readers untrained in social scientific theory and philosophical reasoning, the quarrels of experts can be intimidating. We are tempted to take the easy way out, to turn our backs on their worries and squabbles. But before we shy away, we should remember how very crucial the issue is; it has been said, arguably, that our survival as a people depends on our finding good answers to rural America's problems. It is unarguable that the survival of the family farm depends on our response. So, as difficult as the questions are, we must try to answer them.

The essays collected here provide some answers. Taken as a group, they give a thorough introduction to the facts of the current crisis. They also suggest the major routes that our ethical reflections might follow. While the authors reach different conclusions about what should be done, they agree on the urgency of the question: Is there a moral obligation to save the family farm?

In these pages many will agree with Bishop Maurice Dingman: "If we lose the farms, then we have lost the soul of our nation." Dingman sees "a frightening correlation between what is happening here and what has happened in Central America," and he warns that "if we continue to drift the way we are now [with family farms being sold] we will end up like El Salvador, with ownership of the land in the hands of a few" (Shanley 1985, 5, 62). Call them "populists," "farm fundamentalists," "Jeffersonian agrarians," or "traditionalists." By whatever name, agrarianizers such as Wendell Berry, Jim Hightower, and Tom Harkin all side with Dingman (I prefer the term *agrarianizer* to the more familiar *agrarian* because *agrarian* connotes unprogressive and traditionalist. Rather than speaking as Kirkendall does of agrarians and modernizers, I wish to recognize that *agrarianizers* like Wes Jackson are not only progressive, but also scientific, even post modern.) Farms like my aunt and uncle's are essential to our cultural identity, supplying us with the type of people we need to make our country thrive. Thomas Jefferson believed that lots of small- to medium-sized farms were indispensable to American democracy; they distributed landownership widely while training young men and women in the virtues of independence, hard work, respect for animal and human life, and love of family and nation.

Not everyone in these pages will agree with the agrarianizers. Many are convinced that agriculture should be assimilated completely into the international market economy. These authors will show some skepticism toward what they regard as inflated claims for family farms. Call them "democratic capitalists," "agribusiness modernizers," or "free-marketers." By whatever name, those such as Jesse Helms, Michael Novak, Luther Tweeten, and Gregg Easterbrook all think that Bishop Dingman has oversimplified the story. They understand the present rural crisis not so much as a moral dilemma but rather as an unfortunate episode in the continuing saga of economic expansion and contraction. There is an overabundance of corn and wheat these days; the answer is either to find more buyers for our products, lower our prices, or reduce the amount being produced. Since buyers are in short demand, and since the Third World (which would like to increase its caloric intake) has no wealth of its own to spend on grain, the best answer seems to be to lower prices while adjusting supply downward. If this means having fewer farmers to produce fewer bushels of soybeans, this will inevitably cause some suffering in the short-run. But this is evidence, to the modernizer, that America needs fewer farmers; that is what the law of supply and demand is telling us right now.

Both modernizers and agrarianizers agree on many of the facts pertaining to the present situation. It was brought on by a rapid period of economic growth and inflation in the 1970s, followed by a period of disinflation in the early eighties, shrinking export markets, a strong dollar, high interest rates, large production costs, low commodity prices, exorbitant land prices, and oligopolistic practices in the agricultural supply and food manufacturing sectors. Where the two groups differ is on which moral should be taken from this tale. Agrarianizers read the story through Dingman's glasses: as a tragedy, involving the unjust demise of a heroic figure and the imminent collapse of a once great society. Modernizers read it the way Michael Novak does: as a success story in which the hero's hard work is rewarded with profits while society's technological advances reward it with lots of low-priced goods.

This book does not try to argue that one of these interpretations is right and the other wrong. Rather, it sets out to provide the economic and sociological facts of the present crisis without ignoring the philosophical and theological foundations the reader will need in order to make his or her own judgment about the question posed in the book's title. In order to accomplish these goals, several different sorts of articles had to be collected. The central virtue of the anthology is in the diversity of voices found here. The book should be valuable to agricultural economists, who will be familiar with names like Tweeten, Johnson, Harl, and Boehlje. But it should also be of interest to family farmers, political scientists, philosophers, theologians, clergy, and laity, to those directly involved in determining agricultural policy and to those who wonder what sort of nation we ought to become in the future. The plurality of perspectives represented here is the book's primary value. It is an interdisciplinary collection in the best sense, offering discussions of the economic and financial facts, but including as well the stories of those directly affected by the facts.

The Book of Job has a sad resonance today. Job said with confidence that the land was not crying out at him. Can we say the same? Job ate the produce of the land and paid the land back. Can we say the same? Job did not "cause the death" of the land's rightful owners. Have we? When Job worked his land, it produced wheat and barley. When our grandchildren go out to work ours, will it produce only thistles and weeds? We must decide whether our agricultural policies are sound, and whether the family farm as an institution deserves preservation. It is not obvious that it does. In the pages to come, the debate is joined; it is our moral obligation to decide who is right and, on the basis of our decision, to act as wisely as possible. The fate of real farm families — people like Harold and Sandy Pippert — depends, in part, on our response.

A Plan for Discussion

Like good philosophy, good theology begins with the puzzles and problems of everyday life. Starting from the practical concerns of people around us, we move as theologians into theoretical reflection. Here, we try to get clear about just what the puzzle is, and what resources we have to try to solve it: moral intuitions, communal values, theories about human rights and patterns of exchange, traditional wisdom, and so on. When we move to this more abstract level of reasoning, we discover that our intuitions and considered judgments are not always harmonious. When they conflict, we must try to attain some moral balance, a "reflective equilibrium" that brings order to our conflicted state (Rawls 1971, 17–53).

But good theology does not come to rest in an airtight logical system. It always has a practical aim: to help us understand and change ourselves and our world. Inevitably, good theology drives us back to the ordinary world of commerce and family. Sometimes we come back with concrete solutions. More often, we come back with a better sense of direction. That sense of direction we call practical wisdom: a "knowing how to proceed" rather than a "knowing that this is the only right answer."

This book does theology in accordance with this idea. It begins with the facts, moves into philosophical and theological reflection, and returns to the affairs of modern agriculture and politics. In the first chapter we find an analysis of the family farm by Gregg Easterbrook, an essay written with a slant. While laying out the various dimensions of the current problems, Easterbrook also wishes to discredit some popular myths about the family farm. His piece does not pretend to be a wholly neutral presentation; Easterbrook seems to know that that sort of writing is uninteresting. Instead, he presents the facts as carefully as he can while telling his own revisionist story about contemporary American farming. Easterbrook's piece is a good example of what the Protestant ethicist James Gustafson (1981, 1–3) calls an "evaluative description" of a moral situation. Easterbrook assesses what he sees while describing it.

Other observers put the facts together in different ways. Easterbrook's critics have another narrative to tell about our agricultural system; their evaluations appear at the end of Part One.

The analysis of the family farm in this first part gives rise to several questions, not the least important of which is this: who are these family farmers and what are they saying about current affairs? Part Two presents what narrative theologians have taught us to call the story we live out (Hauerwas 1977, 71–81). No interpretation of a moral problem is complete until we know what the agents themselves believe about their actions. In order that we might take seriously the perspective of the ones most affected by the drama in farm country, we listen to the stories of farmers in various financial positions. We hear about farmers who have taken their lives, who are feeling resentment, are barely staying in business, are doing relatively well, and are doing quite well.

Getting the agent's story, however, is not enough. We need to reflect in a critical manner on all of these narratives. Many of the methods of the social sciences help us to do this: psychoanalysis helps to uncover deeper tensions in the individual psyche, history helps to probe the collective past, sociology assists in understanding cultural institutions. Part Three presents the views of a historian, a sociologist, and an economist.

This sort of work is clearly that of academia. If there is anything dearer to the heart of a farmer than questioning the wisdom of experts, it is proving the experts wrong. When it comes to specific studies in history or economics, a farmer could hardly prove an expert wrong except by becoming a better expert. But when it comes to the general political power that experts wield, farmers seem to have few opportunities open to them. Indeed, farm spokespersons have charged that agricultural scientists and economists have great influence in our society, and they have used it against family farmers. Are university researchers only passive onlookers who report and analyze facts? Or are they powerful — if quiet — instruments of social transformation? If they are powerful, have they used their power fairly?

It has been said that colleges of agriculture, originally established to help the masses improve their farm skills and cultural awareness, have actually served the interests of big agribusiness corporations. This charge is made in an essay by Jim Hightower and Susan DeMarco and reprinted in Part Four. After Hightower's initial volley, Glenn Johnson, an agricultural economist, and Tony Smith, a Marxist philosopher, lock horns over this issue.

Part Five raises the question of justice. What is it? Has the family farmer been treated unjustly? Hightower makes this charge in his second essay, "The Case for the Family Farm." Luther Tweeten investigates the charge and denies it, arguing that no particular sized farm has fared better or worse — all things considered — at the hands of federal policy. A stronger claim is made by Catherine Lerza and Michael Jacobson: modern agriculture is dominated by oligopolies that control prices and siphon money from small farmers. A study by Russell Parker and John Connor on the dollars taken from consumers by major food manufacturing companies lends credence to Lerza and Jacobson's claim.

In the final essay in Part Five, Luther Tweeten says that the disjunction in the phrase "food for people *or* profit" is misleading. He suggests that we can have our cake and sell it too. Ultimately, there is no need to choose between bread and capital; we can have food for people *and* profit.

It should be clear that there is a vigorous debate about American agriculture. To oversimplify, there are modernizers on one side agreeing with Tweeten that agriculture can equally serve human and economic needs. But to do that, we need less labor — fewer farmers — in agriculture. On the other side, there are agrarianizers who with Hightower think that agriculture, in the end, always winds up serving either humans *or* profits. The latter group wants farming to serve people first and money second. Their view in support of farmers may remind one of the slightly different argument of the Hebrew prophets; we can either serve mammon or Yahweh, but not both.

By following the above route, our moral reflection eventually winds up with theological concerns. Can religious traditions help us solve the problem of the family farm?

Unfortunately, religious thinkers, Christian ones in this sampling, do not speak with a single voice. The debate between modernizers and agrarianizers is found both within the Roman Catholic church and among Protestants. Michael Novak, a Catholic, tells a parable of Iowa whose message is that free initiative and hard work ultimately pay off. His theology of the corporation, which follows, can be read as a democratic capitalist's defense of agribusiness. The U.S. Catholic bishops' comments suggest a more skeptical view. With Jeremiah, they are suspicious of our capacity to deal adequately with the hungry when agricultural production is made subservient to the profit motive. The bishops simply do not believe that we can encourage rural communities to thrive if we continue to allow vertically integrated, multinational businesses to dominate the farm production, supply, and food manufacturing sectors.

The debate, an intense and important one, is not settled in these pages. Charles Lutz, a Lutheran, does offer something of a mediating position, however. He calls for the church to be primarily an organ of help, consolation, and spiritual strength through all crises and to all people.

As liberation theologians constantly remind us, good theology should issue not only in useful knowledge but in committed action, *praxis* (Metz 1980, 49–60). What should we do in the economic and political realms knowing what we now know about the plight of the family farm? In part, the answer to this question hinges on whether life on the medium-sized farm is really superior to life on any other kind of farm. Appealing to spiritual and moral values, Wendell Berry maintains that it is. Michael Boehlje contends that social scientists are not able to find many distinct advantages to privately owned, family-run farms. Berry answers with a comment about the "neutrality" of social scientists.

If good theology ought to end in action, the last question is the one raised above: which action? Congressional action is only one sort, but it may be the one that can affect us most. For this reason, the final part asks what we, through Congress, ought to do. Two major proposals were before the House and Senate at the end of 1985. In early 1986 a version of a bill written in part by Sen. Jesse Helms—and pushed heavily by President Reagan—was passed into law. A summary of this law can be found in Chapter 30.

As several commentators have noted, the law does not seem to have solved agriculture's problems. While its defenders plead that it must have time to work, its critics insist that there is no time to waste. The 1985 farm bill puts all its eggs in one basket: lowering product prices so as to regain large portions of a lost international market. But some feel this may never occur (Food and Agricultural Policy Research Institute

1986). Hence the other agrarianizing plan was still being debated in 1987. As an alternative to the so-called free-market approach, the Harkin bill promises to raise farmers' incomes by limiting production. A referendum in August of 1986 showed that many wheat farmers approved of such an idea. The Save the Family Farm Act is included in Part Eight as an alternative direction.

Before starting on either path, however, there is a prior question to be answered.

What Is a Family Farm?

Getting an accurate definition of the family farm is difficult for two reasons. First, farms across the United States are very diverse. One often thinks of the traditional Iowa homestead farm — 160 acres of corn, beans, wheat, hay, hogs, chickens, and cows — as the quintessential family farm. But not many farms in Iowa are like that any more. And across the nation there are many family farms that never were. Should we include subsistence farmers in Appalachia in our definition of family farms? What about sheep ranchers in Wyoming and Montana? Dairy farmers in Wisconsin? Large-scale irrigation operations in California where oranges and vegetables are grown? Peanut growers in Georgia? Cotton farmers in Alabama?

To the extent that any farm is owned and worked by a family, it is a family farm. And yet the farms commonly regarded as family operations tend not to be the large or technologically sophisticated ones. Even though a large, technically sophisticated farm may be owned, managed, and worked by four families all bearing the same name, somehow it does not seem to be what we mean by a "family farm." By the same token, the husband and wife who work full-time in Atlanta and grow pecans on the side in their grove of ten trees do not seem to be family farmers in the classic sense.

The problem is that very large and very small family operations are not the farms usually considered as family farms. Farms with gross annual sales of twenty thousand dollars a year are not generally capable of supporting a family. These hobby farmers almost always have a substantial amount of off-farm income. On the whole, the current crisis has not severely affected them. On the other end of the spectrum, farms with gross annual sales of over two hundred thousand dollars are not usually owned and operated by a single family; they are normally controlled by a corporation, operated by a farm management firm, and worked by hired labor. Even though the owners may be an extended family that once lived on the farm, it is often the case that none of the current owners are now actually farming the land.

For these reasons, I suggest that a definition focus on the medium-sized farm that is both worked and owned by a family. Luther Tweeten has suggested a definition that I have modified: a family farm is an agricultural operation that is owned by a family or family corporation, has gross annual sales of between forty thousand dollars and two hundred thousand dollars per year, and does not hire more than 1.5 person-years of labor (Tweeten 1984, 1, 10). In 1985, there were only about five hundred thousand farms of this type, accounting for less than 25 percent of the total number of all farms (USDA 1985). This definition was suggested to most of the authors of essays in this volume; for the most part, they have stuck to it.

But not all of them are happy with it, and their discontent deserves explanation. The definition above uses only economic measures as its criteria. Wendell Berry and others argue here that the essence of the family farm is not its capacity to make money but its capacity to connect people with land, nature, and the past. Many of the agrarianizers make this point eloquently; we should not think of family farms in terms of gross sales and hired labor but in terms of the relationships such farms foster between people, past and future generations, animals, and soil.

This is the second problem one confronts in trying to define the family farm. It is a very difficult issue because it introduces criteria that are very hard—perhaps impossible—to measure. It would have us switch our entire frame of reference, focusing not on acres and profit ratios but on the care and love farmers have for their land.

The first problem in defining the subject is that people disagree about how large or small a family farm can be. The second problem is that some people think largeness and smallness are the wrong terms in which to think about the matter. The second problem shows that we are dealing with what philosophers call an "essentially contested concept" (Gallie 1964, 157–91). The experts are at odds not only about what size farm qualifies, but about whether size ought to be the deciding criterion.

I do not think we should throw out the definition from Tweeten; it gives us an empirical tool with which to start. On the other hand, I find the agrarianizers' argument convincing. Can we find a compromise that would satisfy each of the two approaches? Perhaps. If Tweeten's definition is modified one more time, it becomes a gangly, but not unworkable, definition; a family farm is an agricultural operation loved, worked, and owned by a family or family corporation, with gross annual sales of forty thousand dollars to two hundred thousand dollars, hiring less than 1.5 person-years of labor.

Whatever its faults, this definition provides a starting place. Its virtues are that it excludes many large corporate farms and most small hobby farms while accommodating both the social scientists' and the poets' interests. But most importantly, it brings into focus just the farms that are currently in most danger. For it is exactly those farms that most urgently need attention.

● ● ●

There are several people who contributed to the appearance of this book. I wish to express my deep gratitude to them. First, thanks to Edna Wiser and Bernie Power, the best secretaries in the world (next to my mother, of course). Second, thanks to Bill Silag and Carol Kromminga for excellent editorial advice. Third, thanks to Jim Hildreth and Farm Foundation of Illinois for their support of the third annual conference on religious ethics and technological change at Iowa State University. That conference was called "Is There a Moral Obligation to Save the Family Farm?" and many of the essays found in this volume were first presented there. Fourth, thanks to my students in "Religious Ethics" who, in the fall of 1986, read parts of the book in manuscript form and responded with insight and imagination. Fifth, thanks to Kay Silet for composing the index.

Finally, thanks to all those who helped give birth to the Iowa State conference series on religious ethics: Brent Waters and United Ministries in Higher Education, John Elrod, Paul Hollenbach, David Kline, Pat Miller, and the Iowa State University Committee on Lectures. The series has quickly become a place where politicians and businesspersons meet with academics and religious leaders to discuss issues of common concern. The broad range of participants attracted to this forum attests to its unique character. While many meetings on college campuses encourage conversation between scientists and humanists, this one insists on having diverse religious traditions represented. Baptists, Lutherans, Jews, Catholics, and Methodists have all appeared on the program since its inception in 1984. The result has been lively and constructive dialogue. For this reason, the conference series has been praised as progressive. It is unusual, being one of a very few such forums sponsored by a state-funded university.

Religion is extremely influential in American life. Implicitly or explicitly, Jewish and Christian values inform many legislative and judicial decisions. The organizers of the ISU conference believe that the ethical values of our various religious faiths should not be discussed only in the congressional cloakroom or Presbyterian meeting hall. They should be brought out into the open: the floor of the Senate, the university classroom, the town caucus. Given such a hearing, those values might influence policies in a more public, rational manner. We should not let religion do its work behind our backs. It should happen in front of us, where we can see it, and where we can use our minds and hearts to reason with one another. By the same token, we should not adhere blindly to religious dogmas; often, our most deeply rooted values need to be put into question.

Forums such as the one at Iowa State provide an opportunity to put religion and public policy into dialectical relationship where each can challenge and transform the other. Religion can be brought out of its modern, privatized existence, giving it what it ought to have: a healthy, carefully circumscribed, role in American civil life. The essays that follow offer a model for this type of conversation.

References

Collum, Danny. "The Forgotten Farmer." *Sojourners,* October 1986, 14.

Food and Agricultural Policy Research Institute. "An Analysis of the Food Security Act of 1985." FAPRI Staff Report #1-86. Ames: Iowa State University.

Gallie, W. B. *Philosophy and the Historical Understanding.* New York: Schocken, 1964.

Gustafson, James M. *Ethics from a Theocentric Perspective: Volume I, Theology and Ethics.* Chicago: The University of Chicago Press, 1981.

Hauerwas, Stanley. *Truthfulness and Tragedy: Further Investigations into Christian Ethics.* Notre Dame: University of Notre Dame Press, 1977.

Metz, Johann Baptist. *Faith in History and Society: Toward a Practical Fundamental Theology.* New York: Seabury, 1980.

Rawls, John. *A Theory of Justice.* Cambridge, Mass.: Harvard University Press, 1971.

Shanley, Mary Kay. "Meet the Bishop." *The Iowan,* Fall 1985, 5, 62.

Tweeten, Luther. *Causes and Consequences of Structural Change in the Farming Industry.* Washington, D.C.: National Planning Association, 1984.

U.S. Department of Agriculture, Economic Research Service. *Economic Indicators of the Farm Sector,* 1985.

Making Sense of the Family Farm

MANY A COMMENTATOR has been struck by the irony of the current agricultural scene. While Willie Nelson praises the virtues of the proud, independent, American farmer, that same farmer drives a tractor into Washington, D.C., demanding a handout. Throughout our history, rural folk have cursed the weather and the government. Farmers usually admit that they can do little about the first, but if only Uncle Sam would stay out of the way, they would be able to survive.

Or so the story goes. Urban folk have not been hesitant to point out that the farmer's much-touted independence is really supported by a vast network of federal support programs. If it were not for millions of dollars in congressional spending, the fiercely antigovernment farmer would not be able to pay bills. In such a situation, recent appeals for help have sounded self-serving to some. Exhortations to "save the family farm" begin to sound like only so much romanticism.

If we are to discuss the ethical dimensions of the current farm crisis intelligently, we must first understand the mechanisms that support and threaten the family farm. The following article by Gregg Easterbrook provides a controversial introduction to this matter. Easterbrook begins by questioning some of the common myths about farmers. The media often make farmers out as poor, struggling tenants who are not being allowed to continue their humble way of life. Easterbrook contends that this is not true; as a group, he argues, farmers are not poor, not being driven from the land, not dominated by agribusiness, not involved in an unprofitable business, and not overwhelmed with burdensome debt. These are contentious claims; many of them will be directly challenged by those responding to Easterbrook, and several of them will be indirectly refined by others writing in this volume.

The significance of Easterbrook's article is that it gets us quickly into the facts of the matter. For in defending his "revisionist" claims, Easterbrook introduces us to the mysterious inner workings of the American agricultural economy. He touches on the operation of such agencies as the Farmers Home Administration, the Farm Credit Administration, and the Commodity Credit Corporation, and he helps us understand the meaning of such phrases as deficiency payments, target and market prices, and crop subsidy programs.

Easterbrook also discusses the history of agriculture in Iowa and California. Since California farms tend to be larger and more scattered than Iowa farms, Easterbrook's discussion of the Midwest seems most relevant to our concern with family farms.

GREGG EASTERBROOK

Making Sense of Agriculture: A Revisionist Look at Farm Policy

AMERICANS ARE INCLINED TO THINK of a crisis as a shortage, but agriculture is in crisis because of surpluses—too much of a good thing. Farmers in the United States produce more food than their countrymen need or want; and although they do not produce as much as the hungry world needs, they produce far more than it can afford to buy. This has been the case ever since the depression. Indeed, overproduction was a problem even during the depression, and it has become unusually pronounced in the 1980s, as many Western countries and even a few developing nations have joined the United States in growing more food than anyone knows what to do with. The result is a tragedy of plenty, which offends many of our deepest convictions about resources, virtue, and the soil. One way or another, most of the confusion in American agricultural policy today arises from our reluctance to accept the idea that growing food is sometimes the wrong thing to do.

Few economic endeavors have any aura of romance and tradition. We don't get misty at the sight of a chain store framed against a prairie landscape or take comfort in knowing that each morning thousands of lawyers head out into the predawn darkness to tend their lawsuits. Farming, though, occupies an honored place in our culture. Even big-city sophisticates who would sooner die than attend a Grange Hall dance find it reassuring to know that somewhere out there honest folk are working the earth much as it has been worked for centuries.

Agricultural industries, from farming itself to the retailing of farm products, constitute the largest sector of the American economy, accounting for 20 percent of the GNP amd employing more people than the steel and automobile industries combined. Yet many people find it heartless and somehow unfair for anyone to speak of farming as an

GREGG EASTERBROOK is a contributing editor of *Newsweek* and also of *The Atlantic*. This essay is reprinted with permission of the author from *The Atlantic,* July 1985, 63–78. Copyright 1985 by Gregg Easterbrook.

industry subject to the logic of supply and demand. To this sentimental faction the thought that any farmer should have to go out of business seems intolerable. As the cost of federal agricultural subsidies has risen, there has come into being an opposing faction, which dismisses farmers as spoiled welfare dependents who bilk the public on an unprecedented scale. Last winter's campaign for an emergency farm-credit bill seemed to divide politicians and the press into two camps: those who would "give 'em whatever they want" and those who would "let 'em fry."

The actual circumstances of modern farming conform to few if any of the assumptions that underlie the public debate. In order to see what condition American agriculture is in, one must first dispatch a number of widely held misconceptions.

—Farm families as a group are not poor. Their average income in 1983, one of the worst years in memory for agriculture, was $21,907. The average income for all families was $24,580. If one takes into account the lower cost of living in rural areas, farmers live about as well as other Americans. In fact, in some recent years farmers earned more than the national average.

—Farmers are not being driven from the land. From October of 1984 through January of 1985—when what was said to be a dire emergency for farmers was making news—the Farmers Home Administration actually foreclosed on forty-two farms nationwide. The FmHA provides loans to farmers who can't get credit elsewhere. Over the same four-month period its borrowers who "discontinued farming due to financial difficulties"—a broad category that reflects foreclosures by lienholders, bankruptcies, and voluntary liquidations to avert bankruptcy—totaled 1,249, or .5 percent of the FmHA's 264 thousand clients. From January to September of 1984, production-credit associations and federal land banks under the aegis of the Farm Credit Administration (which is much larger than the FmHA, handling about a third of U.S. agricultural debt) actually foreclosed on 2,908 loans nationwide. If one includes bankruptcies and loans in the process of liquidation, a total of 1.6 percent of FCA-aided farmers were in trouble.

Debt problems are real: early this year the FmHA, FCA-backed institutions, and rural farm banks saw delinquency rates reach record highs. But the incidence of dispossessions has been vastly exaggerated. The number of delinquent loans usually peaks early in the year because most FmHA loans come due in January. Newspapers rarely follow up their winter reports of a "dramatic increase" in the number of farms in trouble with summer reports of a dramatic decline.

—Farming is not a disastrous investment. Farm lobbyists don't like to talk about disposable income, preferring to speak of "profit"—a problematic concept when applied to the self-employed, who can treat as business expenses items like vehicles and real estate, which most others must pay for out of their salaries. In 1983 and 1984 a trillion dollars' worth of farm assets generated a profit of forty-eight billion dollars – an average return of 2.4 percent a year. But the annual net return on all corporate assets for the same period was only 5.5 percent.

– The "farm exodus" has been over for years. Much is made of the fact that the number of U.S. farms declined by thirty-three thousand in 1982 and by thirty-one thousand in 1983. But the declines were far bigger in the 1950s and 1960s – a period enshrined in political mythology as better for farmers. In 1951 the number of farms declined by 220 thousand, in 1956 by 140 thousand, and in 1961 by 138 thousand.

– Agribusiness does not dominate farming. Only 3 percent of all U.S. farms are owned by corporations. Moreover, farm ownership is not becoming increasingly concentrated. About 1 percent of all owners of farmland hold 30 percent of farm acreage, but that ratio is the same as it was in 1946.

Most agricultural products are not eligible for federal support. Federal programs concentrate on what are called basic crops: grains, cotton, rice, and dairy products. Fruits, vegetables, livestock, and specialty crops such as nuts and garlic are not subsidized. In recent years these products have often done better than the ones the government takes an interest in – a circumstance that some commentators view as proof that abolishing government programs would solve agriculture's problems. Right now, however, the two categories of farm products are in about the same depth of trouble.

– Most farmers don't get subsidies. Participation in the basic crop-subsidy programs is voluntary, and most farmers stay away. A study released in 1984 by the Senate Budget Committee found that the major subsidy programs covered only 21 percent of farms and 16.5 percent of farm acreage.

Those whose response to the increasing cost of agricultural programs is that we should continue the subsidies for small farmers but prevent large farmers from enjoying them should take note that the programs already function pretty much that way. The Senate Budget Committee study found that the largest farms were the least likely to be enrolled in subsidy programs. The most direct cash subsidy, called deficiency payments, is capped at $50 thousand per farm per year, which renders the program of little value to large operations. Indeed, the 1 percent of owners controlling 30 percent of American farm acreage received only 7 percent of the deficiency payments in 1983 and almost none of the FmHA and Small Business Administration benefits.

—No one can get rich on federal subsidies. While a very large amount of money is spent subsidizing U.S. agriculture, it is spread so thin that few individual farmers receive significant amounts. Indiana farmers averaged $1,323 in federal cash payments in 1982; Kansas farmers averaged $1,577; figures in the rest of the heartland were about the same. In Arizona, where farmers receive by far the highest subsidies in the country, direct payments averaged $27,040. Farmers benefit from a variety of other subsidies, the per capita amounts of which are difficult to calculate but clearly not lavish.

—Most farmers don't have burdensome debt. President Reagan was wrong to say that "around 4 percent at best" of farmers need credit help, but the actual figure is not much higher. According to a study by the U.S. Department of Agriculture (USDA), which farm-spending advocates often cite, only 6.5 percent of all farmers are actually insolvent or on the verge of being so. The Federal Reserve System estimates that 8 percent of farmers have debt-asset ratios over 70 percent, and another 11 percent have debt-asset ratios of 41 to 70 percent. But nearly 58 percent of all farmers are well in the clear with debt-asset ratios of 10 percent or less.

From the impersonal standpoint of economics, the 19 percent of farmers who, according to the Federal Reserve, are in credit trouble might be viewed as representing agriculture's excess production capacity. Last year 81 percent of the production capacity of all U.S. industries was in use, leaving 19 percent idle. Looked at this way, the share of borderline cases in farming is not particularly different from that of other industries.

—Debt has hit farmers hard but not that hard. From 1974 to 1984 outstanding agricultural debt rose 193 percent. Through the same period consumer credit rose 172 percent, mortgage debt rose 167 percent, and all commercial bank debt rose 153 percent.

—The embargo on grain sales to the Soviet Union did not clobber wheat farmers. In 1980, the year the embargo was in full effect, agricultural exports jumped from thirty-one billion dollars to forty billion dollars—the largest increase ever. Wheat exports increased from 1,375 million bushels in 1979 to 1,514 million in 1980 and increased again in 1981, to 1,771 million bushels. In 1982 and 1983, after the embargo was lifted, wheat exports declined.

Just Enough to Be Miserable

Farm groups from the radical American Agriculture Movement to the rock-ribbed conservative American Farm Bureau Federation, condemn federal agricultural programs as a matter of ritual, and nearly every congressional hearing on agriculture commences with a rendition of how horribly the government treats the farmer. This is partly because agricultural programs are a philosophical jumble. Containing elements of free-market risk and federal bailouts, capitalist entrepreneurship and socialist central planning, they do not reinforce anyone's world view.

Consumers might say don't knock success. U.S. agriculture not only produces an abundance of nutritious food but does so at low consumer prices. Americans spend a smaller percentage of their disposable income on food than do the citizens of any other industrial nation; in the past ten years, as the Consumer Price Index has risen by 114 percent, super-market prices have risen by only 96 percent. Even when the cost of subsidies—which consumers must pay too, through their taxes—is taken into account, food remains cheap. If the roughly twenty-one billion dollars being spent in this fiscal year to subsidize agriculture were re-flected directly in consumer costs, supermarket prices would be only 6 percent higher than they are.

Some farmers dislike federal programs because they don't want the government interfering with their lives, and others dislike the programs because they wish the government would interfere more. The range of opinion is not hard to understand. Farm subsidies provide just enough money to keep nearly every farmer in business producing just enough excess supply to hold prices down. This means that farmers who depend on government subsidies will find it hard to become prosperous enough to do without them. It also means that farmers who aren't subsidized get lower prices than they would if no one was subsidized. Everybody works, but everybody is miserable.

The main subsidy programs are deficiency payments and loans, both of which are administered by the Commodity Credit Corporation. Deficiency payments are straightfoward. The farmer simply gets in the mail a check for the difference between the market price and a target price, which is set by Congress. A CCC loan is more involved. A farmer borrows a sum calculated by the government to reflect the value of his or her crop, and puts up the crop itself as collateral. The loan is like a salary. If the market price turns out to be higher than the loan rate, the farmer can sell the crop, satisfy the loan, and keep the premium. If the market price is not higher than the loan rate, the farmer activates a "nonrecourse clause" and turns over the crop—the collateral—to close out the obligation. In effect the farmer has forced the government to buy the farmer's product.

The cost of agricultural subsidies is difficult to predict. For example when the farm bill expiring in 1985 (which covered almost all agricultural subsidies) was passed in 1981, its cost was estimated at $11 billion. Because lawmakers assumed that inflation, running hot in 1981, would continue through the lifetime of the bill, they progammed in large annual increases in loan rates and target prices. Instead inflation cooled, the recession held agricultural exports down, market prices wavered, and loan rates and target prices ended up being higher in relation to market prices than expected. As a result, instead of a total of $11 billion since 1981 the legislation has cost $53 billion. In the 1985 fiscal year alone the outlay for CCC purchases and deficiency payments was expected to reach $14.2 billion.

In theory the subsidy system combats overproduction because farmers who "seal" their crops with the CCC must agree to leave idle a percentage of their acreage. But in practice it can backfire. Eligibility is based on acreage: the more acres a farmer has available for production, the more generous the subsidy can be. Thus farmers sodbust land neither needed nor efficient, solely to get credit for the acreage. And nothing in the program prevents farmers from increasing production by using more fertilizers, pesticides, and machinery on land they are not required to idle. It is not unusual for a farm's output to go up in a year in which acreage has been set aside to please the CCC.

Next in dollar value are subsidized loans from the Farmers Home Administration. In the 1985 year the FmHA was spending about $3 billion for below-market operating loans (loans that pay for seed, fertilizer, and so on). Orginally the FmHA was to provide only temporary assistance, but now many recipients stay in the system year after year. (When the first federal price-support program was created in 1933, the secretary of agriculture, Henry Wallace, called it a "temporary method of dealing with an emergency." The emergency farm-credit bill that monopolized the attention of Congress early in 1985 was, like most agricultural assistance, portrayed as a one-time extraordinary measure. There was also an emergency farm-credit bill in 1984, and barring some miracle, there will have to be another next year.)

FmHA assistance is available only to farmers who have already been turned down by commercial banks or production-credit associations. The 264 thousand farmers kept in business by the FmHA represent only 11 percent of a total 2.4 million. Yet 11 percent can make all the difference in the marketplace. For example, in commodity markets like those for grain, oil, and gold, where products have no distinguishing features and only numbers count, often the final few percentage points — what economists call the marginal supply — sway the price for all producers. OPEC's Arab members set world oil prices for almost a decade though they produced about a third of the world's oil. Oil was in tight supply, it was a seller's market, and all oil became worth what OPEC was charging, because any seller knew that a buyer willing to pay the marginal price could be found. Now there is a buyer's market for oil, and the OPEC price can no longer be enforced.

Likewise, in 1972, the year of the first major Russian grain purchase (the "Great Grain Robbery"), U.S. wheat prices zoomed from $1.76 a bushel to $3.95. The Russian acquisition totaled just 22 percent of the wheat available on U.S. markets that year, and by December there was more surplus grain in our stockpiles than the Soviets had bought. But the marginal supply had been carried away and the market converted from a buyer's to a seller's domain. By the same token, a moderate percentage of extra supply can depress prices by robbing sellers (farmers) of their bargaining leverage. In 1981, for example, the corn harvest increased by 22 percent while domestic consumption and exports — or, as the USDA calls it, total "disappearance" — declined slightly. As a result, corn prices fell from $3.11 a bushel to $2.50.

Overproduction becomes acute in the basic crop categories, because one year's mistakes are stored and added to those of the next. In 1981 U.S. fields produced a record 15.6 million bales of cotton, but only 11.8 million "disappeared." Leftovers from 1981 have been plaguing the cotton market ever since. In 1983, for example, cotton production was down to 7.7 million bales and demand was up to 12.7 million, but because the year began with 8 million bales in storage, cotton supply still exceeded demand by almost 25 percent. Today the price of cotton is about what it was in 1976, with no adjustment for inflation, and it's below what some growers spend on production.

When foreign demand falls off, the cumulative effects of overproduction become especially painful. Roughly four out of ten acres are planted for foreign sale. Since the early 1970s, wheat producers have been serving the foreign market first and the U.S. market as a sideline; in the record years of 1980 and 1981, twice as many bushels were shipped overseas as were used at home. If foreign customers fail to buy, as has lately been the case, there's nowhere for the crops to go but into storage, because the United States already has all the food it needs. Even drastic price cuts would produce at best only a slight increase in U.S. consumption (food generally being subject to what economists call inelastic demand: most buyers want about the same amount regardless of variations in price). Soybean growers produced 2,035 million bushels last year and sold 1,880 million, leaving 155 million in silos. Export demand, which had peaked at 929 million bushels in 1981, was down to 800 bushels. Had export demand stayed at the 1981 level, 1984 would have been a banner year for soybean farming; instead the price of soybeans fell from

$7.75 to $6.60 a bushel, and concern about the surplus hangs over the 1985 year's planting.

Next on the list of federal subsidies is $954 million for soil and water conservation. The same federal policies that encourage sodbusting of questionable land and boosting production by means of chemicals are to blame for a significant portion of the erosion that this subsidy is meant to control. Another $421 million goes to subsidize federal crop insurance and about $350 million is spent on emergency loans to help FmHA farmers whose crops have been damaged by weather.

An extra $24.3 million in direct subsidies to farmers was provided in 1984 by the Small Business Administration under a program that issues cut-rate loans to those hit by natural disasters. (In 1984 SBA "nonphysical-disaster" loans were granted to fertilizer companies in states where commodities released under the payment-in-kind program had reduced demand for fertilizer—one federal subsidy chasing another.) Agriculture also benefits indirectly from the $1.1 billion spent on farm research and extension services, the $18.2 billion spent on food stamps and child nutrition programs, and the $1.8 billion in Food for Peace aid to poor countries, all of which shore up crop demand. Finally, many billions have been spent on federally subsidized irrigation and electrical power in the West. For a variety of reasons, the exact cost of these subsidies— beyond $80 million being spent this year to subsidize 2 and 5 percent loans for power and telephone lines to rural communities—defies calculation. Much of the expense is coming to an end, however, because of a law passed in 1982 that requires western growers to begin paying the full cost of their water.

It is common for farmers—and reporters—to speak as if in a just society virtuous products like food and fiber would only increase in value. Yet we find nothing amiss when the price of computers or eyeglasses falls, and we're upset when the prices of energy and housing climb. A successful economy is supposed to drive down the prices of goods, especially manufactured goods—and the advent of fertilizers, pesticides, self-propelled combines, and large tractors has made agriculture one of the least labor intensive of industries. Each year the USDA charts farming "inputs" for capital and labor. The 1980 input for farm labor was a fifth that of 1930, while the input for machinery was three times greater, and the input for chemicals was twenty times greater. Farm groups say that there is something wrong with the fact that wheat costs less in real terms today than it did in 1870. There would be something wrong if it *didn't* cost less.

Farm-state congressmen often cite the index of prices for farm products from the Producer Price Index kept by the Bureau of Labor

Statistics. The PPI, like the Consumer Price Index, uses 1967 as its base year. Whereas the CPI rose to 311 in 1984, the index of prices for farm products has risen to only 256. The congressmen never mention that the indexes for almost *every* commodity within the PPI are behind or only equal to the CPI. Textile products and apparel are at 210, furniture and household appliances at 219, and rubber and plastic products at 247. Only nonmetallic mineral products at 337 and energy at 657 are significantly ahead of the CPI. For that matter, low producer prices keep the cost of running a farm down. Indeed, according to the USDA, the index of what machines, supplies, interest, taxes, and wages cost farmers runs about 10 percent below the rate of inflation.

No one likes to be thought of as being on the federal dole, least of all farmers, who put a premium on self-reliance. Farm groups across the spectrum invariably say, "We don't want subsidies, we just want a price," meaning higher market prices, and they note that higher prices would result in less federal spending, because deficiency payments would decline. The next logical step is usually not taken. Absent increased demand, higher prices can be realized only if excess production is controlled, either by cutting subsidies and letting some farmers fail (anathema to the left) or by imposing fierce restrictions on how much and what a farmer may plant (anathema to the right).

Farmers find it difficult to face the overproduction issue, mainly because of the nature of rural life. One of the salient cultural differences between farmers and city folk is that farmers live in places where everybody is in pretty much the same line of work. Everybody either is a farmer or provides a service that farmers need. Imagine if advertising executives had to live in complexes populated entirely by other advertising executives and could have only advertising executives for friends. Would they be so aggressive about stealing business? To be true capitalists, farmers would have to view their neighbors as their archenemies. So they compensate by viewing farming itself—the act of working the fields, not of selling the finished product—as their purpose and keeping everybody going as their political challenge. This thinking reflects the kindness and communal purpose we admire in rural life. It also makes for too many farmers.

The regular experience of shared achievement and sorrow in a common pursuit is among the most appealing aspects of rural tradition. Indeed, farm advocates often argue that the communal quality of rural America should be preserved for its own sake, even if economics has passed it by. They say that farm living sets a spiritual example whose worth thus transcends cost-benefit analysis. But when farmers say that their way of life should be preserved for its own sake, inevitably they must argue that all farmers are equally deserving of protection—that farmers have a right to remain farmers.

Iowa: Fellowship Replaced by Machinery

In an industry as large as American agriculture, nothing is typical. There are the livestock pens of Texas, the vast irrigation networks of Nebraska and Arizona, the hog pens of Illinois and Indiana, the pastoral dairies of Wisconsin and Minnesota, the tiny tobacco plots of the mid-South, the citrus orchards of Florida, the uninterrupted wheat fields of Kansas. Nowhere is contrast more distinct between the number-one and number-two agricultural states—California and Iowa.

California is a high-tech paradise. Farms are majestic; the climate is blissful. Californians produce some two hundred commodities including milk (the dairy business, surprisingly, is the state's biggest agricultural concern), cotton, rice, cattle, grapes, vegetables, plums, oranges, and almonds. Many farms are diversified: designed to produce several categories of crops and to shift from one to another as rapidly as demand changes.

In Iowa, farming is practiced more or less the way it always has been. Nearly all Iowa farms are family enterprises, and nearly all raise corn, soybeans, hogs, and cattle. The Iowa earth can be harsh, the weather cruel.

Just what constitutes a family farm of the Iowa variety depends on the beholder. A growing number of the nation's 2.4 million farmers—right now, roughly half—sell less than ten thousand dollars worth of crops a year. Generally they work full-time jobs and farm on the side. According to Luther Tweeten, an agricultural economist at Oklahoma State University, most of these part-time farmers sell what they raise for less than they spend on production, leaving them several hundred to a thousand dollars in the hole each year, at least on paper. Such part-time farmers, Tweeten says, enjoy farming as an avocation and seek to qualify for farm tax breaks.

Those who sell from forty thousand dollars to one hundred thousand dollars worth of crops a year form the group commonly called family farmers. There are 381 thousand of them, making up about 16 percent of the total farm population, and they are the most troubled—holding a disproportionate share of farm debt and typically having a lower disposable income than part-time farmers, because the farm is their sole source of income. Iowa is this group's stronghold.

The typical Iowa farmhouse is weathered beyond its years with paint in various stages of peeling and cracks in interior walls. Visiting two dozen farms in various parts of the state last winter, I saw only one house that could qualify for teleportation to the middle-class suburbs of Atlanta or Portland – or Des Moines, for that matter. None of the farms I saw had paved driveways; several farmers who described themselves as successful mentioned as evidence of their good fortune gravel-covered drives. Owing to the lack of asphalt and concrete, mud was everywhere, deep enough in spots to make walking a trick.

Yet the people on these farms did not live in poverty. Nearly all the farmhouses I visited contained microwave ovens, color TVs, videocassette recorders, and other tokens of consumer culture. No farmer I met drove a fancy car, but none lacked a car, either. All were well clothed and well fed. And the farm equipment that some possessed was a sight to behold: combines with wingspans like those of aircraft; four-wheel-drive tractors that could pull a Greyhound bus from a ditch; a few Steigers, the Corvettes of tractors, fitted out with air conditioners and tapedecks. This gleaming machinery, more than anything else, represents a profound change in the way farmers live – in the debt burdens they bear and in their relationships with their communities.

"Too many tractors with too much horsepower" is how Helen Lester of Milo, who has been farming with her husband, Guy, since the depression, summarized Iowa's predicament. Academics and journalists are not the only ones who began to believe around the mid-1960s that the small farm was doomed. Farmers believed it too. Awesome new tractors and combines would enable a family farmer to cultivate more land than ever. Farmers would almost have to buy more acres in order to spread those capital investments over a larger income base. Big new machines on bigger farms held out the promise that the family farmer could achieve the touted economies of scale enjoyed by sprawling ranches in Texas and the Southwest.

These machines also held out the promise of a more pleasant life, free of tedium and backbreaking labor. A Steiger with its immense power could plow more land in one day than a conventional tractor could plow in a week; the big combines would harvest crops with much less need for manpower. Improved seed varieties and chemicals were also coming into play, and were expected to diminish the demands that raising crops made on a farmer's time. What a dream began to emerge: a bigger farm, a higher income, and less physical work. The race was on. Sales of heavy machinery soared, and farms, even family farms, expanded in size.

The value of agricultural real estate escalated from $216 billion in 1970 to $767 billion in 1980 and then crashed. In 1984 it was $765 billion – a decline of about 23 percent when inflation is taken into account. In all the commentary on this fabulous rise and fall, it is rarely noted that farmers themselves were the driving force behind the price changes. Urban growth, often presented as the culprit, exerted a minor influence at best: only 3 percent of all the land in the United States is built on for cities, suburbs, or highways. When farmers sell, it is almost always to other farmers: the most prized farmland is that which adjoins an existing farm. Through the 1970s farmers sought more land, and farm size increased by an average of 13 percent.

Traditionally farmers have been frugal people, fearful of debt and wary of promises of quick wealth. But they did not respond to the economic developments of the 1970s with characteristic reserve. (In fairness, neither did millions of other Americans. Major corporations lost billions of dollars in sure-thing energy investments, and all sorts of people bought real estate as if the prices could never break, poising their loans on the assumption of perpetual inflation.) Using machines, chemicals, and new land, farmers both here and abroad expanded their productive capacity by so much in the 1970s that a fall became all but certain. Production through the decade rose at 3 percent a year in the United States—a record pace for annual growth. Meanwhile, food demand was not increasing as fast as production, and demand for some foods was falling.

Many farmers who bought large machinery told themselves that they would cover the payments by doing "custom work"—tilling and harvesting for other farms. As ownership of combines and four-hundred-horsepower tractors became unremarkable, more farmers were offering custom work than needing it done. One farmer I visited, Pat Meade in Milo, said, "You can pretty much tell which farms are in trouble by whether they're four-wheel or two-wheel." Meade uses a 1976 one-hundred-horsepower tractor that he maintains himself. He said, "When farmers discovered that a big machine would enable them to do all their plowing in a single day and spend the rest of the week at the coffee shop, a lot of them couldn't resist, even if they couldn't afford it. Combines were the worst. The average family farmer actually uses his combine less than thirty days a year. The rest of the time it sits in a shed. Seventy-five to a hundred thousand dollars sitting in your shed, doing nothing." Farmers who went heavily into debt to buy big machines and more acres were dubbed "plungers."

A century ago the free-silver movement led by William Jennings Bryan centered on the desire for a liberal money supply so that there would be more for farmers to borrow and more for consumers to spend. In the 1970s inflation, though less glorious in conception than free silver, provided the financial climate many farmers had always thought they wanted. Plungers became community heroes. Farm journals were filled with stories of "young tigers" who were not afraid to take on staggering debt loads. "Those who dive deepest will come out on top" was a common saying. When briefly during the 1970s interest rates were below the inflation rate, borrowers came out ahead merely by borrowing. Farmers are dependent on credit even in good times, because they must pay production expenses months before they have crops to sell. The economic conditions of the 1970s seemed to say that it had actually become smart to pile loans on top of loans. But if inflation stopped, the loans would smash into each other like race cars trying to avoid a wreck.

The FmHA, production-credit associations, and banks were just as much to blame for plunging as farmers. With acreage values rising up to 20 percent a year, farmland was an investment that was staying ahead of inflation. Many banks, needing borrowers to generate income on their inflation-pumped deposits, encouraged farmers to leverage themselves to the limit. According to Meade, "During the 1970s there were times when lenders quite literally drove up and down the road, knocked on people's doors, and asked them if they could use more credit." Philip Lehman, a farmer in Slater, Iowa, and an official of the Iowa Farmers Union, an organization that lobbies for increased federal farm aid, sat on the loan-approval board of a production-credit association in the early 1970s. "Things got to where it was difficult for me to put my initials on the applications," Lehman told me. "It was like granting people licenses to go under."

As inflation and gleaming supertractors dispelled farmers' qualms about extravagant spending, so they altered the spirit of farm communities. Farmers who drove Steigers didn't have to call on their neighbors for help when a wagon got stuck in the mud. Those with combines didn't have to wait for harvesting crews, nor did they have to offer to join in harvesting a neighbor's land. Many farmers borrowed to build their own silos and storage facilities during the 1970s, loosening their dependence on town silos and diminishing their obligation to attend co-op meetings. Farmers began to feel guilty of what they disliked most about city dwellers—an absence of community spirit.

"Because farmers want to be self-reliant, the combines and the tractors had an extremely seductive appeal," James Schutter, the pastor of the United Methodist Church in Tingley, Iowa, told me in the winter of 1985. "Machines made you really self-reliant. People didn't realize the machines would also make you isolated. As soon as it became technologically possible to farm independently, everybody wanted that." Milton Henderson of Mt. Ayr, Iowa, who is retired after working for the Iowa State University extension service for thirty years, said, "Threshing and harvesting parties were wonderful events—much warmer and more human community events than the kind we have now, like high school basketball games." He added, "I would never want to go back to the past, plowing four acres a day with a team of horses. I'm just saying the sense of fellowship is gone, replaced by machinery."

Dairy, livestock, and poultry farmers still work year-round, because animals must be tended continually, but well-equipped crop farmers have it easier. They face three months of heavy work during planting and two more at harvest. Farm-crisis stories tend to appear in winter partly because it is then that crop farmers, without daily work to do in their fields, do their lobbying. When the American Agricultural Movement staged its tractorcade in 1979 bringing hundreds of farm machines to Washington to block traffic, AAM members made speech after speech about how their backs were to the wall. Yet many had arrived in brand-new Steigers equipped with stereo systems, and some spent several months in the capital in no apparent hurry to attend to their businesses. William Olmsted, a United Methodist minister in Greenfield, Iowa, has noticed a subtle change in farm sociology. He told me, "Making your rounds in winter, you can knock on doors at five farmhouses in a row and find no one home. They've gone to town, or are on vacation. Farmers used to *always* be home."

Fencerow to Fencerow

Farmers say they expanded in the 1970s because they were sent a signal from the highest levels of government instructing them to do so. There can be no doubt that they were.

In the 1970s the USDA predicted that food production would fall behind world demand. A book called *Famine 1975* attracted considerable attention when it was published in 1967. When the 1972 Russian grain deal caused wheat prices to rise and poor harvests caused food in general to become costlier, many thought that the pessimistic forecasts had been confirmed, and they began to cry shortage. Richard Nixon's secretary of agriculture, Earl Butz, is said to have advised farmers to plant "from fencerow to fencerow." That phrase is now fixed in the heartland's litany of woes as firmly as the Russian grain embargo: farmers refer to it again and again. In 1973 the headline of an article in the *New York Times* declared, "Days of Cheap Food May Be Over." As the decade passed and world harvests remained poor, the idea that the Russians and others would soon be begging us for grain caught hold. Oil prices were rising and the market for gasohol, made with corn alcohol, seemed about to take off. Declarations that agriculture was "the bulwark of democracy," "America's answer to OPEC," and so on became political cliches. When exports were booming in 1980, President Jimmy Carter's secretary of agriculture, Bob Bergland, declared, "The era of chronic overproduction . . . is over."

"About ten years ago it looked like you just couldn't go wrong by expanding," Jim White, a farmer in Pleasantville, Iowa, told me. White is operating his farm under Chapter 11 of the federal bankruptcy code. In the late 1970s White bought land and equipment on credit and also cosigned notes so that two of his sons would be able to enter farming. In White's house a stack of foreclosure documents sits by trophies that White and his father won for being Polk County Corn Champions in 1963 and 1964. "My dad taught me to be cautious, but everybody — I mean everybody — was saying that biggest had become best," White said.

Who in the 1970s could have predicted that by 1985 rising inflation would be only a memory, that the value of the dollar would rise so dramatically on the world exchange market, that developing countries would begin producing food for export, that gasohol would flop (some gas stations in Iowa now post signs proclaiming "No Alcohol in Our Fuel"), and that consumption of dairy products and red meat would decline? This web of events surprised even smooth-talking experts. The average farmer could not possibly have foreseen it. White's production-credit-association loans contained variable-interest-rate clauses. "They said I had nothing to worry about — that rates had varied only a fraction of a point since 1970," White told me. "My rate went from 7 percent to 18.5 percent."

Does the fact that the government has misled farmers confer on them a right to special compensation? The frequency with which the national winds shift leads politicians routinely to give industries bad advice, and part of being a businessperson is knowing what to ignore and what to take seriously. More important, though Earl Butz's advice certainly sounds foolish in today's economic climate, it was delivered thirteen years ago in a different climate. How many businesses could survive by clinging to strategies thirteen years out of date? There would be scant public sympathy for an automobile company still making cars that get ten miles to a gallon. Because of the nature of agriculture, farmers have a more difficult time responding to economic changes than people in other industries. But this does not mean that they should be exempt from having to respond. In February, the entire South Dakota state legislature traveled to Washington to lobby for more farm aid. When South Dakota's governor, William Janklow, appeared on the ABC News program "Nightline" to state his case, he absolved farmers of blame because, he said, "Earl Butz told them to plant fencerow to fencerow" — as though this had happened the year before.

It could be that farmers attach almost religious significance to what Butz said because it was the one time they were told exactly what they wanted to hear. Agriculture, a quiet line of work in which it was nearly impossible to get rich, was going to take off. Thanks to technology, incomes would rise and workloads would fall; farmers would rescue the country from foreign debt and become national heroes. With crops in demand, farmers could sell them on the market, not to the government, and make their money fair and square. Everything was finally going to be all right. It is human nature to cling to the moment when things were going to be all right, and so it is natural that farmers should cling to the vision of the early 1970s.

For farmers, government policy is like the weather. It's good for a spell, and there's just no predicting. In addition to making business planning difficult, flip-flops of policy inevitably build resentment against government, even when the subsidies are flowing.

One farmer in Creston, Iowa, told me, "I know for a fact that agriculture is controlled by a special committee of bankers and manufacturing interests and that no one is allowed to use the committee's name." Another in Corydon, Iowa, said, "There was a secret meeting in 1947 at which a plan was laid out to destroy the family farm, and everything that has happened since comes directly from the plan." The two speakers were not kooks, and I heard similar sentiments from others during my travels in Iowa.

A fair reading of recent history is that the federal government would have to form a firing squad just to shoot itself in the foot. Nevertheless, some farmers do believe that their troubles must have a nefarious source. Many look outside the federal government. A popular target of suspicion is the Chicago Board of Trade, which most farmers resent as a rich man's plaything designed to make easy profits from the workingman's toil, and which some are convinced was invented by the antifarm conspiracy in order to cause chaos in farm prices. (Some farmers now use the commodities exchange to hedge their crops, but the Iowa farmers I spoke to said they simply couldn't bring themselves to do that, even if it does make business sense.) Conspiracy theories in themselves are probably harmless, but their prevalence in rural communities suggests that farmers wish to fix the blame as far away from home as possible — to dwell in an unreal world. That wish helps to explain why farmers seem so bitter in television news footage.

Dissatisfaction rules life on the farm today, and the unhappiness farmers vent in the media or at political rallies probably does more to advance the idea of an endless crisis in farming than does the actual incidence of foreclosures. The unhappiness stems in part from raised and then dashed expectations. "Every year for the last fifty years had been a little better than the year before, until the 1980s," Carolyn Erb, a farmer in Ackworth, Iowa, said when I met with her in the winter of 1985. At some point having a better year begins to seem like a right. That the Washington, D.C., tractorcade was staged in 1979, a year that turned out to be the second best on record for farm profit, shows how consuming a force self-pity can be.

The structure of farm economics guarantees that farmers will be frustrated. Suppose your year's salary would be twenty thousand dollars, forty thousand dollars, or sixty thousand dollars with the amount determined by a lottery based on what day you asked for your pay. That is the arrangement that most farmers have to live with. The advent of international blockbuster deals adds turbulence to the market. "Now our decision on exactly what day to sell can make or break us for the entire year," Nancy Meade told me.

Many businesses, of course, must deal with macroeconomic unknowns, but how many individual workers have to? Auto workers are not expected each time they report for their shifts to perform an analysis of international sales patterns, according to which their wages will later be calculated. Family farmers must be laborers, market analysts, and financial managers all at once. Considering the modest track record of specialists who do nothing but predict agricultural markets, it is unreasonable to expect the records of individual farmers to be any better. The anxiety over when to sell can place extreme stress on a farm marriage: the husband may feel obliged to dictate the selling strategy, which offends the wife, who in turn blames the husband for market changes he couldn't possibly have predicted. The sheer uncertainty is oppressive. To return to the salary-lottery analogy, even if you picked the sixty thousand dollar day you would be upset about the wringer you'd been put through.

Endless waiting for Washington to make up its mind about farm programs produces more unhappiness. Ronald Reagan announced the latest emergency credit program for farmers in September 1984 but as of February 1985 with spring planting approaching, the FmHA had made decisions on only about 2 percent of the applications it had received. Farmers who had asked for aid had to sit and stew all fall and winter wondering whether they would be able to stay in business and powerless to find out. Many of those who ultimately got extra help were inclined to be resentful rather than grateful.

Dissatisfaction extends even to successful farmers. Because federal subsidies have the effect of keeping everybody at least barely in business, overproduction prevents successful farmers from realizing the profits they otherwise might. Since farm commodities are basically interchangeable, farmers cannot compete with one another by offering different features or higher quality as manufacturers can. They can compete only by undercutting already depressed market prices. Those who resist taking the debt plunge or who run their operations unusually well say they feel pressured to pay for the errors of the careless.

Suppose the government stepped into the computer industry, which is suffering from oversupply, to make sure no manufacturer went out of business. Successful companies like IBM and Apple would be unhappy, because the artificial stimulation of supply would prevent them from getting full value for their products. Unsuccessful companies would find themselves in the debilitating position of being dependent on Uncle Sam and ridden with anxiety over whether their handouts would continue. Everyone would be working, yet no one would be happy. There would be a "computer crisis."

In farm communities across the heartland there is one more level of anxiety. Farmers have never been able to look forward to wealth, but they have had a satisfaction that city workers with better pay cannot hope for—the moment when they hand the farm over to their children. For many farmers that moment is the culmination of a responsible life; for the younger generation it is the moment when the world recognizes that they have done what was expected of them. "Right now it's not a responsible act to bring a son into the business, and you don't know what that does to the farmer's mind," Rob Erb, Carolyn's husband, told me. Transferring a highly capitalized farm from parent to child entails leveraging to the hilt. Under these circumstances parents consider themselves failures, and children, unable to do what generations before them have done, feel they have compromised their entire family histories.

After I left Iowa, there was one farm couple that I couldn't get out of my mind: Dennis and Patricia Eddy, who live in Stuart and have three young sons. The Eddys are existing on six subsidized loans. They had qualified for the Reagan emergency credit aid, which would save them about twenty-three thousand dollars in interest payments overall. They hadn't paid income taxes in five years. The kitchen of their farmhouse had been remodeled and was strikingly attractive; I noticed a microwave oven and other minor luxuries. They had a purebred Doberman puppy. Here was a couple who could easily be portrayed as hooked on handouts. "Just once in my life I would like to live at the poverty line," Patricia complained, yet her federal aid is more than many city families with greater need receive—to say nothing of the fact that she lives in her own home. Most urban welfare recipients, I felt sure, would exchange places with the Eddys in the blink of an eye.

Except for the kitchen, however, the Eddy's home was modest. By no stretch of the imagination was the family living indulgently. There didn't seem to be any chance that they would buy a sports car, eat in three-star restaurants, or enjoy many other luxuries that young professionals who went to graduate school on government-subsidized loans consider their due. I felt sure that none of the conservative theoreticians who rail against subsidies in the fastness of paneled libraries would exchange places with the Eddys.

The Eddys said that they were dismayed by having had six different FmHA loan supervisors in eighteen months, each of whom had issued a new set of instructions on how the couple ought to run their lives; by unending anxiety about whether their crops would grow and sell; and by the public perception that those in debt to the FmHA are "bad farmers," which surely is not true.

For a while policymakers diagnosed increasing federal expenditures as a problem caused by "bad farmers" and "bad managers," who squandered their aid. The label has stuck — one farmer introduced himself to me as "just another bad manager" — although there is nearly universal agreement among agricultural observers that truly lazy or incompetent farmers are rare. If anything, farmers are too good at what they do. They produce too much, and they are baffled and anguished by the fact that bringing food out of the ground as they were taught was right does not invariably lead to success. The Eddys were working as hard as they knew how, and that was what stayed with me.

California: MBAs in Overalls

Iowa seems like an outpost on the moon by comparison with the San Joaquin Valley in central California. Nearly anything will grow in the valley's benign climate. The land is flat, which is perfect for farming, and cows graze in the nearby foothills of the Sierra Nevada as if they were in the Alps. In Iowa some farmers had plowed and planted their front yards, but here in Fresno and Tulare counties, one of the world's most productive agricultural regions, every square inch seemed to have been tilled. Farmhouses are modern ranch-style buildings with carports. Machinery doesn't rust. In the Midwest, farms are discrete places because farmers generally buy only land that adjoins their own; in California, however, farms tend to consist of scattered parcels. The scattering lowers the sentimental value of each farm, and it also works against any sense that farms are fortresses to be defended against an encroaching outside world. Farms in California seem much more like businesses. One sees few four-wheel-drive tractors with balloon tires. "Farmers like to buy tractors, managers like to make money," a Californian told me when I visited farms in the state last winter.

One of California's largest operations is Harris Farms, Inc., of Coalinga. The average U.S. farm has 437 acres; Harris has 17 thousand, and it's as diversified as can be—cotton, tomatoes, garlic, almonds, onions, grapes, potatoes, wheat, and cattle. It is not a corporate farm; John Harris, the only child of the founder, is the sole owner. During the 1970s, when the land frenzy drove prices to $3,500 an acre in the valley and as high as $15 thousand an acre in the prestigious vineyard counties, Harris didn't buy. "We could never figure out how to make buying land at those prices pay," he told me.

Harris has 125 full-time workers, whom he calls "employees," and one hundred units of housing for them of a quality better than that of a trailer park but not as good as that of a subdivision. Unskilled seasonal laborers—migrant workers—make $4 an hour, full-time tractor operators make about $14 thousand a year, and foremen make up to $65 thousand. Two foremen, Richard Lobmeyer and Juan Barrera, spend an increasing amount of time with computers: Harris Farms has a large mainframe computer for accounting and half a dozen personal computers for other tasks.

Barrera, for example, uses software designed for tomato growing to track temperature, humidity, and the rate of transpiration and to predict exactly how much irrigation the vines will need. Harris has been switching to less wasteful means of irrigation such as drip pipes, which apply small doses of water directly to a plant's roots. (Before the 1982 law was passed, many heartland farmers complained bitterly that California water was subsidized, ignoring the fact that their own water falls from the sky free of charge.) Much of Harris's production is sold to food processors by a standard supplier's contract—a promise to deliver a certain tonnage of, say, tomatoes on a given day at an agreed price. This reduces the potential for a windfall, but it also eliminates anxiety over fluttering prices.

In rural communities it seems to be considered bad taste to build a luxurious house, and Harris does not flout that convention. He lives in the modest house where he was raised, which sits directly across the parking lot from the company's headquarters, concealed behind hedges. His office, however, would make any lawyer proud; it is large and wood paneled with orginal art on the walls and a commodity-price monitor on the credenza.

Diversification as practiced at Harris Farms is said to be the future for profitable agriculture, because it allows the clever manager to stay a step ahead of demand. Midwestern farmers have some ability to shift among the products their environment will support: wheat, corn, soybeans, sorghum, and livestock. For example, when grain prices fall and livestock prices rise, farmers can feed their grain to the stock and sell it ultimately as meat. Trends in the supply of these commodities tend to be parallel, however, and so the degree to which midwestern farmers can diversify often does not amount to much. Being able to shift among unrelated crops—cotton to onions to lettuce—makes for better protection against the erratic market. Working with small specialty crops also enables growers to get an idea of where they stand relative to the competition. Lobmeyer explained, "There are 11 million acres of cotton fields scattered around the United States. Realistically, there's no way we can get a sense of what other cotton growers are doing. But there are just 294 thousand acres of process tomatoes in the country, and 84 percent of them are here in California."

Harris Farms operates almost entirely without federal subsidies. Most of its crops aren't eligible for support. Harris does not enroll his cotton and wheat acreage with the CCC, because, he said, the maximum loan is inconsequential for a seventeen thousand-acre enterprise and not worth the bother of going after.

Despite his computers, his economies of scale, and his lack of dependence on the government, Harris is pessimistic. He told me, "At this point farmers have become capable of producing just about everything. The regulated programs get all the attention because they involve federal deficits and family farms, but profit margins in specialty crops are becoming almost as bad." The market for grapes, once a glamour crop, is more depressed than the market for corn. Citrus fruits are about the only crop widely agreed to be profitable this year—profitable, that is, for growers who weren't hit by a series of frosts in Florida and Texas that cut supply. Harris observed that if the country had not experienced a cycle of poor weather for crops in three of the past five years, which led to disappointing yields, surpluses would be even greater than they are, and prices lower.

The weather of late has been unusually severe in southern Iowa, where the gullied bottomland is in any case ill-suited to farming. In 1983 the rain was so relentless that farms were flooded with mud. Pat and Nancy Meade, who live in southern Iowa, lost twenty of the hundred calves born on their farm that year because they drowned in the mud. A high percentage of the hard-luck farm stories presented as typical on television and in newsmagazines have southern Iowa datelines.

To some extent a bad year can be offset by the generous tax treatment accorded to agriculture. For example, many types of farm buildings and fruit trees can be depreciated, and deductions can be taken for supplies that will not actually be used until the following year. Played properly, the game allows the payment of taxes to be postponed almost indefinitely.

Investors who buy into farms do not, as popular lore would have it, "profit from losing money": the best investment is always one that makes money. The situation of a "tax farmer" is roughly like this. Say that a doctor in the 50 percent tax bracket has one-hundred dollars in marginal income. If he or she does nothing, the government takes fifty dollars and he or she keeps fifty dollars. If the doctor shelters the money in a farm generating a one-hundred dollar deduction, the tax bill is reduced by fifty dollars. In both cases the doctor walks away with fifty dollars. But in the second case he or she also has some equity in the farm.

Since the doctor keeps fifty dollars no matter what, all the investment has to do is make a dollar and the doctor will come out ahead. Thus a commercial farmer who needs a decent income to care for a family is at a disadvantage with respect to a tax farmer for whom any income is pure gravy.

In the long run, agriculture's favored tax status may do farm families more harm than good, because shelters encourage the overproduction that is at the core of agriculture's problems. From one pocket the government hands out tax breaks to encourage more farm investment and from the other it hands out subsidies to compensate for the depressed prices that investment causes. Nevertheless, farm-state legislators have extreme difficulty opposing any measure presented as a "farm tax break."

Surrounded by Surplus

Agricultural exports have stopped increasing over the past few years after having grown steadily through the 1970s. The strength of the dollar is usually cited as the reason, and without doubt the strong dollar is a factor. But the growth trend in exports probably would have stalled regardless, because many countries are generating food surpluses of their own. Moreover, almost all of the industrialized countries have higher internal subsidies and more-restrictive trade laws than we do.

In the late 1960s the countries in the European Economic Community met up to 90 percent of their demand for grain. In 1984, a bumper crop year, EEC farmers met 125 percent of that demand. U.S. agricultural sales to the USSR make headlines, but sales to Western Europe have been far more significant in volume. In 1983, for example, Europe bought about $10 billion worth of U.S. food—ten times as much as the Soviet Union did. In 1984 Soviet grain purchases hit a record high, and yet total sales to "centrally planned countries" accounted for just 10 percent of U.S. exports. Asian nations are our best customers. In 1983 they bought $13.5 billion worth of U.S. agricultural products. But since then sales to Asia have been stable or falling.

World wheat production, 447 million metric tons in 1978, is expected to be 505 million metric tons this year—a 13 percent increase. Also since 1978 world rice production has increased by 20 percent and world cotton production by some 35 percent. In every part of the world except Africa and Japan, food production is increasing faster than population.

The increased production, particularly in developing countries, can be attributed partly to the spread of modern technology and farming methods—progress that the United States has encouraged and in many cases has funded and supervised. We have a moral obligation to share our agricultural secrets with the world and should be proud that we have done so, but the cost has turned out to be greater than we expected. The world's increased production can also be attributed to U.S. farm-subsidy programs. For many commodities American support prices become the world floor prices, because whenever the going rate falls below what the CCC will pay, a significant portion of the supply is withdrawn from the market. Equilibrium between supply and demand is restored and the decline in the market price is halted. Nonrecourse loan programs, intended to limit the risk to American farmers, have the effect of limiting the risk to farmers in other countries as well. For example, Argentina knew in 1981, courtesy of an act of Congress, approximately what the floor price for grain would be, and thus could plan accordingly.

The agricultural programs of other countries are far more protectionist than those of the United States. Within the EEC agricultural commodities move freely, but what are in effect tariffs prevent food from being brought into the community at less than the regulated price. Thus American farmers are not permitted to undercut the market price in Europe, whereas any shipper may sell cheaply here. When the EEC has surplus grain, exporters receive direct subsidies to sell their overstock at below-market prices. The European subsidies are threatening to break the bank; several times in recent years the EEC Commission has recommended that the subsidies be cut only to have farm interests in member nations beat the effort back. In 1986 Spain and Portugal—two potentially major agricultural producers—were expected to join the community, hide behind its tariff curtain, and use its subsidies to boost production.

Japan's regulations are even more stringent. Various restrictions limit the sale in Japan of U.S. commodities for which there might be higher demand—particularly beef and fruits. According to Yujiro Hayami and Masayoshi Honma, economists at Tokyo Metropolitan University, Switzerland has the world's most protectionist agricultural system with domestic prices half again as high as the world average. Japan is next with domestic prices 45 percent higher than the world average. The EEC's domestic prices are 25 percent higher, and the United States, where domestic prices are slightly below the world average, ranks last in protectionism.

It is perhaps understandable that Europe and Japan, which have known famines in this century, should be anxious to keep farm production up. It's harder to understand the Reagan administration's behavior. In March, President Reagan withdrew the pressure on the Japanese to observe voluntary auto export quotas and asked nothing in return. If the administration continues to push for cutbacks in U.S. agricultural production without insisting that members of the EEC cut back as well, subsidized European farmers will rush in to grab U.S. export markets.

We should be grateful that Japan has so little land, else it would surely make life as hard for American farmers as it has for American manufacturers. Another Asian country, however, has lots of land and a strong agricultural tradition: China. In 1980 China bought $2.3 billion worth of U.S. food and fiber. Now its purchases are down to $500 million and the country has begun to export cotton. Chinese officials have been stepping back from the commune system and introducing a limited form of capitalism in their fields; the results have been dramatic. Since 1978 China's cotton production has risen by 150 percent, its soybean production by 84 percent, its wheat production by 58 percent, and its rice production by 27 percent. Many production techniques used in China are backward (for example, cotton is shipped in nonstandard-sized bales, small enough for two men to lift manually), but Western methods and machines are being introduced. In China workers are accustomed to strict discipline and low wages; modern production could transform the country into the world's breadbasket. "They are much closer to it already than we like to think about," Kenneth Billings, the president of Fresno County's Federal Land Bank, says.

U.S. farm productivity by acre, which has declined slightly from its peak in 1982 can be expected to rise again soon, as new hybrids, growth hormones, and chemicals come onto the market. A little further down the road gene splicing might make possible entirely new strains of plants that could spark another Green Revolution. The problems of oversupply in the future could make those today seem like warm-ups.

The Annual Crisis

Senate Majority Leader Robert Dole said recently that in every one of his twenty-five years in Congress there has been a farm problem. During the late 1940s President Harry S Truman tried to cut back many farm subsidies and was rebuffed by a livid Democratic Congress. The Senate majority leader at that time, Scott Lucas of Illinois, sounded very much like Dole when he complained of senators who "are constantly talking about economy in government . . . who have no hesitancy in getting off the economy bandwagon . . . to take care of their own communities . . . regardless of what the cost may be in the future." The Eisenhower administration pushed through some farm support cuts in 1953, but when it tried to go further in 1958, it was defeated. That year Senator Allen Ellender of Louisiana argued, "This is not the time to lower prices farmers receive, particularly when farm income is already at an all-time low." New Deal descendant Lyndon Johnson moved to cut farm subsidies by $540 million in the budget he submitted in 1969, just before he left office. Nixon angered farmers with his attempts to change farm programs. Carter, a farmer, not only asked that subsidies be cut but stood fast during the 1979 tractorcade, refusing to make significant concessions.

The job of secretary of agriculture has chewed up most of those who have held it. Truman's "Brannan plan," named for Agriculture Secretary Charles Brannan, became a euphemism for political lunacy, the racetrack MX of its day. Radical farmers like some members of the American Agricultural Movement point to Eisenhower's Ezra Taft Benson as the first mastermind of the antifarm conspiracy. LBJ's Orville Freeman, whose affiliation with the Minnesota Democratic Farmer Labor party was beyond reproach, got into trouble by suggesting that cotton growers cut their prices in order to compete in the world market. Carter's secretary, Bob Bergland, who himself had been a farmer, ended his term amid acrimony. Reagan's first secretary, John Block, formerly an Illinois hog farmer, was an object of scorn for having gone along with Reagan in the effort to cut farm subsidies. Block was hardly helped by the fact that his former business partner, John Curry, was a plunger who bought up land with FmHA assistance and then defaulted. Every farmer I met in the Midwest knew about this case; some spoke of Curry as if he were a poisonous snake.

Partly because of the continuous anxiety to which they are subject, farmers have a long history of crying wolf. The agriculture committees, staffed by farm-bloc congressmen, do too. In 1965 a report of the House Agriculture Committee warned, "Hundreds of thousands of our most progressive farmers will find their debt positions intolerable and will be forced into bankruptcy"—language almost identical to that used last winter.

Checking old newspapers, I found that farmers have been proclaiming the "worst year since the depression" regularly since the early 1950s. Reporting that placed the claim in historical perspective was rare, except for the occasional "down with farmers" piece that exaggerated in the other direction. Covering agriculture is a delicate matter for journalists and especially for television crews. Someone who loses a farm loses job, home, and way of life all on the same day. It would seem heartless of a reporter to mention that dispossessions are the exception or that thousands of other people experience tragedies that are of equal weight but that simply don't fit a news peg.

"They're taking away my land" is a plaintive cry of farmers, and when true, it's terrible to hear. Often it isn't true. A farmer who defaults on a loan is not losing the land so much as some part of the farmer's investment in it. The bank is the true owner of the farm, just as ultimately it is the owner of a mortgaged house.

Through the 1970s the farm bloc complained that farmers were being destroyed by inflation. Senator Jesse Helms of North Carolina, who is the chairman of the Senate Agriculture Committee, said in 1981, "Farmers understand that unless this inflation is cured, they don't stand a chance." Now that the cure has taken effect, the *lack* of inflation is said to be a special hardship for farmers. Similarly, cries of "foul" were heard in response to the escalation in farmland prices while it was taking place, even though it was making holdings more valuable. Four years ago Catherine Lerza, the head of the National Family Farm Coalition, declared that high land prices had created a crisis for farmers. At about the same time Representative Berkley Bedell of Iowa proposed a $250 million Beginning Farmers Assistance Act to subsidize purchases of farmland. Now, of course, it is the decline in land prices that is said to pose a crisis for family farmers.

Using the Department of Agriculture's parity tables, some farm advocates claim that farmers are worse off today than ever. Parity tables are supposed to show the buying power of farmers; the period of 1910 to 1914, when farm prices were strong, is taken as the base period and given a value of one hundred. In 1984 parity hovered around fifty-eight—that is, farmers had 58 percent of the buying power of their forerunners decades earlier. Figures like these are extremely misleading, however, because calculations of parity treat the present as if it were 1914. For example, they take into account the fact that tractors cost more but not that they do more. Parity makes an effective round of political ammunition, but it is not a reliable indicator on which to base policy. One farm-state congressional staff member told me, "Any farmer who seriously believes he would be living twice as well if it were 1914 again is crazy."

At a House Agriculture Subcommittee hearing in February 1985, I gave up trying to count the number of times words like *desperate, disaster, unprecedented,* and *dying* were used. Representative Steve Gunderson of Wisconsin asserted, "Clearly, within the credit crisis confronting agriculture we are literally facing the fundamental destruction of rural society." Representative E. Thomas Coleman of Missouri declared that "farmers are faced with such high interest rates, declining land values, and low return for their products that they are reliving the Great Depression of the 1930s." At this hearing and throughout the battle over emergency aid, farm-bloc congressmen repeatedly predicted that as many as 10 percent of the nation's farmers could go bankrupt by March 1 — the traditional deadline of banks for issuing spring planting loans — if substantially more than $650 million was not provided. A bonus-credit-aid bill was passed but Reagan vetoed it. The first of March came and went, the wave of foreclosures that was supposed to sweep the country did not occur, and the story vanished.

Resisting pressure for spending is a recurring challenge in Washington, and the ability of even conservative Republicans from farm country to forget their speeches about deficits and call for more agricultural subsidies has been remarkable for years. Farmers make powerful claims on the legislature in part because they make up a high percentage of the population in rural communities. In heartland congressional districts almost everyone is tied to the farm economy. No other constituent group — not auto workers in Detroit or gas producers in Louisiana — places such a statistical lock on its representatives.

Another running theme in Washington is the demand for a crackdown on rampant waste in somebody else's program. Farm groups are among those who have most vociferously protested the federal debt, which contributes to high interest rates and the strength of the dollar. But they also want more spending for farms, which would drive the deficit up. Last winter five of the country's largest agricultural associations formed a lobby called the Balanced Budget Brigade. While some spokesmen were making the rounds on Capitol Hill demanding that the deficit be slashed, others were demanding extra credit aid. During his appearance on "Nightline," Governor Janklow sneered at Congress's failure to balance the budget — a feat that even New Right thinkers admit will be out of the question for many years — and boasted that states balance their budgets. Janklow didn't add that the main reason why state budgets are balanced is that federal grants supply 18.5 percent, on average, of state revenue. If a cosmic philanthropist made an 18.5 percent contribution to the federal budget, then the federal budget would also balance.

Finding solutions for a chronic problem like agricultural overproduction can vex democratic systems, which almost inevitably focus their attention on the demands of the moment. "Every time we do a farm bill, the short-term outlook — what happens one year down the road — overwhelms all other considerations," William Hoagland, the deputy staff director of the Senate Budget Committee, who was raised on an Indiana farm, told me recently. "In 1981 all we talked about was inflation, which turned out to be a moot point almost immediately after we finished debating it. This year it's debt. What we want from agriculture, what our long-term goals ought to be, stands no chance [of consideration], compared to whatever was in that morning's paper."

Eventually Congress will have to face the fact that there are too many farmers. [See Easterbrook's response to critics for a comment on this sentence.] The farm bill that Reagan has proposed, which in effect would abandon those parts of the federal program that subsidize the least successful farmers, may not be perfect, but so far it is the only one to concentrate on the problem of overproduction. The solutions to that problem do not lie solely in the realm of economic abstraction. They will involve a painful human cost. If the Reagan plan or something like it is enacted, Dennis and Patricia Eddy, for example, may lose their farm. That would not be a happy day for them or for any caring citizen. But the Eddys are young, responsible, bright, and eager to work. If they can't land on their feet, who can?

Early this year Representative E. Kika de la Garza of Texas, the chairman of the House Agriculture Committee, said he might support changes in federal agriculture programs, but only if they could be achieved "without sacrificing one single farmer." This is like saying let's cut back that bloated defense budget — as long as no contractors lose work. There can still be family farms. It's just that not every person who wants a farm can have one, no matter how fervently we might wish he or she could.

Responses to "Making Sense of Agriculture"

IF EASTERBROOK IS RIGHT, we need go no further; we already have a good answer to our question. The family farms currently in trouble represent agriculture's "excess production capacity." Culling them will be painful, but it must be done. Hence there is no reason—much less a "moral obligation"—to save them. The problem is overproduction, not loss of family farms. Indeed there is too much farming for Easterbrook.

Easterbrook has been praised for writing what one *Atlantic* reader, Joseph Elvove of Charleston, South Carolina, called "the most lucid, well-balanced account of American farm programs [written] since their inception." But when Easterbrook's article appeared in *The Atlantic,* not everyone who read it agreed. Many took offense at Easterbrook's conclusion, which he identifies here as a misprint, "that there are too many farmers." Some felt he failed to appreciate the magnitude of the problems in the Corn Belt. Others believed his assessment of the Reagan plan to move agriculture toward the free market was overly optimistic.

In the October 1985 issue of *The Atlantic,* letters to the editor appeared. Four of them—by Dennis Harbaugh, Tim Carter, Jeffrey Ostler, and Craig Cramer—are reprinted here. Also included is a commentary by Marty Strange, which criticizes Easterbrook's praise of "MBAs in overalls" and suggests that we should not encourage the formation of large agribusinesses. Claiming "bigger is not better," he says that family farms are more efficient, flexible, and better stewards of the land than corporate farms. He plays the opening notes of a tune we will hear throughout this volume: family farms are worth saving because they have emotional value to us, are the backbone of our democratic system, make better use of our natural resources, guard our land from erosion and chemical abuse, and distribute landownership in a fair way.

Easterbrook's response to his critics follows Strange's piece.

DENNIS HARBAUGH, TIM A. CARTER,
JEFFREY OSTLER, CRAIG CRAMER,
AND MARTY STRANGE

What Easterbrook Fails
to Make Sense Of

DENNIS HARBAUGH. Gregg Easterbrook says after visiting twenty-four Iowa farms that "the people on these farms did not live in poverty." In fact U.S. Department of Agriculture statistics show that the average Iowa farmer recorded a net income *loss* of $1,891 during 1983. The average net farm income for all fifty states dropped to $6,793 during the same year, which represents an income $3,407 below the federally established poverty level for a family of four. There *is* poverty in the heartland.

And although Easterbrook tries to dispel common misconceptions surrounding American agriculture, he repeats one of the most common myths of all: that there are too many farmers. Easterbrook correctly lists overproduction as one of several problems troubling agriculture, but the assumption that the problem of overproduction can somehow be corrected by reducing the number of farmers is faulty and misleading. Indeed, the past thirty years has demonstrated an inverse relationship between the number of farmers and agricultural production.

Easterbrook claims that Reagan's proposed farm bill "is the only one to concentrate on the problem of overproduction." The opposite is actually true. The problem of overproduction will be solved by reducing the number of acres tilled, not by reducing the number of farmers. Agricultural economists project that production levels of major commodities would increase under Reagan's plan as farmers strove to produce more in order to compensate for the lower prices that Under-Secretary of International Affairs and Commodity Programs Daniel Amstutz recently warned farmers to expect.

DENNIS HARBAUGH lives in Des Moines, Iowa. TIM A. CARTER is a farmer in Bethel, Maine. JEFFREY OSTLER is a graduate student in the Department of History at the University of Iowa. CRAIG CRAMER is an editor of an agricultural journal. MARTY STRANGE is codirector of the Center for Rural Affairs in Walthill, Nebraska and serves on the Board of Directors of Rural Coalition, Inc., and the National Catholic Rural Life Conference.

For readers sincerely interested in the problem of overproduction and in the future of American agriculture, I suggest for consideration the Save the Family Farm Act sponsored by Iowa Senator Tom Harkin and a host of rural and urban legislators. This bill offers farmers a referendum on a mandatory supply-management program that, if implemented, would reduce production of most commodities currently covered by federal support programs including corn, wheat, barley, cotton, oats, rice, and soybeans.

The primary criticism of this proposal is that commodity prices will rise and consequently U.S. exports will decrease. However, reduction in production is projected to keep commodity prices – and farm income – high enough to withstand a decrease in exports. In any event, exports are only one factor in the farm-income equation, and high levels of exports do not guarantee correspondingly high levels of agricultural income. That's just another of those misconceptions.

TIM A. CARTER. Gregg Easterbrook suggests that farming is not a bad investment and that farmers make good money. His figures don't tell the real truth, because the farmers now receiving a high income have a large investment with a low debt load and the return on this investment in agricultural land is part of the reported income. This return on agricultural property has been caused by inflation's increasing land values, a return that is really of little benefit if you do not wish to sell your property. Easterbrook points out that farmers are emotionally attached to their property but dismisses this as not significant. I wonder if his attitude would be the same if he lived in a house his great-great-great-grandfather built. He points out that farmers "cried wolf" this spring and that even though President Reagan vetoed the emergency farm bill very few farmers folded. The president may have vetoed the bill, but then to prevent disaster, money beyond what was promised from other government sources was pumped into agriculture.

Easterbrook suggests that President Reagan's plan to put agriculture completely on its own is probably the best course of action. This would be like kicking a man who is down. For one thing, President Reagan's plan will pit an American farm in an open market against subsidized foreign farms. A shock wave would flow through agriculture and rural America. The weaker farms would go first; healthier farms would follow. In the end most of the agricultural resources would wind up belonging to big corporations. The consumer would have to pay more for food. Rural communities dependent on local farms for their existence would become ghost towns. The corporations would raise the food from which they could profit the most. We could even become a nation dependent on foreign food sources.

There are alternatives. The dairy industry just finished a very successful experiment with a diversion program that reduced milk surpluses and saved taxpayers billions of dollars. The diversion program assessed (taxed) all dairy farmers and then used the revenues to pay those farmers who could and would reduce their proven capacity to produce. It was a supply-control program in which each farmer shared the cost. The National Milk Producers Federation, which represents about 80 percent of milk producers in the country, wants the diversion program reinstated. I think this approach deserves support and could also be implemented in other commodities with a surplus problem.

JEFFREY OSTLER. Gregg Easterbrook concludes his "revisionist view of farm policy" with the observation that "Congress will have to face the fact that there are too many farmers." He endorses the Reagan administration's free-market approach of eliminating farm subsidies for the least successful farmers while recognizing "the painful human cost" that would result. There is a major problem, however, in Easterbrook's assumption that forcing marginal farmers out of business will solve the problem of overproduction. After foreclosure an Iowa family farm continues to grow corn, but it does so for the profit of new owners — usually corporations or wealthy investors. Reduction of total acreage in production, not in the number of farmers, is the key to reducing overproduction.

The Save the Family Farm Act would reduce total acreage in production by establishing a conservation reserve for highly erodible farmland. This program, of course, is not a panacea, but it addresses the problem of overproduction more correctly than free-market approaches.

CRAIG CRAMER. The trouble with agriculture is that we are trying to impose inadequate economic-decision-making criteria on a biological system.

Much of the California paradise Easterbrook describes is actually a desert, made to bloom only by enormous infusions of federal subsidies to supply irrigation water. The vegetable farms there are diversified only in the economic sense. Biologically they are unstable monocultures propped up by costly inputs of agricultural chemicals. The harsh Iowa landscape is actually some of the most productive and fertile cropland in the world, even though economics are driving us to send two bushels of Iowa topsoil down the Mississippi for every bushel of Iowa corn we harvest.

The loss of soil is analogous to our crumbling industrial infrastructure. Soil is our capital, our resource base, which supports current income and ensures future production. But unlike a factory, soil, given proper care, can be not only maintained but improved in fertility and in ability to produce food. Soil erosion causes billions of dollars of damage annually by clogging waterways and decreasing the quality of drinking water. But, according to an Iowa State University study, erosion costs the farmer only $1.60 to $2.00 per acre—less than the price of a bushel of corn even in today's depressed market. Although it is clearly in our national interest to reduce erosion in both the short and the long run, we don't have an economic system that takes these externalized costs into account.

MARTY STRANGE. Gregg Easterbrook's article brings to mind Mark Twain's essay, "How I Edited an Agricultural Paper." In the regular editor's absence, Twain serves as stand-in and deluges his readers with nonsense about harvesting turnips from trees, "setting out" cornstalks in July, and ganders that spawn. In his defense, Twain says he never heard of the notion that you had to know something about agriculture to write about it. Somehow, Easterbrook manages to cram more open nonsense into one serious article than even Twain could muster in humor.

Easterbrook fails to make sense of farm policy, using data carelessly and sometimes misleadingly. Asserting, for example, that the farm exodus has been over for years, he points out that the number of farms declined by only thirty-one thousand in 1983, while in earlier years the number was much higher. He does not note that this attrition is from a progressively smaller base each year and that the *rate* of decline in 1985 is as high as any time since the depression. Nor does he comment on the fact that most of those leaving agriculture before 1980 *chose* to do so; the recent *exodus* has been forced; farmers have had to leave against their will. And his comment that the high rate of decline in the 1950s and 1960s occurred during "a period enshrined in political mythology as better for farmers," is pure hyperbole. The fifties and sixties were decades of discontent almost equal to the eighties, as even a casual observer of agriculture would know. One of my earliest memories of social protest is televised scenes of farmers dumping milk in the early 1960s.

Easterbrook says agribusiness does not dominate farming, using simplistic statistics about corporate farming. But he ignores the larger issue of concentration in food production. Fewer than 5 percent of the farms produce half the food in America.

Furthermore, Easterbrook states correctly that most eligible farmers do not participate in crop programs, but fails to advise his readers that the programs he is referring to are designed to raise market prices through price-support loans and production cutbacks. Those who choose not to participate are shirking the responsibility of joining with the participants in voluntary production cutbacks. But they benefit from the higher prices that result from others having done so. He is right in pointing out that these tend to be larger farmers—precisely the kind of farmers he fawningly describes as "MBAs in overalls." They are generally referred to as "free-riders" because they benefit from the sacrifices of others, and leave naive observers with the mistaken impression that there is something virtuous about not participating in federal farm programs.

Again, Easterbrook argues that farming is not a "disastrous investment" because farmers earned an average annual rate of return of 2.4 percent on farm assets in 1983–84, comparable to the annual net return on all corporate assets of "only" 5.5 percent for the same period. But 5.5 percent is 129 percent larger than 2.4 percent, and if farmers had had such a rate of return, net farm income for those years would have been $110 billion instead of $48 billion, and there would be no talk of a farm crisis today. Farmers are not just slightly behind corporations in profit making; they are way behind. In this case, Easterbrook's use of the modifier "only" can only be viewed as deceptive. Using facts in this way, Easterbrook might convince the uninformed that there is no difference between rats and rabbits, since both are only rodents.

But more disappointing than such abuses of data is that Easterbrook seems to have missed the significance of much of what he does reveal about the paradoxes of agriculture. After citing a litany of examples of overexpansion, tax-motivated investment by nonfarmers, and aggressive lending by bankers during the boom 1970s, he can reach no deeper conclusion than that there are too many farmers. He seems entirely oblivious to the fact that he has built the case for concluding quite the opposite: there are not too many farmers, but there is too much money invested in agriculture.

This is particularly unfortunate because properly treated, the information Easterbrook provides might have produced an informed contribution to farm-policy debate. If the "excess resource" contributing to overproduction is capital, not people, then *whose capital* will be squeezed out during the crisis? Will it be farmers' capital, investors', or bankers'? Had this question been posed, it might have led to the revelation that most of the emergency credit proposals debated in Congress this year were designed to save bankers, not farmers.

Contrary to Easterbrook's conclusion, there are not too many farmers in America. Most food is produced on farms already large enough to capture nearly all economies of size, according to a host of studies on the subject. We have simply reached the point of diminishing return on the substitution of capital for labor. The issue before the nation is whether those who farm will be able to own the land and the machinery with which they farm.

That issue is complicated by evidence from the current crisis that the larger the farm, the more the likelihood of financial stress. According to USDA data, the probability of being highly leveraged with debt increases as farm size increases. In fact, over one-third of the "troubled" debt is held by 2 percent of all farms. These are large farmers with over $250 thousand in sales who borrowed aggressively to expand, bidding too much for land, spending too much on machinery, overbuilding production capacity, and soaking up available credit. This small group of pacesetters put all farmers on the same treadmill of higher costs (especially for land) and lower prices for crops. And when they fail, these pacesetters take a lot of people with them. Both politicians and bankers are therefore afraid to let them fail. The challenge of farm policy now is to find ways to help the majority who are victims of this minority, without validating the expansionist behavior of the few.

This is no small chore, and Easterbrook does not make the burden any lighter. Instead, he paints all farmers in trouble as greedy expansionists while praising the "MBAs in overalls" whom they emulate.

Easterbrook's admiration for these big farms based on the notion they don't receive any federal subsidies is uninformed. He says, for example, that big irrigated farms pay "full cost" for federal irrigation project water. The fact is that for the first 960 acres, they pay none of the interest cost of constructing the project, and no more than what they can afford, in any case. Above 960 acres, they only pay "full cost" for water used on land leased from others. Out of ignorance or dishonesty, Easterbrook ignores these facts and admonishes the rest of us not to complain about such subsidies because we get our rainfall "free." I've never read a statement that does as much self-inflicted intellectual damage.

Easterbrook also misses the fact that these big farms that do not participate in the federal programs for eligible crops are "free-riders" on the rest. And in his enthusiasm for big fruit and vegetable farms, Easterbrook misses entirely the price-manipulating effects of state and federal marketing orders for most of the very crops he claims grow "without federal subsidies."

In seeking to demythologize the family farm, he feeds an even larger myth that bigger is better. That is what got us where we are in farming today.

A Response to My Critics

FIRST I WOULD LIKE to point out that the version of my article appearing in *The Atlantic* and reprinted here contains a typographical error that I wish to correct. I did not intend to say that "there are too many farmers." I wrote instead that "there is too much farming." It's total production, not numbers of farmers, that matters.

Harbaugh speaks of agricultural "net income," meaning profit. Calculating profits for the self-employed — especially farmers who reside at their places of business — is an inexact science, since living expenses blur into production costs. For example, farmers can deduct most fuel bills and can depreciate most pickup trucks and vans, though no one could seriously claim that such vehicles are never used for personal trips. If an urban worker subtracted living expenses from his or her salary, he or she wouldn't show much of a "profit" either.

Without question there are farmers who are poor, but my point was that the *typical* farmer does not live in poverty. If during 1983 each Iowa farmer had really "lost" $1,891 — in the commonsense meaning of receiving nothing while paying $1,891 out of pocket — there wouldn't be a single person left in Hawkeye agriculture by now. The modest 1983 average profit contrasts with another Department of Agriculture figure, "total income per farm for family personal spending and investment." In 1983 that figure was $24,092.

Harbaugh notes that strict federal acreage controls (production limitations that exist today are easily circumvented) might combat overproduction just as well as reduced price supports — a point the article acknowledged. He is pulling a fast one, though, by implying that such regulation offers a simple solution. Enacting meaningful limits would entail at least as much political furor as enacting price cuts, and enforcing them by routine physical inspection of 2.4 million American farms (the only way to give strict limits teeth) would be a bureaucratic nightmare, assuming it could be done at all. This doesn't make production controls a bad idea. But I feel that on balance reduced price supports are a better idea, especially since they promise lower rather than higher federal spending.

Ostler makes the related point that a cut in the number of farmers does not necessarily mean a drop in overproduction. Strictly speaking, he's got me. But does the point stand scrutiny? If federal price supports were reduced, causing some farmers to quit and sell their land to other farmers who didn't quit — yes, those other farmers might keep the land in production. But now, because of the reduced federal price supports, they are doing so at their own risk. If they produce too much, they must bear the cost, just as suppliers in other businesses must bear the cost of any overproduction. It can't be that farm prices are too low for the little guy to get by and high enough for the big guy to make a killing on the same land at the same time. Provided that tax-shelter laws are successful at preventing the wealthy investors Ostler cautions against indirectly billing the public, overproduction under the circumstances described above should be less likely. And if it does continue, consumers will reap low food prices at the wealthy investors' expense.

Strange's claim about the rate of farm decline is not supported by fact. The 1983 decline of 31 thousand farms with 2.3 million remaining represents a loss of 1.3 percent. During 1960, 138 thousand farms shut down leaving 3.8 million at year's end, a decline of 3.6 percent. In 1950, 220 thousand farms went out of business and 5.4 million were left, a decline of 4 percent. Figures for most of the 1950s and 1960s are similar. As for his other claim about the 1950s and 1960s, my article makes very clear that I was not discussing what actually happened, but the *mythology* of the period.

Hard after accusing me of "using simplistic statistics" about corporate presence in farming, Strange declares "Fewer than 5 percent of the farms produce half the food in America." That's a nice simple statistic, but what does it prove? I can't think of another major industry where half the production is still in the hands of the little guy; Strange seems to be making my point, not his.

His contention that farmers who stay out of support programs are secretly taking advantage of them does not persuade me. Driving up prices may be the official purposes of federal programs, but that purpose ceased to be more than a polite fiction many years ago. Where are these "higher prices that result [from supports]" to which Strange alludes? And if one can make a bundle by dropping out of support programs, why don't more farmers do so? No doubt many farmers who seal their crops resent those who don't. Let's bear in mind that many farmers who don't seal likewise begrudge those who do — feeling that if federal spending did not keep acreage in production artificially, prices would increase through market forces. The notion that those refusing subsidies enjoy a "free ride" while those who take the money bear a weighty burden is hard to swallow even by the standards of agricultural paradox.

My point about return on investment was not that 2.4 percent is the same as 5.5 percent — one surely is twice as good as the other — only that corporations do not make the huge returns some people imagine. The gap between farm and corporate returns is real, but not enormous.

Mr. Strange's final and central point about capital leaves me scratching my head. If too much capital is the problem, aren't farmers themselves to blame? Nobody made them buy combines, big tractors, or extra land. And if excessive investment is the ultimate problem, there are two things we can do about it. We can substitute labor for capital — in other words, farmers could work for less and accept a lower standard of living. (Somehow I suspect Strange would object to this approach.) Or, we can shift the capital to other economic sectors — in other words, have less farming.

SUGGESTED READINGS

Foreign Affairs and National Defense Division, Congressional Research Service, Library of Congress. *Feeding the World's Population: Developments in the Decade Following the World Food Conference of 1974.* Report prepared for the Committee on Foreign Affairs, U.S. House of Representatives. Washington, D.C.: Government Printing Office, 1984.

Marion, Bruce W. *The Organization and Performance of the U.S. Food System.* NC 117 Committee. Lexington, Mass.: Lexington Books, 1986.

Solkoff, Joel. *The Politics of Food.* San Francisco: Sierra Club, 1985.

World Food Institute. *World Food Trade and U.S. Agriculture, 1960–1984.* Ames: Iowa State University, October, 1985.

Farmers' Stories

DO WE HAVE too many farmers? Should we be encouraging people to get out of, not into, this line of work?

These are important questions. In order to answer them we must learn as much as we can about the history and economy of U.S. farming and about the values and principles by which Americans live. We will pursue these questions in Part Three. But there is a prior concern, an even more pressing need: to hear the stories of those farmers currently going through these severe times. In the course of reading statistics, percentages, and price differentials, it is easy for us to lose touch with the lives represented by these numbers. The graphs and percentages are important; they give us objective knowledge about the extent of the problem. But what the data do not carry with them is the constant reminder that these figures signify the losses and tragedies of real women, men, and children. Before getting too mired in the debates of social scientists, we need first to recognize the subjective dimensions of the problem. For behind every statistic is a human being, someone like us forced to cope with the prospect of unemployment, the closing of a main-street barbershop, the sale of a hay baler, the auction of a piece of land first owned by one's great grandfather. Before getting caught up in the abstract figures of economists and sociologists, we need to hear the voices of farmers themselves, for listening to them tells us how it feels to live with their fears, losses, hopes, and dreams.

For the most part, family farmers are not authors. The men and women whose writings we find in these chapters are not intellectuals. They are workers, and we should not expect to find tightly argued academic treatises here. But this is not to say that these people work "only with their hands." Contrary to some common myths about farmers, the fact is that farmers on the whole are well educated, thoughtful, and informed. What follows is evidence; these stories of success and failure, grief and happiness are eloquent witness to the human stuff of which the farm crisis is made.

In the past decade, many farmers have been forced out of agriculture. They have had to find work in other places. But some farmers have not made it that far. Some, apparently depressed by their inability to master cruel circumstances, have committed suicide. Others have found ways to cope and, for now, to hang on. Still others, some 30 or 40 percent of all farmers, have not only survived but have flourished, even in the past few years.

It is important for us, having been introduced to the farm economy, now to learn how it feels to live in rural areas. The next five articles give us a snapshot of life in Iowa farm country in the mid-1980s. The first to speak is Paul Hendrickson. Hendrickson is not a farmer, and he does not speak for himself. In "Those Who Are No Longer with Us" he speaks on behalf of farmers and specifically those farmers who can no longer speak for themselves. These are the ones who have lost not only their farms but their lives. These are the ones who have killed themselves and sometimes others: the most visible victims of the crisis.

Carol Hodne is a farmer, or as she clarifies the point, the daughter of a farmer. She may never get a chance to farm. Her own future on the land, she fears, "has been stolen by deliberate government policies" designed to uproot today's farmers. Hodne's bitterness is representative of a sentiment common in rural areas today. But it is neither thoroughgoing nor unmatched by hope. She suggests several things farmers might do to vent their emotions, take control of their lives, and insure the future of family farming.

Denise O'Brien and her husband have taken over his parent's farm. They are, as she admits, not doing well financially; they use "food stamps, heating assistance, and any other program" for which they are eligible. They continue to "hang in there" because of their devotion to family farming.

Erwin Johnson is another farmer from Iowa. He represents that silent minority of farmers who, through their present hardships, continue to turn a profit. Johnson writes in a personal and open tone about his situation, something he admits is difficult for successful farmers to do these days. He is sensitive to the problems of those around him, but he speaks optimistically about the future. He questions the ability of the corporately managed farm to compete with family farmers and believes that the future lies with family farms, farms that will have to be larger than they have been in the past.

Finally, C. L. William Haw, the only author in this chapter not from Iowa, speaks as a corporate farmer. When he wrote his paper, Haw was the president of National Alfalfa Dehydrating and Milling Co., a business that owned fifty thousand acres of irrigated land. Haw argues that firms such as his have better access to capital, marketing techniques, and new technologies than do family farmers. Agribusinesses can also provide a steadier income to their farmer-employees. Moreover, corporate farmers are not the villains they are often made out to be; he cites statistics that (as of 1978, when his essay first appeared) no more than 2 percent of agricultural land is corporately owned and of that small figure fully "92 percent are family farm corporations."

It is interesting to note that the problem of interpreting statistics, a problem that appears repeatedly in the following pages and contributes to the fog surrounding many farm issues, rears up here. Richard Rodefeld of Michigan State University has criticized the USDA report on which Haw's figures are based, saying that it vastly understates "the size and impact of corporate farming" (Blobaum 1975, 83–84).

From bankrupt suicide victims to corporate managers making a profit, the voices of American farmers are diverse. It appears that there is no single story we can tell that will adequately represent every "family farmer." But this fact, of course, makes it all that much more important to listen to the different autobiographies here recorded.

Reference

Blobaum, Roger. "Wresting Agriculture from Corporate Farmers." In *Food for People Not for Profit,* pp. 83–84, edited by Catherine Lerza and Michael Jacobson. New York: Ballantine, 1975.

PAUL HENDRICKSON

Those Who Are No Longer with Us

DATELINE: Ames, Iowa. . . . It's May of 1985. Beauty is a cruel mask when the earth rolls right up to the edge of the interstate freshly turned. When the redbud trees are bleeding into pinks and magentas. When the evening rain is soft as lanolin.

And yet . . . five students from Iowa State University killed themselves during the 1985 academic year. Why? Nobody really knows. It's almost as if acute stress were an infectious disease in Iowa, like pinkeye in cows. A blooded humming presence in the sweet-smelling air. People can just reach up now and touch it.

In March of 1985 in a place called Strawberry Point (population 1,463), men with mud on their boots sat in St. Mary's Catholic Church and wrote names on small pieces of paper. Maybe it was the name of the person in the Federal Land Bank who killed their loan. Maybe it was the auctioneer who sold off the family possessions as if they were bingo cards. Maybe it was the smart-ass from John Deere who had said, sorry, this time he'd just have to have cash. How are you supposed to get your corn in when they won't give you credit?

One by one, these proud, humiliated men got up from their pews and walked to the altar and deposited their slips of paper into a coffee can wrapped in tinfoil. Then they set it on fire. What they were trying to do was burn away their bitterness and anger before something worse happened. Yes, it was symbolism, but it was also an expression of community grief. The priest who ran it said it was an effort to find a spiritual dimension to so much suffering and loss.

A month later, a man near the town of Osage told his wife he'd be back by supper. He had recently sold out, and the sale didn't go well. He

PAUL HENDRICKSON is a staff writer with the *Washington Post*. This essay appeared originally as "The Fields of Fear" in the *Washington Post,* 30 May 1985, and is reprinted with permission of the publisher and author.

and his wife were renters on the land, and the land had turned sour as gall. All five of his children were dead. (Four of them were killed in the same car crash years ago.) Maybe it was the lousy sale; maybe it was the lousy world. A priest said he just walked out into an open field and shot himself. He was in his sixties. There was no note.

A farmer up near Mason City was digging a coffin-sized hole behind his house a while back. His wife rushed up.

"Oh, my God," she said, putting her hands up to her mouth. "What are you doing?"

"It's not for me," he said and kept on digging. "It's for our banker."

They got him psychiatric help.

All over the state, it is happening, and has been happening, and few want to talk of it. Neighbors avert their eyes. But it isn't only suicide and murder or the threat of it. Less savage gods are loose here, too: wife beating, alcoholism, child abuse. All of it is alarmingly up, say social workers and psychologists and ministers.

What is the explanation? A desperate economy is much of it, of course. The rest of it is eerily devoid of logic. But violence, self-directed or otherwise, isn't chained to reason. Maybe we've been living too long off old myths. Maybe Meredith Willson and "The Music Man" died a long time ago.

Statistics won't tell this sad story, but here are several chilling ones:

• A farm goes down in America every six minutes.
• In Iowa, according to a poll in *Farm Journal* in 1985, 42 percent of all farmers are thought to be "sliding toward insolvency."
• One-third of all Iowa farmers are facing foreclosure. What this means in the jargon of agricultural economists is that their debt-to-asset ratio is seventy or higher: seventy cents or more on every dollar they're worth is owed to a pale figure in a slack suit behind a big desk in a bank.
• According to a sociologist at the University of Missouri, the suicide rate among midwestern farmers is 30 to 40 percent above the national nonfarm rate—and rising.

And yet, farmers have always had a way of defying odds. Many have found credit this year when all the betting went the other way. The spring crop has gotten in the ground, after all.

Says Paul Lasley, director of the Iowa Farm and Rural Life Poll at Iowa State: "We know suicide is happening. All the signals I get tell me it's happening. There's a lot of despair. I don't know—mark it from the fall of 1983. But how do you count? Let me give you an example of the problem: Occupation is listed on Iowa death certificates, but they don't

put them in the computer that keeps track of vital statistics. You'd have
to go through by hand and try to figure out which ones were farmers."

The first Iowa State suicide occurred in the fall of 1984. He was a
good kid from good German stock. He shot himself on his parents' farm
in a bucolic little spot at the top of the state named Buffalo Center. He
was in love with things coming up out of the ground. He left his dorm
one night, rented a room in a motel, drove home several days later. The
neighbors spotted his car by the side of the road.

"He was just lying out there in the corn," says his ag-ed adviser,
pulling lint off his sock, unable to look up.

But here is the mystery: his parents' farm wasn't going under. It is
doing fine, in fact. The flash point was elsewhere.

In the 1984–85 academic year, three teenagers in Storm Lake committed suicide. One was the basketball coach's son. They say he just
walked past his parents into his bedroom and shot himself. A psychia-
trist was brought in from the Menninger Foundation in Topeka, Kansas.
He talked to a lot of people and ran some tests and told the town its
problems were pretty normal. What happened might have been just a
statistical quirk.

In Harlan (population 5,357), on the western edge of the state, three
farmers killed themselves in eighteen months. The American Psychological Association sent a writer out earlier this year, and residents hinted
darkly that the actual number was higher than that.

These are random headlines from the suicide file at the *Des Moines
Register,* the state's largest daily: "Man Slays Woman, Then Himself,"
"Sioux City Boy, 10, Dies of Hanging," "Officers Say Farmer Left Notes
Planning Suicide," "Jobless Clinton Plumber Shoots, Kills Wife, Then
Self," "Farmer Slays Ex-Fiancee, Then Turns Shotgun on Self," "De-
spondent Woman Commits Fiery Suicide," "Minister Who Drowned
Despaired Over Losing Job."

"We had to start a separate file for the first quarter of the year," says
the paper's librarian, matter-of-factly. She kindly does a computer run
on suicides in the last nine months, and the list is as long as her leg. The
file from ten years ago is scant by comparison, but the comparison isn't
really fair. For one thing, until recently medical examiners in Iowa have
been reluctant to put the word "suicide" on death certificates. They're
still reluctant. The word cancels life insurance policies. But it is more
than that. People want to respect their neighbors. The word was and is
such a taboo.

So a man is found hanging one morning from a rope in his machine
shed. So somebody puts a .410 shotgun in his mouth under a beech tree
one evening. So somebody is found crushed beneath his International

Harvester tractor. Well, you put it down as "accidental." Either way, somebody didn't have to watch the sheriff come out and nail signs on his land. Everybody can tell these stories now.

In the town of Spencer, up near the Minnesota border, it is not uncommon for seventy-five families a week to walk into the Northwest Iowa Mental Health Clinic. One of the common thoughts that come out is suicide. Four separate crisis intervention support groups are currently operating in Spencer. Breaking through the resistance is the hardest part. "We're past the point of positive pretending," says Joan Blundell, a social worker at the clinic.

Life in rural America was never easy. If it wasn't the ruinous weather, it was the harsh loneliness. And still, the outside world wants to romanticize rural life — or mock it.

To the naked eye, much of Iowa is doing just fine. Many farms are fat and healthy. So are some city dwellers. The temptation is to say the problem doesn't exist. It's the hard-to-see bottom tier that's hurting. The gap between rich and poor widens in Iowa, as it does everywhere else in America. The midwestern state that helped vote in Ronald Reagan in the last two elections now gives him, in a recent poll, a 24 percent approval rating. In February 1985, the president had a 42 percent approval rating in Iowa. But then in March he vetoed farm emergency credit relief legislation.

You land at Des Moines and cows are grazing behind a fence at the end of the runway. You drive out into the countryside and the quiet crashes all around you. It is the contrast between what pleases the eye and what burdens the soul, but you don't have to live in Iowa to feel that. People have long suffered at the hands of their place's beauty, whether in northern Ireland or South Africa or Appalachia.

There are no easy answers to any of this. The five Iowa State suicides in 1984–85, none of which occurred on campus, are thought to be the highest number of self-inflicted deaths of any college in the country. That report went out on National Public Radio. The university would like to think it is an isolated phenomenon, freakish as lightning in a rainless summer sky. And, in fact, maybe it is.

But then the Office of Student Life at ISU released the final results of a March 1985 student stress poll. More than half of the 212 students polled — 54 percent — said their anxiety was indeed related to the state's farm crisis. One in every four polled said they felt that "life was not worth living." Extrapolated to the total campus enrollment of twenty-six thousand, that would mean more than six thousand students are walking around with suicidal feelings. That number blew the lid off the poll.

Maybe the questions were phrased wrong. Maybe you'd get that response on any campus in the seemingly placid eighties.

An outsider dare not make something more than it is. The five ISU students (two from the same dorm) conform to no particular profile. One was in preveterinary. One was a new transfer into entymology. Two had no farm background at all. Overall, the links in these five student deaths to the state's rural stress seem tenuous. And yet one would be a fool not to think something is terribly awry in the middle of the country.

"It would be a mistake to think that ISU is an island of psychological depression," University President Robert Parks told the *Des Moines Register*. The *Register* broke the suicide story. "This may be part of a sad nationwide condition."

The university is in the town of Ames, in Story County, in the richest agricultural belt in one of the richest agricultural states. The land is so black it almost hurts your eyes. You stand on the steps of the massive student union and watch kids fishing in Laverne Lake, right on campus. The bells in the carillon toll every fifteen minutes. Lovers drift head to head. It feels like a 1940s movie starring William Holden as Biff Baker with June Allyson on his handlebars. In the library, students sit near a mural engraved with words from Daniel Webster: "When Tillage Begins, Other Arts Follow."

"I wouldn't argue the right to cover the story, no, no," says Dave Lendt, the university's public information director, a nice man who of course wants to put the lid on. "We're all First Amendment freaks around here. I do know that these are not flaky people. I do know that these are people who have values, who do not come here without anchors, without family ties. These are people who do not believe they can take up a space in the world without giving something back. What you've got here are center-of-the-road, corn-fed, land grant type of students."

Lisa Birnbach, author of *The Official Preppy Handbook,* came to Ames a while back and called it Silo Tech and Mule U. Har-har. In Agronomy 600, there are lectures entitled "Water Relationships in Alfalfa" and "Effect of Residues on Maize Growth." The University of Iowa is two hours away in Iowa City, and over there they like to style themselves as the Left Bank of the Mississippi. Iowa City is where the artists are, Ames is where the hayseeds are—never mind that Iowa State has a huge engineering college, that it did some of the earliest atomic energy research in the nation, that its National Public Radio affiliate plays Liszt and Mozart. Things are never quite what they seem.

In ISU faculty mailboxes these days are memos about detecting "warning signs."

His name is Pete Kapustka, and his voice is high and reedy, his body trim and muscular. He wears cowboy boots and walks with one hand in his back pocket. He's thought about suicide, yes, driving his car into a bridge, but he's past that. He's only twenty but now he's head of the family. His father, at forty-six, is on his back in an Omaha hospital: bone cancer. He may never farm again. In fact, there may be nothing left to farm. The family is threatened with foreclosure — not just to the bank, that's the half of it, but to relatives who hold the notes.

"I found out last night my dad's family is serving us with papers of forfeiture," he says. "I sat up half the night trying to figure out what to do."

He came home from Iowa State on 1 April 1985 to get the spring crops in. College has to wait. He brought home with him his favorite poster and put it on a new wall: "They that sow in tears shall reap in joy."

"Now my dad, without ever telling me to my face has entrusted me with putting this year's crops in the ground," he says, and there is a sonorous, almost dreamy, quality about the sentence. "From his bed in Omaha, my dad figured out how much seed corn we need. He's been farming down in Omaha for a couple of weeks now."

He went to see his dad in the hospital. He had just come out of surgery. There he was, full of tubes and dangling things, with nurses and funny-smelling stuff all around, and all he could fix on was his calluses. He kept saying, "Pete, Pete, come over here and look at my calluses. They're peeling."

The son who has become the man is telling you this in a huge equipment shed on a suddenly glowering afternoon in Barnum, which is a wide spot in the road west of Fort Dodge. His eldest sister is in the house making supper; she has become the mom, because the real mom spends a couple of days every week in Omaha. The wind is racking fiercely at an aluminum roof. Pete's little brother, Tony, who is ten, is skimming stones on a dirt floor. Tony is chubby and in a red satin school jacket. The difference between ten and twenty, between the big brother and little brother, seems as vast as the tilled fields spreading beyond the house.

"I've been to Chicago once, for about twenty hours," Pete says, laughing at himself.

He slaps at the fender of a mud-caked tractor. "This is the staple of the American family farm, a 4020 John Deere."

"I can drive 'er," Tony calls over.

Pete walks around a four-wheel drive with an enclosed cab.

"This is our 8640. We bought it when things were good. It cost eighty thousand dollars. You couldn't sell it now for twenty thousand dollars."

"Nah, we don't use 'er much," says Tony.

Pete squints over at his little brother. "I'm about numb to all this stuff. I keep a journal. It's hard to remember when things were normal. But I've always heard, if you think you're crazy, you're probably not."

In the late fifties his dad came out of the Army and had no choices but to take over the family farm. His own father, Pete's grandpa, had walked out into a winter field just a little while before and split his head open on a patch of ice. He just bled out. By the time they got him into the house, he was almost gone. So the son, just sprung from the service, had all his options foreclosed.

"And in a way, I feel it's kind of happening all over again."

Nights are the roughest. You go in and try to sleep. Often it doesn't come. You're up anyway as soon as the light is pale.

One bad night a couple of months ago, Pete Kapustka wrote a letter to the president of the United States. He wrote it in longhand on Alpha Gamma Rho stationery. Alpha Gamma Rho is the big ag house at Iowa State. The stationery was a kind of classy yellow parchment; he figured that might catch the president's eye.

"I said, 'Sir, we're just a family farm. My dad's dying of cancer, we're losing our land, and you made a spectacle of it. Please do something, sir, do anything.' "

He never heard back.

In Nora Springs, Mary Beth Janssen and her husband Gary run farm survival meetings. They've set up a computer in their living room. They meet with fellow farmers in local churches. The first thing they do is pray. There is no particular agenda—people just stand up and talk, get it out. Sometimes Gary accompanies bankrupt friends to the bank.

Mary Beth is telling you this on the phone and she sounds as if she's talking from Ethiopia. "We've been through bankruptcy. We're still farming and sometimes I ask myself, What the hell for? Gary and I lost a farm that had been in the family eighty years. 'We were failures.' You hear a lot of that. When somebody goes down, you hear, 'Oh he deserved it.' I think people are so afraid that it's going to happen to them next. But Gary and I are through rolling over and giving in. For years we sat on the dumb side of the desk when it came to bankers. What Gary and I are trying to get people to realize is that it's not their fault. It's been coming on for years. That's what happened in the thirties."

Gary takes the line. "I have no degree in anything. Yeah, I've probably saved some lives. To say who, I couldn't give you a name. I don't know who's in the back row. The people in the front row, you don't worry about."

CAROL HODNE

We Whose Future Has Been Stolen

I AM NOT NOW A FARMER. I am one of the many children of the farm who always thought that somehow we would be able to carry on the tradition of family farming. I am no longer certain that I will be able to see that dream fulfilled. My future on the land has been stolen, I believe, by deliberate government policies designed to keep people like myself off the land. As a consequence, one of my most deeply held wishes — to return to the land — may never come true.

I represent our family's fourth generation of farmers, farmers who came from Scandinavia and Germany. Decades ago, my ancestors left their farms in Europe because they were poor and starving. They sought a new land here. Ironically, after four generations of family farming in the United States, my generation is seeing the end of family-type agriculture in this country. My neighbors are being starved out by conscious government policies of the last thirty years. My home town of Irwin, Iowa, is drying up as businesses are closing and people are leaving. The only visible growth is the number of elderly residents and number of huge grain elevators where "they pump the blood out of our community," as my mother explained two harvests ago.

Family Farmers Are the Best Stewards of the Soil

Based on my experiences on our western Iowa farm, I feel that perhaps our greatest moral obligation to save family farms stems from the moral imperative of preserving our soil and water resources. Our farm is one of the few diversified cattle operations left in our part of the state. Extensive terracing, crop rotation, and other conservation practices have been part of our family for decades. We have loved and cared

---

CAROL HODNE is executive director of the North American Farm Alliance.

for the land as much as we have loved and cared for any family member. The hands and hearts of my family have been devoted to this land; we want to pass it along to the hands and hearts of future farmers in better condition than it was in when we received it.

I fear for our land now and the land of farmers around the globe as the ever-deepening farm crisis surrounds us and them with deepening debts, fewer neighbors to care for the land, and the greed of insurance companies, land management companies, and wealthy individuals. These groups and individuals see in today's devalued farmland a new source of short-term profits, not a heritage and gift to be passed along to tomorrow's children.

I believe that this stewardship ethic is being destroyed today as land management companies are buying up more land and as foreclosures, contract hog feeding, and other mechanisms of corporate, industrialized agriculture are pushing families off their land. Our children's futures, along with our precious topsoil, is being washed down ever-deepening gullies.

The destruction of the cattle industry in Iowa is a particularly ominous trend for land stewardship. The destruction of diversified family-owned livestock farms and ranches through low livestock prices and financial and tax advantages to corporate feedlots is a major cause of increasing soil and water erosion. Family producers can't compete with the huge vertically integrated corporate feedlots that can buy our corn and other feed more cheaply than we can produce it. As cattle feeding has disappeared from Iowa, farmers have had to plow up fragile hillsides and convert them to cash crops, greatly increasing erosion of some of our nation's finest soil.

While taking an annual first of spring drive with my father in 1984, I was heartbroken as he pointed out some of the worst water and soil erosion ever seen on the rich loess hillsides of our neighborhood. Gone were the usual new calves bounding along the hillsides. In their place were gullies filled with unprecedented layers of silt, stalks, and debris.

Family Farms Are Not Lost; They Are Stolen

This destruction of family farms and deterioration of our environment is not the result of "natural economic forces." It is not the result of good policies gone bad. Today's farm crisis is rooted in the conscious dismantling of the cost-effective, prosperity-generating federal farm policies initiated during the 1930s. In contrast, farm programs of the last three decades have been designed to enforce low farm prices. These policies designed to ensure cheap raw materials here and abroad for multinational, monopoly grain traders and food processors are rapidly destroying family farms and rural communities throughout our nation.

Clear solutions do exist to the farm and rural crisis. Farmers around the country are building widespread agreement on the following demands:

1. We want farm prices that will return costs of production and a reasonable profit coupled with supply management. Using the nonrecourse loan program the way it was intended to work and worked successfully from 1933 through the early 1950s, we can raise farm prices in the marketplace with minimal government expense.

2. We need farm debt adjustment with a moratorium on foreclosures and repossessions.

3. We also need emergency survival assistance for those who need food, clothing, shelter, and health care.

Farmers Want the Right to Vote on Their Own Destinies

A producer referendum allowing farmers to vote for higher prices through controlling how much they produce is supported by increasing numbers of farmers. For example, recent polls showed that 81 percent of Kansas farmers and 73 percent of the Nebraska general public favored a producer referendum. Of Iowa farmers, 66 percent supported letting farmers set mandatory production controls.

So far the Congress has only begun to listen to us. A producer referendum with prices near 70 percent of parity and mandatory supply management are the basic principles of the Save the Family Farm Act introduced by Iowa's Senator Tom Harkin. Although it did not win in 1985, we gained an incredible amount of support and we will be back to force Congress to rewrite the disastrous 1985 farm bill. This so-called Food Security Act ought to be called the Farmer Reduction Act because it lowers farm prices even further. According to a new Food and Agricultural Policy Research Institute analysis (FAPRI 1986), these lower prices mean that export values will rise only slightly. In the meantime, direct government costs through deficiency payments will be almost double the export value of the crops. For example, for the 1986 crop year, export value of corn is estimated to be $3.44 billion while the direct cost to the government is estimated to be $5.932 billion.

We gained many friends in our fight in 1984 for the Save the Family Farm Act: farmers across the country and friends in the churches, union halls, and unemployment lines. Well over one hundred major groups endorsed and fought for it. We gained consistent support from the Congressional Black Caucus, Rev. Jesse Jackson, the Conference of Black Mayors, and many other national leaders.

We also found we had support for higher U.S. prices from farmers in other countries who recognize that the United States sets the world prices for many major farm commodities. For example, in the late days of the farm bill debate, representatives of Argentina's and Brazil's governments came to plead for us to raise farm prices. They knew that lower U.S. prices would mean lower prices for them, which means they may have to default on their debt payments.

The North American Farm Alliance has worked closely with farmers in Europe, Canada, and Central America. Through our direct contact, we know the impact of our low prices on other countries. Our government's goals of lowering U.S. farm prices, increasing exports at any cost, and weakening European food security systems are leading to a dangerous escalation of U.S. initiated trade wars with the European Economic Community. Scapegoating the EEC and European farmers diverts us from real solutions.

Another moral obligation to save family farms in the United States and to raise farm prices relates to the especially devastating impact of our low farm prices on the Third World. Cheap imports from the United States make it hard for Third World farmers to compete. They are discouraged from producing food for domestic consumption and are forced to grow more nonfood export cash crops. Increasing dependency on food imports is increasing hunger in these countries as mounting debt and interest payments use up much of their foreign earnings, making it hard to buy food from other countries no matter how low the price goes.

Those of us in the United States who are concerned about ending hunger must first take a look at what our cheap raw material prices are doing to farmers and hungry people around the globe. One of the best ways to end hunger is to increase the self-sufficiency of farmers everywhere. This means giving farmers a reasonable profit in the United States and abroad. Raising our prices will not translate into lower export sales; other countries will follow suit, raising their prices as well.

Opponents to Economic Justice for Farmers Are Clear

During the 1985 farm bill debate, we also learned more clearly who opposed our goals of higher farm prices, lower government costs, and production controls. For example, production controls for wheat were opposed by the so-called Farm Coalition Group, which includes the American Bakers Association, the American Meat Institute, the Chamber of Commerce of the United States, Ralston Purina, the National Broiler Council, the National Fertilizer Solutions Association, and so on.

Why would such groups oppose farmers' desires for the government to assist us in controlling our production and getting a price that would cover our costs of production? Most simply, their profits depend on our cheap raw materials here and abroad. The more we produce, the more cheaply they can buy our raw materials, and the more money they can make.

We were also sorry to see the American Farm Bureau Federation join with such large corporate interests to oppose farmers' right to control our own destinies. The Farm Bureau opposed the idea of a farmer's referendum whereby we could limit production and get a better price for our goods. I say this as a former "Farm Bureau daughter"; my father served on the Iowa Farm Bureau board of directors for twelve years trying to fight for farmers' interests over the interests of Farm Bureau's businesses and insurance company interests. I was appalled to see the Farm Bureau speak so forcefully and consistently against higher price supports and a referendum.

In closing, I ask you to take a stand for family farms over corporate agriculture, for sustainable agriculture over global environmental destruction, and for food security over increasing world hunger. If you do, there is still the chance that we whose future has been stolen may find our hope suddenly restored.

Reference

"An Analysis of the Food Security Act of 1985." FAPRI Staff Report #1-86. Ames, Iowa: Food and Agricultural Policy Research Institute, February 1986.

DENISE O'BRIEN

We Who Are Hanging on the Edge

TEN YEARS AGO my husband, Larry Harris, and I were married. We started farming. Little did we know what we were getting into. Even if we had known, however, we still would have made the decision to farm, as Larry was born of the land and we are both now married to the land. We have chosen to farm somewhat differently than most. Preferring not to harm our valuable air, water, and land resources, we decided, early on, to farm nonchemically. Our labor and diversity replace the harmful chemicals most people in agriculture are tied to. Conservation has played an important part in our decision making, moneywise and landwise. Neither one of us has had an off-farm job; we still remain committed to the idea of a family deriving its entire income from the resources of a farm.

Ours was a traditional takeover of the family farm. Larry's parents decided to leave farming and move to town. We moved into the home and onto the farm where Larry was raised. As the years have passed, we have developed very strong ties with our parents as have our children with their grandparents. We believe this is important to the health and welfare of our families and our communities. During the past several years as the economic situation has changed, farming has become more stressful. Both families, father's and son's, have had to make sacrifices to stay in farming. This is not unusual among modern farm families; we have been fortunate that with the passing of farmland from father to son there has been no extra burden of land or machinery purchases. We have been able to "get by" on very little capital investment and therefore have kept costs down. This has *not* kept us immune to the hardships others in agriculture are suffering in this crisis. We are getting by with the aid of

---

DENISE O'BRIEN is a dairy farmer in Atlantic, Iowa, a founding member of the Iowa Farm Unity Coalition, and a board member of Prairiefire Rural Action.

food stamps, heating assistance, and any other program that we are eligible for. It hurts our pride and injures our dignity; but we are willing to suffer those indignities in order to survive. Our commitment to farming, and to raising a family on that farm, is that strong. Few children these days are raised by both parents and grandparents. It is a challenge to be able to do this.

As I mentioned before, our farming operation is done without the use of harmful chemicals. We raise hogs, milk cows, and have a cow/calf beef herd, operate a U-PICK strawberry business, and have eight hundred apple trees that will start producing in three more years. Three hundred acres on a good solid crop rotation that includes oats, alfalfa, wheat, and corn is essential to our existence. There is more than enough ground and livestock for two full-time people to handle. Unfortunately there is not enough profit in farming to support a family. So what are we to do? If we do what is necessary to farm properly, nonchemically, then there is no time for off-farm jobs. We feel certain that if we took jobs off the farm our children and our land would be neglected and we would no longer be good stewards of the soil. But prices are so low now that we can only make enough money to barely hang in there. What will happen to us if prices drop further?

Four years ago our commitment to farming took on a new dimension. My husband and I became involved with a number of people who were beginning to see what was happening to the future of agriculture. It was not a picture of comfort and satisfaction. Land prices were starting to fall and prices for our crops were low. The future of agriculture was beginning to look dim for a number of people. You might say that those first knocked out of farming were the "bad managers." But even that is debatable. What is no longer debatable is that good farmers who are good managers are being displaced. The magnitude of this problem now encompasses everyone. There are few who have not been touched by it.

A conference sponsored by the National Catholic Rural Life and the National Council of Churches in which Larry and I participated changed our perception of our position. Our experience led us to become active in trying to change the direction agriculture has taken since the 1950s. Our commitment led to my involvement with the Iowa Farm Unity Coalition and Larry's intensified involvement with child rearing and farming. We are willing to invest our time on this level and in these ways only because we feel so strongly about stopping the direction of agriculture toward large corporate farming.

The work I have done as the vice-chairperson of the Iowa Farm Unity Coalition has given me contact with farmers all over the state. The pain and suffering I see has made me both angry and frustrated. To hear the news every day that our country is experiencing a recovery makes me ask questions — questions like who is in recovery and why are *we* missing it? To have a recovery based on defense contracts and displacement of thousands of farm families does not seem to be sound economic practice. It is hard not to draw comparisons between the haves and the have-nots. To have a recovery at the sacrifice of the hardworking, honest, proud people of the land is wrong and unjust!

It is easy to read statistics and not relate to them in human terms. When one reads that one-third of our farmers are in serious trouble, what does that mean? In personal terms it means that payments on the land or machinery are not being made. It means that someone has had to leave the farm for a job, usually the woman, but sometimes both husband and wife. It means children left at home alone or at a sitter. Sleepless nights, alcohol or drug abuse, domestic violence, and the ultimate violence — suicide — are all taking place in households in rural America. Never before has there been such a need for mental health facilities.

Is this what we went into farming for? What is happening to the proud farmer? Why should this happen to the foundation of our country? We should ask ourselves these questions and especially think about one: should the economic recovery take place on the backs of farmers? President Reagan promised everyone a safety net as we embarked on economic recovery. That safety net seems to have many large holes through which innocent victims have fallen.

Recently, there was a large rally in Ames, Iowa, that brought out fifteen thousand farmers. All of them wished to send a single message to Washington: something needs to be done. My husband and I attended that rally hoping to find the needed encouragement to continue with our work. What has happened since that day in February 1985? The farm economy has been hit with blow after blow of falling land prices and extremely low commodity prices. The final devastating blow was the 1985 farm bill that passed Congress and was signed by the president. This farm bill was supported by Farm Bureau and other agribusiness-oriented organizations but not by grass roots farmers who understand they cannot grow corn for less than two dollars per bushel. The 1985 farm bill, along with the provisions of the Gramm-Rudman Act, will have terrible effects, not only in this country but internationally. The United States sets the world price on most commodities, so this country is in essence knocking out its competitors by setting the price so low. Third World countries will have much trouble competing with our price and will be forced to heavily subsidize their farmers. So the United States bankrupts Third World farmers at the same time it bankrupts its own.

Who wins? The big grain cartels that deal in quantity and not price. Who is going to pay? The American taxpayers. As the Third World countries find it increasingly hard to pay their debt, the American taxpayer will have to provide a bailout of some kind. We cannot depend on the export market as the answer to our problems as the 1985 farm bill recommends. There are more and more countries such as China, India, and countries in Europe that now export rather than import grain. We face stiffer and stiffer competition.

What is the answer if not the 1985 farm bill? Senator Tom Harkin offered an alternative to the administration's bill. It was a farm bill written in part by farmers that would have helped our farm economy tremendously. Supported by Iowa Senator Charles Grassley, the Save the Family Farm Act would have brought up farm prices and returned profit to agriculture. The bill would have given farmers the opportunity to decide for themselves whether or not to have mandatory controls on their commodities. The act would have set the direction of agriculture policy towards medium-sized, more efficient farms. This would help us to keep people on the land and to keep our valuable resources of land, water, and air in the hands of farmers: in the hands of people who care about them.

There are some basic precepts upon which our country was founded that are being threatened by the transition in agriculture. Jeffersonian democracy advocates that land be owned by many and not concentrated in the hands of few. As we stand witness to what is happening in Central America where much land is owned by few people, we can understand how that can happen here in America. In many countries in Central America there is widespread poverty, high infant mortality rates, and hunger. It is not because there is not enough land; it is because too few people own it.

My view of the rural crisis is not an objective view. My family and I are fighting for survival along with the many other citizens that make up rural America. I see pain and suffering in the eyes of many when they ask, What are we going to do? There is no social justice in what is happening right now in our country. Farmers are victims of economic policy that has been handed down from the current administration. It would not, of course, be fair to put the entire blame on this administration. What is happening is the fruition of forty years of bad farm policy. We farmers are being treated as second-class citizens, as people without rights in this time of economic change.

We farmers need a fair price for our products, we need conservation incentives, and we need emergency support to pull through this time of need. We do not need welfare programs nor do we need to undercut farmers around the world. We who are hanging in there continue to hang in there because we want to work cooperatively with one another, doing what we do best: raising crops to feed a hungry world!

ERWIN JOHNSON

We Who Are Doing All Right

I AM A FARMER, son, husband, father of two daughters, community volunteer, friend, veterinarian, consultant, Iowan, accountant, Christian, lobbyist, leader, borrower, renter, plumber, idealist, American, lover, shepherd, salesman, market analyst, Charles Citian, ISU fan, bank board member, farm manager, landowner, corporate executive, agronomist, advocate of the free enterprise system, soil conservationist, grandson, extension service associate, world citizen, electrician, family man, businessman, co-op board member, Floyd Countian, carpenter, boss, housekeeper, and steward of the land. I list these roles not to stroke my ego but to let you know that a family farmer is much more than just a fellow sitting on a tractor with a seedcorn hat on his head advocating the purchase of some pesticide as seen on TV or heard on radio commercials.

The beliefs that I have are to a great extent the product of my past. I grew up in a rural community centered around a Methodist church near Charles City. I am second-generation American with a long history of German blood flowing through my veins. My grandfather and grandmother were poor German farm people who had the desire to escape the bondage of poverty and saw the opportunity for a new start in life in the United States. They came to the United States in the early 1900s with my father in tow and began our present farming operation in Floyd County.

My neighbors are named Schmidt, Heitz, Fluhrer, Koehler, Pruessner, Forsyth, Hirsch, and Wulff: all of German ancestry. Many of them live on Century Farms passed on from generation to generation. They were my Sunday school teachers, 4-H leaders, and Methodist Youth Fellowship chaperones. They might even have flushed me out of a favorite "parking spot" a time or two to keep me on the straight and

ERWIN JOHNSON is a farmer in Charles City, Iowa, and a member of the Farm Bureau.

narrow path. We share joy, sorrow, laughter, prayer, marriage, and funerals together. My neighbors are part of who I am.

As a child, 4-H, Future Farmers of America, music, hoeing thistles, skinny-dipping in our creek, playing football between the "mudpies" in the cow pasture, and learning a love for nature, God, and humankind were important in my development. I was loved and loved in return.

After high school, I attended an excellent university, Iowa State, and graduated after living and working in South America for several months. I then joined the world community by serving in Southeast Asia for two years as an agricultural development volunteer. Employment with the U.S. government then beckoned me to continue my work in Laos working with the United States Agency for International Development. During this five-year period living in Venezuela, Laos, and Thailand, I shared my life with compesinos, peasants, and *kung u bawnok*. I touched the lives of the poor and rich, refugees of the war and royalty, ministers of government and village dwellers. These experiences left a profound and indelible imprint on my personality, ideals, and outlook on life. I do not view my world as just farming, or just Charles City, Iowa, or just the United States. It is a big world, a huge planet, inhabited by diverse people of diverse languages, religions, cultures, and lifestyles. Through this diversity I observe common threads that bind us all together: love, desire for peace, friendship, loyalty, and a need for hope.

Working, living, and traveling overseas gave me the opportunity to observe different political and agricultural systems. In my opinion, there is no superior system or general policy than what exists in the United States. I believe this to be true, even as I concede we need to make improvements.

In 1970 my wife Yoshiko and I came to Iowa and started to farm with my parents in the same community in which I grew up. We farm close to six hundred acres of cropland. My wife and I incorporated our farming operation with my parents' in 1979. We have one full-time employee who helps us raise hogs, sheep, corn, soybeans, and alfalfa. As a corporate officer and farm manager, I attempt to maximize profits by using computerized farm records, hedging or utilizing options trading on the Chicago Board of Trade and Mercantile Exchange, keeping detailed enterprise analyses of the various commodities we produce, and keeping informed of events around the United States and world that influence the prices of the products I produce and buy.

I am a businessman! I farm to make a living for my family. I farm to make profits. Whatever skills, techniques, or resources that are available to help me achieve those goals I will take advantage of. I employ a marketing advisory service to help. I study commodity charts. I consult with specialists in the area of management, markets, animal health, farm policy, tax codes, and computers. I try to surround myself with people who are positive, creative, and active.

In my opinion we farmers have many common traits in our personalities. One trait is our tendency to tell each other of our failures. Too seldom do we tell each other of our successes. A farmer who verbalizes his or her successes is considered boastful, egocentric, and a person to shun. It is with this knowledge in mind that I am reluctant to tell you about our financial successes.

We made profits from our farm business in 1985 and, as I write, project some profit for our business in 1986. Our hog and sheep enterprises generated good income in 1985. The corn, soybean, and alfalfa enterprises enjoyed average to poor years, but we are diversified and the pluses definitely were good to us. Every decision we have made was not good. We are purchasing some land at 1981 peak prices and are now scrambling to make that purchase a workable situation. Our debt-to-asset ratio is in the 40 percent range. This is livable, but it continues to increase each year as land values decrease. We are worried about the future, especially as it relates to our debt load. But we have not stopped looking for opportunities that might help brighten that future.

Even in today's highly negative atmosphere in the farming community, many farmers are making money, generating positive cash flows, and paying off debt. Please realize that these same family farmers are not going to stand up before a TV camera and tell others of their success. You will hear farmers saying that they are in trouble, making adjustments, quitting, or sliding into insolvency. These are difficult times for people in agriculture. But the majority are surviving and many are prospering.

I decided to farm because it was one vocation in which I could be my own boss. If I succeed using my skills, talents, and intelligence, it gives me great satisfaction. I desire to live and work within a system where I have the "opportunity to make a profit." Notice that I did not say a "guarantee" but an "opportunity." In any environment where there is opportunity for success, the opportunity for failure also exists. Farm people, for the most part, are independent sorts and understand this environment. It helps to make us, on the whole, conservative, determined, and aggressive.

I am an advocate of the system where supply and demand determines the price of our outputs and inputs. I do not desire to be part of an American agriculture treated as a public utility with mandated governmental or quasi-governmental controls. Government intervention has crippled agriculture and helped drive people off the land. Producers who hope to survive this current upheaval must look to the business side of their operations and not to government. I want to produce for the market, not government programs. To be a vital sector of the U.S. economy, farmers must be competitive, progressive, aggressive, and able to adapt to a changing world environment. We need to grasp ahold of our costs of production, cut back where needed, and achieve greater efficiencies. Not everyone will survive nor should everyone survive. We are part of an Iowa, U.S., and world economy.

I would like to address the moral question in the title of this volume. Two paragraphs from the book *Beyond Power* written by Marilyn French and published by Summit Books in 1985 came to me as I considered this question.

> Morality is a neutral term; it has no specific content. It refers to the set of values by which we judge, which guide our behavior, and even our emotions. Morals are our real values—not qualities we claim to revere, to which we give lip service, and not much more. Our moral system is a system of priorities, or rather a texture, an interlocking set of shifting, ambivalent, and conditional goods and ills. Our morality manifests itself in our choices—how we live, to what we devote our time and money, the kind of people we want to be. The morality of a society also manifests itself in its choices, how much its people spend on what, its language and art, its mode of production, its mode of ordering itself and the kind of order deemed desirable.
>
> Morality is a personal and communal affair; when it reaches the public realm, it is called politics. The relation of the two is not like two sides of a coin, but the inner and outer sides of a balloon; that is, it is the same thing, seen from the inside and outside.

Can the family farm survive and is there a moral obligation to save it? The answer to both questions is yes. I hold close those values or morals that were ingrained in me as a young person by my ancestry, my religion, the environment of rural Iowa, and nature. I desire for my daughters to possess these same values. The family farm is the place where I know the opportunity still exists for this transfer from generation to generation. We need to save this business, social, and ethical environment. Most family farms are surviving but we need to change the role of government. We need to pare back the influence of government in the areas of farm policy and taxation. Almost every government farm policy that I have observed in the past seventeen years has always helped the small farmer a little and the large farmer a lot. We have subsidized the growth in farm size with government programs designed to help the small- or average-sized farm.

Through tax laws we have encouraged or directly aided the growth in farm size by the use of investment credit, fast depreciation, cash accounting, capital gains, and tax-loss farming by outside investors. Again, these tax shelters have aided the small farmer a little and the large farmer a lot. I believe that I can compete with the so-called super-farms or large corporate farms if I am competing on a level field. That means I do not want the government providing disproportionate subsidies to the large operations. American laws have encouraged farmers to overinvest in machinery, tile, buildings, and other capital investments. Such investments are depreciated out for tax purposes long before their real usefulness is finished. These same laws have discouraged the farmer from investing in labor. As a result, we are labor efficient and capital-intensive even though there is no shortage of men or women who would prefer to live and work on the farm. Therein lies the risk that has come home to roost today. We could lay off extra labor in bad times. But today we cannot put off the bank or lender from whom all that capital investment is borrowed.

Is it possible to maintain the family farm "way of life" without maintaining the family farm business? To various degrees I believe it is an alternative to maintain the farm "way of life" without the business. A large number of people who live near Charles City have for several decades held jobs at Oliver-White Farm Equipment or Salsbury Laboratories. These people have also farmed, raised their families on the farm, and carried on a way of life that rural farm life offers. Depending on off-farm income has complemented their desire to live in the country. It is working for them.

In 1968 a devastating killer tornado roared through the center of Charles City destroying hundreds of homes, many businesses, and thirteen lives. Charles City could have died then. But it didn't. People joined together to repair, renew, restore, and rebuild. Our town is more attractive now and the will of our people is as strong as ever. As a former Iowa farm boy, Rev. Robert Schuller wrote *Tough Times Never Last, But Tough People Do.*

I must interject my personal opinion about the general state of our U.S. economy and the present role of government via the executive branch, Gramm-Rudman, and tax reform. I think that the present monstrous buildup in our defense spending (using the federal government's ability to deficit spend or borrow against the future) is morally bankrupt. I do not view the strength of America as being in tanks, guns, nerve gas, Star Wars, or space weapons. I view our strength as being in the will of our people, in hard work, productivity, education, and attending to the spiritual and material needs of our people. To be spending increased billions of dollars each year for weapons of destruction while the human needs are so great is dead wrong.

We in agriculture need the federal government's help in this transition period from government dependence to markets. We need a period of four to five years as legislated in the 1985 farm bill to gradually change. To cut this aid quickly and indiscriminately each time the administration sees an opportunity to do so is wrong. To eliminate the agencies of government, the Extension Service, Soil Conservation Service, Environmental Protection Agency, and Department of Education that can help us survive educationally, environmentally, and financially, is wrong.

I envision the future family farm as being a larger family unit. The large corporate farm is not efficient enough to compete in a market-oriented system, but there will be a tendency towards fewer and bigger family farms. There will also be many more smaller farm operations that depend on off-farm income for their survival. We cannot live in the past with 160-acre farm operations. We must learn to adapt to changing times and economic environments. Our future as family farmers lies with brains, hard work, and diligence. Our future depends on education, intelligence, technical competence, and self-discipline. These qualities do not come easily. Business-minded family farmers can survive in today's environment. Our hope for the future springs from personal action and trust in ourselves.

C. L. WILLIAM HAW

We Who Are Corporate Farmers

EVERYBODY LOVES THE FAMILY FARM and so do I. But most people think they must hate the corporate farm in order to continue to love the family farm. This is the farthest thing in the world from the truth.

I am a corporate farmer. A real-life, corporation-type farmer. But I am not the family farmer's greatest enemy. The farmer's greatest enemies are the other family farmers down the road who need to grow and expand to become viable economic units. These units seem to be becoming larger all the time, for the only way to become large enough to survive these days is to buy out neighbors.

Let me tell you what a corporate farm is. Our company, National Alfalfa, owns and farms fifty thousand acres of irrigated land in northern Nebraska, western Kansas, and the panhandle of Texas. It is hard country. Before we farmed this land, it was virtually nonproductive, semiarid rangeland. Before we developed that land by spending ten million dollars for irrigation equipment and another ten million dollars for harvesting and storage and cultivating equipment, it would barely support a cow and a calf on forty acres. Its economic value was such that much of it had been leased for twenty-five cents per acre until the irrigation development was completed. That land now produces over three million bushels of grain every year. The feedlots on our facilities, which are leased to others but which consume a portion of the grain and alfalfa that we grow, are finishing forty thousand cattle per year. Two hundred fifty people in our company at the maximum time of employment produce enough food to feed, in most parts of the world, a city of five hundred thousand people. We are proud of what we do. Again, our company is virtually unique, to my knowledge, in that we are farming on that scale as a corporation. Until we made the company private, we were

---

C. L. WILLIAM HAW was president of the National Alfalfa Dehydrating and Milling Company when he wrote this essay.

listed on the American Stock Exchange with fourteen hundred stock-holders. It was a public company in every sense of the word.

What are the things that we as a corporation can do well and, perhaps, even better than the family farmer? We have access to capital. My background is in the banking business. The company several years ago was a public company managed by a farmer, a lifelong farmer, who farmed well but made some bad financial and marketing mistakes. Maybe this is the way that marriages between family farmers and busi-nesses can occur. We now have access to capital. We have lines of credit from the largest and the best banks in the country. We can get the money when we need it. We can get it at the rates comparable to the most creditworthy borrowers in the country. Maybe the corporation has a larger capital base to start with. It takes a capital base to operate a farm. Even a smaller farm by today's standards would take half a million dollars invested to get going. What is the interest on five hundred thou-sand dollars, the minimum investment on a fair-sized farm, at 8 percent? It is forty thousand dollars. It will take all of the gross revenues of that farm just to pay the interest if it is all borrowed money.

Maybe something else that a corporation or a business underpin-ning for a farm could do is provide better marketing. We seem to have been able to do that well. We have a lot of products to sell. We have some technical expertise to do it. Because we have such a large base, we have people who can devote most of their time to do that marketing job. A corporate farming structure (if that is what a large structure is) can also take advantage of the latest and most efficient equipment available on the market today. The sum of fifty thousand dollars for a tractor is a lot of money. If it will extend the productivity of its operator, it is a wonderful investment.

What are the disadvantages of being a corporate farmer? There is only one and it is a big one. The American farmer, or the family farmer, is the backbone of American agriculture. He or she has a great degree of dedication, selflessness, drive, and ability unequaled by any other group of working people that I know of in the world. The degree of drive and efficiency is impossible to compete with. Fortunately, the corporate farm does not necessarily have to compete. Virtually all employees of our farm division were born and raised on family farms. They chose to work with us. We do not make them work for us. They were family farmers who chose to work with us for a variety of reasons. Maybe they do not want to risk that half million dollars of capital or maybe they do not have it. Maybe, having been family farmers all their lives, they know that the hard years, the tough years, have far outnumbered the good ones for family farmers in this country. We give them the opportunity to make a good living and good salary in the bad years and we have a bonus program, an incentive program, so that they are directly rewarded. This program is based on the production that comes from the units of farming they are responsible for.

So maybe the corporate farm does not need to be in opposition to the family farm. Maybe there are some advantages to people with that background of working in a corporate structure if they are given the incentive of being paid commensurately with their production. The people who work for us do not come to work at 8:00 A.M. and punch out at 5:00 P.M. Their pay is based on the number of bushels of corn that comes out of the fields they are responsible for. Their working day during busy seasons such as harvesttime starts at about 7:00 A.M. and ends at 10:00 P.M. with a sandwich on the combine at noon.

Whatever the reasons, we have been able to succeed. We have been an extremely high-production company. Our farm productivity, our yields, have been better than the local averages by a large margin in every area that we have farmed. Two reasons for our success have been (1) the land, but more important, (2) those who farm the land. The corporation does not make the farming decisions; those who farm the land make the decisions. The corporation cannot farm the land; people farm the land.

One further point. The great scapegoat of the problems of the family farm in recent years has been corporate farming. Nineteen of our states have either passed or proposed anticorporate farm legislation. Who is this great villain that everybody wants to legislate out of business? How much influence do we corporate farmers really have? Corporate farms today own 2 percent of the agricultural land in this country. Of that 2 percent, 92 percent are family farm corporations. So the public corporate farm is not a major factor in the agricultural setup in this country. And it certainly is not a villain. It may help over a period of time to bring some business and marketing and financial abilities together with this rare and wonderful thing that is the American farmer. Somehow the two of them will probably come together to constitute American agriculture in the future.

SUGGESTED READINGS

"Family Farms Or Corporate Agriculture?" *North American Farmer,* 25 Jan. 1985, 3.

Harrison, Fraser. *Strange Land: The Countryside, Myth and Reality.* London: Sidgwick and Jackson, 1982.

Ketchum, Richard M. *Second Cutting: Letters from the Country.* New York: Viking, 1981.

League of Rural Voters. "U.S. Farm Policy and World Hunger: The Deadly Connection." Minneapolis, Minn., 1986.

Logan, Ben. *The Land Remembers: The Story of a Farm and Its People.* New York: Viking, 1975.

Ritchie, Mark. *The Loss of Our Family Farms.* San Francisco: Center for Rural Studies and U.S. Farmers' Association, 1979.

With These Hands. Photographs by Ken Light. Text by Paula DiPerna. New York: Pilgrim, 1986.

History of the Family Farm

IN AN ESSAY on the differences between types of moral language, Henry David Aiken (1962, chap. 4) calls attention to the complexity of ethical discourse. He points out that when we talk about issues like abortion, euthanasia, and distributing wealth, we naturally use a diverse array of expressions. It is important to get clear about these various levels. Aiken suggests that there are four different levels of moral language: the premoral, the moral, the ethical, and the postethical.

At the first, premoral level, we have not yet reached truly ethical discourse. Sometimes we just vent our emotions saying, "Hurrah for the family farm!" or "A pox on all agricultural fundamentalists!" These are examples of expressive language; they communicate raw feelings. As Aiken suggests, this is not immoral or unethical language, it is simply premoral. In it we have not yet reached the level of ethical reflection. We should not ask whether this or that emotion is morally justifiable; our emotions are not things we should blame or praise ourselves for. My mother used to say that you cannot help how you feel. This is not only good psychological advice; it is good moral theory.

We enter the sphere of moral discourse when we begin to reason about actions. Here we enter the second level. Should we try to save the family farm? Should we do away with investment tax credits favoring large corporate farms? Do smaller farms continue to serve the common good in the modern world? These questions invite us to think about historical developments, the choices those developments force upon us, and the values we should use in making those choices. At this, the moral level, we must describe what is taking place, ask what solutions are possible, and figure out which one to select. At this level there are two subtypes of language: (1) the appraisal of actions and facts and (2) the appeal to rules that may guide our decisions. Thus we might decide (at the level of the first subtype) that family farms are in fact endangered, and that this is unfortunate since they represent the most equitable system of landownership. At the level of the second subtype, we might agree that society should always pursue a fair system of distributing resources. If all of these judgments were correct, then we would have a moral argument for saving the family farm.

Of course, we could always ask whether we reasoned correctly. Did we get all the facts straight? Are family farms really endangered? Are they really the most equitable way to allocate landownership in the 1980s? Then again we might ask whether the moral rules that guided our reasoning were really the best ones. Should a society always try to spread landownership as evenly as possible? What if the vast majority of citizens do not care to own farmland?

When conflicts arise at this second level, we are forced to move to Aiken's third level, that of ethical principles. Which of our values take precedence? Which rules are more important than the others? Should we ever alter the rules, change our moral code? Such questions rarely arise in actual conversation, but they become very important when two rules clash, or when one of our values tells us to do something our conscience knows is wrong.

The fourth type of discourse moves us beyond moral language strictly understood. Occasionally we ask ourselves why we should be moral in the first place. Why should I do what the moral code says? Aiken calls this the postethical level, because it really concerns existential or even religious matters, not ethics.

We can see that understanding moral discourse is a very complex matter, perhaps as complex as understanding the American agricultural economy! But it does simplify things a bit to observe that the questions having to do with our moral obligations occur primarily at the second and third levels. In turn, this directs our attention to such things as the facts about ourselves, our neighbors, and our society, and the rules and principles that are generally accepted in our culture. Thus our first task in deciding whether there are ethical reasons for saving the family farm is simply to find out what is going on. Then we can ask, what ought we to do about it?

The following three essays take us a long way toward answering the first question. The opening piece, "A History of the Family Farm," is written by one of this country's foremost agricultural historians, Richard S. Kirkendall. Tracing the history of the Jeffersonian agrarian tradition as it meets up with the modern technological tradition, Kirkendall shows us what is going on; the family farm "is not nearly as important as it once was." Having provided a solid introduction to the facts he directs us towards the third level with a pointed question: "If we are to be persuaded that now when family farms are so few in number we have a moral obligation to save them, advocates must explain why this is so."

With Kirkendall's review of the past two hundred years as background, the facts pertaining specifically to the recent crisis are discussed by Paul Lasley. As the director of the Iowa Farm and Rural Life Poll, Lasley has had extensive contact with Iowa family farmers the past five years. His analysis of the changing economic, sociological, and psychological conditions of small farming communities gives us a closer look at the most up-to-date changes occurring in the Midwest. Lasley concludes that if family farm supporters are going to convince non-farm people to help preserve their institution, they should not argue that small farms represent a unique system of values. Rather, he says, they represent one part of our society that might be able to help foster the development of a new ideology, a new system of values and beliefs including community, family, environmental quality, and cooperation.

Of most immediate concern to family farmers themselves is the mounting debt that looms over thousands in the Corn Belt. Neil Harl's analysis, "The Financial Crisis in the United States," addresses this issue. He urges the formation of a national agricultural financing corporation, which would have the mandate to buy farmland from troubled owners in the short term, rent it back immediately to its present occupants, and encourage them to repurchase it as soon as possible. Harl seems to think that it is possible to save many—although certainly not all—family farms. But this will only prove feasible if the country as a whole demonstrates the will to do so.

Once we have all of the relevant facts before us we can move on to consider questions of social justice, fairness, and stewardship. But we should be sure we know "what is going on" before delving into matters of ethical theory and theology. So we temporarily postpone the ethical questions, taking them up in the parts that follow.

Reference

Aiken, H. D. *Reason and Conduct: New Bearings in Moral Philosophy.* New York: Knopf, 1962.

RICHARD S. KIRKENDALL

A History of the Family Farm

SINCE THE BEGINNING of the United States as a nation, the family farm has been an institution and an idea of large importance. To its admirers, it has seemed to be a foundation of much that is good in America, above all, democracy. Yet, although for more than a century the number of family farms increased dramatically, the history of their importance has been largely a story of decline; and during the past half century, the number has dropped sharply so that now little more than 1 percent of the American people live and work on such farms. As an institution, the family farm is not nearly as important as it once was. As an idea, it still has numerous adherents, but they have difficulty making the claims for it that once were made.

Historians have their own way of looking at human affairs, one that emphasizes development or change over time. The development of agriculture and rural life is a very big subject, one of the largest in the American story. Once nearly the whole of the nation was rural; the farm population is now only a small part, and that change is a fundamental part of American history. Farm people were the American majority; now they are only a minority, and no other group in American history has experienced such a change—a change from majority to minority status.

Agriculture's people have been dynamic in other ways. They have been significantly involved in two gigantic population movements—the move to the American West and the move to the American city. Rural people have become increasingly productive over the course of our history, doing so by building a large number of farms and then by consolidating many of them and applying technology to them. This productivity has been a problem as well as a great blessing. It has pressed down on farm prices as well as served needs and conferred power.

RICHARD S. KIRKENDALL is Henry A. Wallace Professor of History at Iowa State University.

American agricultural history is not only the story of change. Tradition has also been important. An agarian tradition took shape in the eighteenth and nineteenth centuries that exalted farming and rural life, looking upon them as having great importance for the nation's welfare. That tradition glorified—and still glorifies—the family farmers and portrays independence as one of their greatest qualities. Yet as agriculture and rural life were modernized, some of our farms became large corporations; and even the surviving commercial family farms have not only become large in size but also have become parts of an agribusiness complex in which much of significance in supplying food and fiber takes place off the farm.

Two concepts—tradition and modernization—help us organize and understand American agricultural history. The interplay between them and the eventual triumph of modernization form the central theme of that history.

In the beginning, the United States was almost entirely rural, and agrarian ideas made great sense to many Americans. About 95 percent of the population in 1790 was rural, and nearly all of those rural people worked on farms and plantations. The opportunity to make a farm was one of the great attractions of America, and ideas that had been available since the days of ancient Rome and before encouraged Americans to believe that farming was the best way of life and the most important economic activity. To agrarians, the United States seemed to be a superior place because it supplied more and better opportunities to farm than any other part of the world. Difficult to acquire in Britain and Europe, land was abundant here, and it was rich, and the persons who worked the land, the agrarian nationalist proclaimed, usually owned it and could use it as they wished. This was in contrast to Europe where land use was dictated by other people such as the aristocratic rulers.

The idea of the family farm was a basic part of Jeffersonian agrarianism. Americans had inherited agrarianism from Europe, but they gave it a twist of their own. Greeks and Romans had created that philosophy; it enjoyed a revival during the Renaissance, and it flourished in England and on the European continent in the eighteenth century. But the ideas about superiority of farming and farm people were not then tied to a particular size of farm or political system. The ideas were associated with large as well as small farms and with both aristocracies and monarchies.

American democrats, above all Thomas Jefferson, took old agrarian ideas and gave them a democratic interpretation. Here, the often expressed ideas about the benefits for the human personality of work on the land was very important. To Jefferson and others who thought like he did, such work developed the personality type needed for the success of a democratic political system. But that work must be done on family farms, and family farmers must be the majority of the population.

This democratic version of agrarianism had several major dimensions. It involved a theory of personality formation. It assumed that the nation needed a large number of family farms, each endowed with the resources required for families to lead good lives. It stressed the importance of landownership. It insisted that a family farmer (who in most cases was male) must not be obligated to others such as landlords, must be free to do what he wished with his land and its products but should have no more land than his family could use. And family farmers deserved large roles in government and special attention from it.

Over the years, agrarians would develop other claims for the importance of farmers. They included ideas about the economic value of agriculture and farm people. Agriculture was defined as the most important part of the economy for it produced the essential commodities, and farm people were looked upon as essential consumers as well as producers who must prosper for the nation to do so. Such ideas obviously assumed that farmers participated in the market and were not self-sufficient. Jefferson accepted this assumption, as his belief in free trade and the value for American farmers of the European market indicated. The needs of Europe seemed to him to be one guarantee that there would be many opportunities for American farmers and that thus the nation would succeed.

To the Jeffersonians, abundant land provided the other guarantee of success for the democratic experiment. Western lands seemed especially important; families eager to acquire farms could find many opportunities to do so in the West. It seemed so important to Jefferson that, as president, he brushed aside his own constitutional inhibitions to purchase Louisiana. That acquisition appeared to him to promise that the United States would remain a nation of farmers for many generations. To the agrarians, it seemed necessary not only to acquire western lands but also to remove Indians from them so that American farmers could occupy the acres.

From the beginning, the idea of a democracy based on family farms had to compete against alternative visions of what the nation should be. One challenge came from within American agriculture itself. That was the plantation system that dominated life in the pre–Civil War South. Plantations were customarily owned by individuals, not corporations, but unlike family farms, which were numerous in the South as well as the North, slaves rather than the owners of the land and their families supplied most of the labor. To its defenders, slavery gave planters the

leisure required to be good citizens. Family farmers had to work too
hard.

The antislavery movement, especially its Free-Soil variant, had an
agrarian dimension. Slavery was wrong because it violated the democratic version of the agrarian creed. Above all, it must not be allowed to
migrate into the vitally important West.

Western lands had a crucial role to play in the development of an
agrarian democracy. They guaranteed its prosperity and continuation.
This notion exerted a major influence on the development of national
land policies during the first half of the nineteenth century. In addition
to encouraging the expansion of the nation and the removal of the Indians from land coveted by white farmers, democratic agrarianism promoted a steady lowering of the selling price for land owned by the
federal government. Finally, as a major part of the triumph of the Free-
Soil movement during the Civil War, Congress passed the Homestead
Act in 1862, which offered free family-sized farms (160 acres) to those
who would make farms out of the land. Such gifts seemed justified by
the benefits that would come to the nation, its people, and its land from
the interactions between people and land. People had a natural right to
land, agrarians like George Julian and Horace Greeley argued, and labor
gave land its value. They also maintained that urban workers would
benefit from western land. It would function as a "safety valve" by
offering workers an alternative to their jobs. Because they had the alternative, they would not be subservient like European workers and
would develop the personalities required to make a democratic system
successful.

During the course of the nineteenth century, millions of people
poured into the Middle West, the lower Mississippi valley, and beyond.
They were lured by land and the promise of good markets in the East,
the South, and Europe, and of access to them. They were eager to take
advantage of opportunities and were not tied down to families and
places. They even established farms in an area known as "the Great
American Desert," encouraged by a theory that "rain follows the plow."
From 1860 to 1910, the greatest period of farm building in our history,
the number of rural people increased from eighteen to nearly fifty million, and they enlarged the number of farms from 2 to 6.4 million.

By the beginning of the twentieth century, American farmers had
developed a complex as well as a large agricultural system. Influenced by
varied land, weather, and transportation patterns, the system was
divided into several regions emphasizing different kinds of farming and
ranching. In the Northeast, dairying was most important. Cotton was
the chief cash crop in the deep South. Midwestern farmers raised a wide

variety of crops and livestock but emphasized corn and hogs. On the Great Plains, wheat was dominant, cattle were important. Rural people in the Rocky Mountains focused their energies on livestock, while those in the Far West were diversified.

Farming received a boost with the passage of the Homestead Act and the end of slavery after the Civil War. During Reconstruction, however, family farms did not replace plantations in the South. To be sure, the planters failed in their efforts to reestablish gang labor largely because the freed people were unwilling to work in gangs as they had as slaves on the large plantations. But because the federal government did not seize the land of the planters and redistribute it to black families and because few blacks could afford to buy land, sharecropping rather than family farming succeeded the old system. Whites continued to own the land; they divided the plantations into small units and rented them to black families. Because of their economic weakness, the blacks paid their rent with a portion of the crop and borrowed from their landlords the things they needed to live and farm. This dependent relationship clashed with the philosophy of agrarian democracy. Although passage of the Homestead Act suggests that democratic agrarianism had enormous influence in the 1860s, the fact that southern land was not redistributed indicates that this philosophy was not the only one with persuasive power.

Just as abolition failed to lead to the multiplication of family farms in the South, the Homestead Act failed to fulfill the dreams of its champions. It did increase the number of family farms and contribute to the great era of farm making in the half century after the Civil War; but many new farmers during those years had to pay for their farms, often going into debt to do so. Homesteading was not the only way to acquire land. The federal government continued to sell land, thereby enabling speculators to obtain it. And the federal government also made large grants from its domain for purposes other than homesteading. It granted land for the development of state colleges and for the construction of railroads.

Obviously, the belief in the family farm was not the only influence on land policy in the mid-nineteenth century. Of course, land grants for colleges and railroads were designed, in part, to benefit farming. But the move to establish colleges was influenced by convictions about the defects as well as the importance of farmers and the value of social mobility, including movement off the farm by young people eager to improve their lot in life. Railroads were important for farming as a business, not as the foundation of democracy.

The building of colleges and railroads was part of another vision of

America that conflicted with the Jeffersonian vision. The growth of manufacturing was another part. Jefferson had, late in life, come to accept manufacturing for the United States but had done so only on the assumption that it would never be more than a very small part and that the great majority of Americans would continue to be farmers. In the early days of the building of factories here, it seemed possible to keep them small and thus to keep the urban population small. But with the substitution of steam for water power, factories grew greatly in size and stimulated the growth of cities.

With farming still demanding much work by people, the agricultural population was large, as Jefferson had hoped it would become; but rural society did not conform with his aspirations in every way. Midwestern agriculture, dominated by commercial family farmers, resembled his vision most closely. The Cotton South diverged from it widely. There, a large number of farmers were sharecroppers who farmed very small units, did not own the land on which they worked or the shacks in which they lived, and even depended on their landlords for supplies. Dependence and poverty characterized the lot of these black and white farmers. The scene in places like Appalachia or the Ozarks would also have depressed the great apostle of the American agrarian tradition, for there most farms were largely self-sufficient and very small and the people were poor.

The nation as a whole had moved far from what it had been in Jefferson's day. Although the farm population had grown rapidly, the urban population had grown even faster. Fed by immigrants from abroad and migrants from rural America, the urban population had grown from 22 percent of the total in 1860 to 51.4 percent in 1920. Although many of the places that the census taker called cities were small, the urban trend was clear and strong.

Farmers themselves had contributed significantly to the dashing of the dream of an agrarian nation. By expanding production enormously, they permitted and helped cities and industries to grow. City people could get the food and fiber they needed from American farmers, and industrializers could get capital from abroad, for exports of American wheat, cotton, and other farm products balanced the flow of European funds. By moving away from self-sufficiency, the rapidly expanding farm population gave American manufacturers a large home market. In addition, the children of the often large farm families tended to move to the city, finding its economic and cultural opportunities better than those available in rural America.

By the late nineteenth century, farmers continued to be the largest group in the American population, but they were no longer the majority.

To some people, influenced by Jeffersonianism, the change was troublesome. This was especially so for those who also suffered from low farm prices, as was true of the cotton farmers of the South and the wheat farmers of the recently occupied Great Plains. They joined in a great uprising known as the Populist Revolt, reasserted agrarian values, criticized the recent changes in American life, especially what seemed to them to be an undemocratic concentration of economic and political power. They insisted that family farmers, given their great importance for democracy and economic prosperity, deserved to be helped by government. The Populists did not wish to destroy the new industrial system, for farmers wished to share in its benefits; but Populists did want to make farmers more important in the scheme of things and did not believe that farmers deserved to suffer, as many were. The defects in American life lay beyond the farm fences.

Largely white, middle-class, landowning farmers in the South and on the Great Plains, the Populists produced cotton and wheat, crops that sold in a now overcrowded world market. Served by an inadequate money system, their prices had been falling so that they could not pay their debts or prosper. Influenced by the agrarian tradition, they felt that the America they believed in had been displaced and that they were exploited in the new one by an immoral, unproductive "money power" or "monopoly capital." Convinced that they were virtuous hardworking people, they believed they deserved to flourish, so they tried to change, not farmers and farming, but the system within which they operated. Not seeking a "Golden Age" free of cities and factories, they pressed for political reforms that would destroy monopolies and change the money system and thereby permit family farmers to enjoy good times.

The Populist Revolt failed in its efforts to reassert agrarian values. The Democratic party absorbed much of the movement in 1896, focused attention on only one of its reforms, free silver, and suffered a decisive defeat. Not even all farmers voted for the Democratic candidate for the presidency, William Jennings Bryan of Nebraska. Midwestern and eastern farmers voted Republican in large numbers as did wage earners and the urban middle classes. Many who did so identified with the new urban industrial way of life and saw the Populists and Bryan as a threat to it and the material abundance and opportunities for economic and social advance that it seemed to promise.

In the meantime, another point of view was emerging that emphasized changing—modernizing—farming and rural life. It had a variety of participants including the railroads with their development offices. The United States Department of Agriculture (USDA), the land grant colleges, the experiment stations, and the extension services focused their energies on the modernization of farm people, their institutions, and their practices. The people involved in the establishment and early operations of these agencies believed that farmers and farms were very important; they also believed that farm practices needed to be changed. They must become more scientific.

In their formative years, these agencies moved through three stages. The first emerged in the Civil War years and involved the establishment of the USDA and a number of agricultural colleges. The passage of the Morrill Act allowed land to be given to the states to encourage them to set up such colleges. Then in the 1870s and 1880s, the agricultural scientists and educators and their associates established experiment stations to supplement and strengthen the research in the Department of Agriculture and give the agricultural colleges lessons of practical value to teach. Congress gave this effort a boost with the passage of the Hatch Act in 1887, which made federal funds available for stations attached to land grant colleges. Finally, the extension services took shape in the late nineteenth and early twentieth centuries. They were designed to bring the new scientific knowledge directly to the farmers and to demonstrate to them that they would benefit from becoming more scientific. Again Congress aided the movement, this time with the passage of the Smith-Lever Act in 1914, which gave federal money for the development of extension services everywhere in the nation.

The Smith-Lever Act was a victory for a modernization effort known as the Country Life movement. Not a farmers' movement, it was composed mainly of city people, although many of them had been born and raised on farms and all looked upon agriculture and rural life as very important. Many features of the contemporary rural scene troubled them: the large-scale migration of people to the cities, tenancy and absentee ownership, overworked women, poor health, inadequate roads, poor schools and churches, inefficiency, and so on. They feared that soon the nation would not have the kind of rural population it needed. At work in a period of relatively high farm prices, they feared that the food supply would become inadequate and too costly for the fast-growing urban population.

Thus, they proposed a series of changes (such as the consolidation of rural schools and the development of the extension services) hoping they would hold the best rural people on the land and make agriculture more efficient. They sought a renovated agriculture and rural life that would be a substantial and significant part of the nation and serve its needs efficiently. Neither flattered nor persuaded by the analyses and prescriptions, many rural people resisted the Country Lifers. But Teddy Roosevelt liked them and established the Country Life Commission to push the cause along. Congress, especially during Woodrow Wilson's years as president, passed several laws that incorporated the ideas of the movement and improved rural mail service and rural roads, encouraged the formation of cooperatives, strengthened the Department of Agriculture, boosted vocational agricultural and home economics in the high schools, and expanded agricultural extension.

The American Farm Bureau Federation emerged in these years as another modernizing force, although it was, like the Country Life movement, influenced by the agrarian tradition. It took shape from 1911 to 1920, beginning with the formation of county farm bureaus, moving to the organization of state federations, and then bringing the county and state bureaus together into a national federation. At first the movement was promoted by certain types of business people and public officials in the USDA, the agricultural colleges, and their extension services who feared agrarian radicalism such as the Populists had expressed and hoped to make agriculture more prosperous by making it more scientific. The organization prospered, stabilizing at about three hundred thousand farm families in the 1920s, most of them in the Middle West, especially Iowa and Illinois. The members were the more substantial farmers, the rural businessmen. They seemed most capable of applying science, shared an interest in higher prices, and seemed likely to give the movement a nonradical bent. The organization imitated rather than attacked urban business, organizing farmers as urban businesspeople were organizing, forming organizations for economic action — the marketing cooperatives — and engaging in political action seeking policies favorable to their interests.

Another view of farm life had emerged alongside the tradition of agrarian democracy. The newer view was "modernization." It did not place the same heavy burden on farms and farm people that the tradition imposed on them. According to advocates of modernization, farming was essentially a business and important for economic reasons. It produced essential commodities. The main thrust of modernization was the substitution of technologies for people. As the modernization process moved forward, it enlarged the scale of rural institutions including rural schools as well as farms and developed a vast complex, or agricultural system, now often called "agribusiness," in which farms were only a part. The process formed part of a larger transformation involving the advance of science, technology, industry, the city, and capitalism.

The agrarian tradition and agricultural modernization overlapped in one highly significant way but differed in others. Both preached the great importance of agriculture; but one proclaimed the political importance of farms and farm people, while the other emphasized their eco-

nomic significance. Moreover, traditionalists insisted that the farm population must be very large — a very substantial percentage of the total — if
a democratic system was to function successfully. The modernizers, however, were indifferent to the question of population size, emphasizing
instead the ability to produce and the contributions of technologies.
They believed in a large farm population only when the technology
seemed to demand it.

By 1920 the number of farms in the United States had nearly
reached it peak. By then there were nearly six and a half million farms,
four and a half million more than had existed in 1860. Not all of these
were family farms. Many were operated by sharecroppers — over a half
million in the South. In fact, this and other forms of tenancy were on the
rise. Farms operated by tenants had increased from one million in 1880
to two and a half million in 1920. Furthermore, some of the nation's
farms were large corporate enterprises on which most of the work was
done by wage laborers. This type was especially prominent in the central
valley of California where the prevalence of large farms had long distressed traditionalists, who believed that farming should look like it did
in the Middle West and hoped that irrigation projects would produce a
great increase in the number of family farms in the Golden State. In the
Middle West and East, the family farm, the farm owned and worked by
a family with little or no hired labor, was the dominant type.

Although the farm population was large, it was only a large minority of the American people. For the first time, rural America, which
included people in towns and villages of less than twenty-five hundred as
well as farmers, was smaller than urban America. There were fewer than
fifty-two million rural people in 1920, more than fifty-four million urban ones. Thirty-two million Americans lived on farms, approximately
30 percent of the total, down from nearly 44 percent in 1880. In later
years, however, the percentage and the number of people who lived on
farms in 1920 would seem very large.

Also in 1920, farm prices fell sharply, and the greatest crisis in the
history of American agriculture emerged. The crisis was greater than the
other major ones — the crisis of the late nineteenth century and the crisis
of the 1980s — in part because the farm population was larger, and thus
more people were directly affected by it. Furthermore, this crisis lasted
for an unusually long time. It persisted, with varying degrees of intensity, for two decades, and it had gigantic consequences. It influenced a
substantial expansion of the federal government in agriculture and
helped to initiate the Great American Agricultural Revolution of the
1940s and the years thereafter.

While the crisis strengthened desires among farm people to leave
farms, it also generated schemes to save farmers. In the 1920s, the most

prominent of those schemes, the McNary-Haugen plan, was affected by modernization. It proposed that farmers should adopt rather than rebel against a practice of urban business—the use of the protective tariff to establish a two-price system—and it advocated the establishment of a government corporation to market "surplus" farm products. But the agrarian tradition also influenced the plan. Its champions, including Henry A. Wallace of Iowa, believed that the industrialization and urbanization of America were moving too fast and too far and that the nation needed a government program that would raise farm prices and thus persuade large numbers of people to remain on the farm.

After 1929 the crisis entered its most severe phase; and in 1933 a New Deal for the farmer began to take shape. Both democratic agrarianism and agricultural modernization, as well as the Great Farm Crisis, influenced the agricultural New Deal. Most obviously, it enlarged the role of the federal government in farming. This formed part of the modernization process; the developing agricultural system included much more than farm families working the land. The New Deal also made heavy use of an agency of modernization, the extension services, in the administration of the farm program. The agricultural New Deal stressed the business character of farming, defining the "farm problem" as low prices, seeking to raise them so as to make the business profitable again, encouraging farmers to regulate their production as industrialists did, strengthening the agricultural credit system so as to supply credit to good risks, and making nonrecourse loans on agricultural commodities so as to guarantee higher prices.

The agricultural New Dealers thought in terms of the great importance of farmers but tended to emphasize their economic importance, rather than their political importance, portraying them more often as essential consumers and producers than as the foundation of democracy. The champions of these farm programs drew heavily on the argument about the large value of farm purchasing power for the health of the entire economy to justify federal aid to the farmer. As time passed, these New Dealers made increasing use of the old argument about the crucial nature of the things that farmers produced. From an early point, the New Deal distributed some of the surplus crops to people on relief, influenced in part by an apparent paradox: the coexistence of a highly productive agriculture and hungry people. Soon Wallace made plans for an "Ever Normal Granary," which became a program in 1938. The official line was that government must store farm crops to assure consumers that they would have enough to eat. The New Deal also launched an experiment with food stamps in the late 1930s to make more food available to the urban poor.

As the agricultural New Deal developed, other efforts to change farming and rural life were added. An old idea of the agricultural scientists — soil conservation — became a part of the New Deal in the Soil Conservation Service and other agencies. The Rural Electrification Administration, established in 1935, quickly became a major modernizing force. Its aim was to make farm life more like city life. Before its establishment, only 750 thousand of the nation's farms had electricity. By 1941, 2.3 million did.

These features of the New Deal reflected the influence of agricultural modernization, but the agrarian tradition also exerted influence. Most obviously, the New Dealers — above all, Henry A. Wallace, the secretary of agriculture — hoped to maintain a large farm population composed mainly of family farmers. The Agricultural Adjustment Administration (AAA) tried to strengthen established farmers. In its major program — production control — AAA did urge farmers to copy an urban pattern of behavior rather than try to change it. Yet production control was only a means, and the goal was to protect commercial family farmers and hold them on the land so that the United States would continue to have a substantial farm population. Also, as was indicated by the formation of committees with farmers as members to administer and plan farm programs, agricultural New Dealers assumed that farmers were people who could participate effectively in government.

The existence of extreme poverty in parts of rural America troubled some New Dealers. They made the federal government's first attack on the problem of rural poverty, doing so chiefly through the Resettlement Administration (RA) and the Farm Security Administration (FSA). Over 3.5 million of the more than 6.5 million farms had less than one hundred acres, and farms of that size in most types of agriculture were too small to support a family. Also, many of the small family farmers did not own their land. Reflecting the influence of the agrarian tradition, RA and FSA tried to improve the lot of the rural poor in various ways so as to keep them on the land. They made loans both to tenants to enable them to become farm owners and to small farmers to enable them to enlarge and improve their farms so that they could support families. Some New Dealers preferred the migration of large numbers of rural people to the cities but recognized that there were no job openings there in the 1930s. Also, the Farm Bureau and many congresspeople believed that the federal government should not devote resources to the rural poor and kept RA-FSA appropriations small.

FSA's program to increase the number of family farmers was one of the smaller parts of the New Deal for agriculture, and New Deal benefits were not limited to family farmers. Government benefits flowed also to the corporate farms of California and the large landowners of the South.

In response to critics, Wallace and others insisted that they must have the cooperation of such farmers if the programs, especially production control, were to succeed.

Furthermore, the agricultural New Deal promoted and facilitated technological change. The U.S. Department of Agriculture expanded agricultural research during the 1930s. New Deal payments to farmers and cuts in acreage and higher incomes gave farmers inducements to make changes on their own. Rather than pass on a portion of the payments to their sharecroppers, many southern planters held on to the money, lowered the status of their workers from sharecroppers to wage laborers, and evicted some of their sharecroppers. Some farmers, including cotton planters, reduced their need for sharecroppers and hired labor by investing some of their new resources in tractors, a relatively new technological development.

Tractors had first become important in the 1920s, chiefly in the Middle West and on the Great Plains, and they grew in importance in the late 1930s even becoming numerous in the South for the first time. In 1920, less than 4 percent of the farms had tractors. By 1940, over 23 percent did. The number of tractors on American farms, which had increased only slightly in the first half of the 1930s, jumped from 1 million in 1935 to 1.6 million five years later.

In this fashion, the New Deal for agriculture contributed to a slow decline in the farm population. After falling by more than a million during the 1920s, it grew during the early 1930s, nearly reaching by 1933 its 1916 high of 32.5 million. Then decline began again, and the number fell to 30.5 million in 1940, 23.2 percent of the total American population. The decline would have been greater if there had been more jobs available in the cities.

In the 1940s the pace of change in agricultural America became revolutionary. The Great Farm Crisis of the 1920s and 1930s contributed to the Great American Agricultural Revolution. By 1940 many farm people were tired of the hard work and low rewards of farming; and analysts including Wallace spoke of the "surplus" farm population. Too many people seemed to be working too hard for too little. Farm life was not as good as agrarians insisted.

Other influences also contributed to the revolution. The economic boom ushered in by World War II enlarged urban opportunities. The agencies of modernization, including the agricultural colleges and the corporations involved in the agricultural system such as John Deere and Pioneer, functioned as revolutionaries.

The revolution had a technological dimension. The tractor—the substitution of motor power for animal and human power—was a major feature. Tractors now became much more widely used, larger, faster, and more powerful; and machines were developed and adapted for use with them. The mechanical cotton picker illustrates the change. Responsible for only 10 percent of the cotton harvest in 1949, it harvested 96 percent twenty years later. By then, there were nearly five million tractors on the nation's less than three million farms.

There were other features of the technological revolution. One was more productive crop varieties such as hybrid corn. Henry A. Wallace introduced it commercially in 1926; farmers began to use large amounts of it in the 1930s; and it completely displaced its competitors early in the postwar period. By then other types of hybrids were being introduced including hybrid chickens. Agriculturists developed better breeds of live-stock, and farmers increased dramatically their use of chemical fertilizer and began to use insecticides, herbicides, and fungicides. These chemicals eliminated hard and unpleasant tasks and greatly increased output. Farmers also became more specialized and made increasing use of modern managerial practices including larger farms and computers.

The rapid rate of change suggests that farmers as a group were more receptive than ever before to advice from the modernizers. The industrial corporations that had tractors, seeds, chemicals, and other products to sell farmers supplied some of the advice. The Department of Agriculture, the land grant colleges, the experiment stations, and the extension services also functioned as promoters of the technological revolution and contributed significantly to the development of the new technologies. Farmers who could afford to do so now welcomed what these agencies had to offer, for increased efficiency seemed to provide a way to greater profits.

The new technologies made the farmers who employed them much more productive. Output per farm person increased more rapidly than the output of workers in manufacturing in the years after 1940. Before the days of hybrid corn, yields averaged below forty bushels per acre; now yields average well over one hundred bushels. Before the advent of the mechanical cotton picker, field hands worked 140 hours to produce a bale of cotton; by 1968 it took 25 hours. In 1820 before the first technological revolution had begun, one farm worker could supply the food needs of four people; in 1945 as the second revolution was getting into full swing, one worker served the needs of fifteen people; by 1969 one worker served forty-five consumers; now the ratio is one to more than seventy. The farm person is, of course, helped by scientists, technicians, educators, government officials, manufacturers, salespeople, and others.

Family farmers still produce most of America's farm output, but their numbers have dwindled, and their farms have grown in size. At the end of World War II, the average American farm had 174 acres; by 1960 it had expanded to 303 acres. By 1975 it had 376. The average commercial farm jumped from 404 acres in 1959 to 534 acres in 1974. The new technologies permit families to farm much larger units than they could before World War II. Encouraged to expand, some owners of family farms rented as well as bought additional acres.

As farms have grown in size, corporation farms have increased in number. Some are owned by families who have chosen the corporate farm because of the legal benefits it provides. Other corporate farms exist largely for the tax benefits they give to their nonfarm owners who are not interested in making profits from their farm operations. Most corporate farms, however, are giants employing the latest technologies and people who work for salaries and wages. Some of these are parts of urban-controlled organizations. The large corporate farms are not products of the economies of scale; they are not more efficient than well-run, highly capitalized family farms.

To use a term coined in the 1950s and used frequently since then, American commercial farms of both the family and the corporate types are now parts of a complex agribusiness system. That system includes all of the businesses that participate in the production and distribution of food and fiber. It includes feed, seed, and fertilizer companies, farm machinery manufacturers, food processors, and other businesses. It is in these areas that the most important corporate giants involved with agriculture are located. Some parts of the system are much more important now than they were before 1940 because their products are more important in farm operations than they were or they supply things that farmers once provided for themselves such as energy, seed, and fertilizer. Some firms in the complex are linked more closely with farms (or ranches) than others, some because they own farms or own feedlots, some because a system of contract farming has developed rapidly in some lines. An example is the poultry industry, which has shifted decision-making power off of the farm and into the hands of firms that supply feed or process chickens or have some other reason to tie particular farmers close to them. The off-farm parts of the agricultural system, in other words, have been enlarged, and some agricultural functions have been transferred off of the farm.

In this vast complex, most of the income to farmers goes to the larger farmers, but most of the total income goes to the other parts. In the late 1970s, less than 20 percent of the farms earned more than 75 percent of the farm income. Processors and retailers, however, normally obtain about 60 percent of the income received from the retail sale of agricultural products. Farmers get about 40 percent but must pass on much of that to their suppliers.

The revolution had a demographic dimension as well as a technological one. After 1940 there was a massive migration of people from country to city. In four decades, more than thirty million people made the move, a number greater than the number of people who moved from Europe and Asia to the United States and settled here permanently during the great period of immigration from 1815 to 1914. And the people who moved now had adjustment problems that were at least as large as those faced by immigrants earlier. Neither group had been trained for life in the city. Earlier, nearly all of the migrants from farm to city had been whites; later, many were blacks from the South and victims of poverty and poor schooling.

"The United States was born in the country and has moved to the city," one historian, Richard Hofstadter (1955, 23), wrote three decades ago. His observation is even truer today than when he made it.

What are the factors responsible for this enormously significant change in American life since 1940? One is a constant of American history: the relatively high rural birth rate. But that is not the most significant part of the story. Why didn't people who were born on farms remain in agriculture? A major part of the answer is that they were not needed there but could be put to work in the cities, at least much of the time. With new technologies in use, agriculture did not need as many people as it had employed in the past. Cotton planters, for example, substituted tractors and mechanical pickers for sharecroppers and wage laborers. Farmers who employed the new technologies could operate larger units. To enlarge their farms, they bought neighboring land, often freeing the people who had owned and worked it to seek employment in other places.

The farm population dropped from over 23 percent of the total in 1940 to less than 3 percent today. Clearly modernization has triumphed over agrarian democracy.

Government policies must receive some credit for the transformation of agriculture and rural life. The USDA backed by Congress conducted research that contributed mightily to technological changes. Washington also helped farmers obtain the funds demanded by the modernization process. The federal government continued the price-support program that had been introduced by the New Deal. It was controversial, and critics talked of doing away with it. The Republican secretary of agriculture during the 1950s, Ezra Taft Benson, was especially forceful on this point. But the debate actually focused on the level at which prices should be supported rather than whether or not government should be removed immediately from decisions about agricultural prices; and Washington remained active in the area, even reaching a record level of expenditures in 1983 under a Republican administration that had promised a drastic reduction in the operations of the federal government. Thus, it continued to support prices and to make payments to the farmers who participated in the programs, thereby helping them obtain funds for modernization. The largest payments went to the largest farmers.

That Washington continued to give so much help to a dwindling minority is a problem demanding an explanation. The strength conferred by organization is part of the answer. The vocal and active participation of farmers in politics is another. So are the skills and positions in Congress of the rural representatives there. But the agrarian tradition cannot be ignored. It continued to exert influence, persuading many city people and politicians that they must not let the farm population completely die out even if protecting it meant higher food prices and the expenditure of tax dollars. A development that began in the mid-1960s must also be noted—the reestablishment and vast expansion of the food stamp program. Here was a farm program that helped a large and growing number of city people. Placed and kept in the major farm bills and the Department of Agriculture, it helped to offset urban criticism of government aid to the farmers and to hold on to urban votes for farm legislation.

Washington also contributed to the transformation of agriculture and rural life in negative ways. After the war, it dropped New Deal programs on behalf of the rural poor and never reestablished them. There were, in other words, no programs such as those administered by the FSA that might have encouraged and helped sharecroppers, farm laborers, and small farmers to remain on the land. Such programs seemed to the Farm Bureau and others to violate the logic of business, however desirable they might appear from a democratic agrarian point of view. Thus the Farm Security Administration was destroyed in the 1940s.

Some people did suggest that programs be developed to help the rural poor make the transition to urban life, but here the agrarian tradition got in the way. Americans seemingly could not allow themselves to plan migration out of farming. So the rural exodus took on massive proportions soon after World War II began and moved forward without guidance. Had humane planning been employed, the nation might have avoided the burning cities of the 1960s.

The federal government also failed to enforce the 160-acre limitation in reclamation law, especially in the gigantic Central Valley Project in California. That legal principle was intended to restrict water from federal reclamation projects to family farmers and was designed to guarantee that irrigation would increase the number of them. But foes in

California argued that application would be impractical and undesirable for it would destroy the large corporate farms or force them to use other ways of getting water and would hamper economic growth. This view prevailed.

In spite of efforts to do so, Washington failed to abolish the cost-price squeeze from the lives of farmers. In addition to the price-support system, the government attempted to cut farm production with the soil bank program of the 1950s and the PIK plan of the 1980s. Washington employed and proposed subsidies, as in the Brannan Plan of 1949–1950, and sought to eliminate the squeeze by influencing demand via Public Law 480 and the food stamp program. Nevertheless, farmers continued to be caught in cost-price squeezes, doing so in the 1950s as well as today. Some government policies such as embargoes, deficit financing, and tight money contributed to the squeeze at times.

Washington's failure contributed to the movement out of farming. The technological changes raised costs of farming substantially. Farmers had to pay for things like gasoline and fertilizer that they had not paid for earlier, and the things that they needed to farm in modern ways became increasingly expensive. Thus, farmers burdened with debts encountered price drops that forced them to sell out and move to the city in search of satisfactory employment. Unlike their financially strong neighbors, they could not respond to the problem of low prices by expanding operations to reduce unit costs. Some hard-pressed farmers responded to the cost-price squeeze by joining the National Farmers Organization in the 1960s and the American Agriculture Movement in the late 1970s; they staged demonstrations and threatened farm strikes and holding actions but failed to raise prices. Some farmers caught in the squeeze turned to part-time jobs off the farm in hope of holding on to their land.

The national government affected family farmers in still other ways. For example, it pursued tax policies that encouraged city people to buy farms. Often the result was displacement of farm people, at least as farm owners. Ownership by a farm family, of course, is an essential part of the definition of a family farm.

The family farm has become a rare species, nearly extinct. Many, perhaps most, of our high-tech farms today are family farms. In fact, family farmers are even less dependent on hired labor now than they were in the nineteenth century, for the technology enables family members to do the work that needs to be done even on quite large farms. But the new technology is also used on plantations and corporate farms where the work is done by tenants and people on wages and salaries. Furthermore, the new technology is used on part-time farms by people who depend mainly on city jobs for their income. With hardly more than 1 percent of the people living and working on American family farms today, democracy surely cannot depend upon them for its health as the agrarian democrats desired.

One implication of the historical account is that we must look beyond history for a positive answer to the question posed by the title of this volume. If there is a moral obligation to save the family farm, then Americans have often behaved in immoral ways. This is not to say that belief in the family farm has not exerted an influence on American history. It has done so at a number of points, but other beliefs and other forces have exerted greater influence, eventually overwhelming family farms. Americans had opportunities to preserve the importance of this institution at crucial points when such farms were still numerous but failed to do what the task demanded.

If we are to be persuaded that now when family farms are so few in number we have a moral obligation to save them, advocates must explain why this is so. Is it because human beings live on them and we have obligations to human beings wherever they may be? Or is it because family farms have unique social value no matter what their number? When they are small in number, how can they exert a significant influence on society? As a historian who finds in history no basis for a positive answer to the question of this anthology, I would like to be persuaded by others that such an answer should be given. The most important suggestion I can make as a historian is that the argument must not assume that family farmers still consititute a large part of the American population.

Reference

Hofstadter, Richard. *The Age of Reform: From Bryan to FDR*. New York: Knopf, 1955.

The Crisis in Iowa

THIS PAPER EXPLORES TRENDS in agriculture related to community viability. The final section of the paper focuses on farm families' assessments of these changes, using data from the Iowa Farm and Rural Life Poll. Family farms are by no means representative of *all* types of farms found in the United States, but they are the dominant type of farm in Iowa and the Midwest. For this reason, I will generally confine my discussion to Iowa family farms.

The current set of problems plaguing agriculture results from the inability of institutions and people to adjust to a rapidly changing economy. Agricultural change is not new, but the pace of change has greatly accelerated. This gives the aura of chaos or crisis.

From the colonial period until the mid-1800s the rate of agricultural change was slow, allowing considerable time for community and institutional adjustment. However, the rate of agricultural change increased with the industrial revolution. It brought mechanization to the farm: the mechanical reaper, the steel plow, the threshing machine, and other labor-saving devices. While farm size had been limited previously by the endurance of men, women, and draft animals, it was now possible for farm size to increase by substituting capital investments for human and animal power. The adoption of new technologies from the mechanical revolution brought dramatic changes to the rural social structure. New lands could be settled and brought into production, machinery manufacturing provided new jobs, agricultural finance and credit became more important, and the village blacksmith became a mechanic.

A second wave of agricultural change swept the rural landscape in the mid-1900s. This revolution, which built upon the mechanical revolution, introduced further mechanization and the energy intensification of agriculture. Key developments out of this revolution include commer-

PAUL LASLEY is associate professor of sociology and anthropology at Iowa State University and director of the Iowa Farm and Rural Life Poll.

cially produced fertilizer, agricultural chemicals, hybrid seed corn, and further refinements in agricultural equipment. This wave of change led to further energy and capital intensification of farming. The social and economic adjustments that were required by this wave of change were profound. New fertility practices, the complete mechanization of farming replacing draft animals, the enormous gains in productivity, and the capacity for larger farms are just a few of the important societal adjustments springing from this revolution.

A third agricultural revolution began to develop in the mid-1970s, which many predict will have even more profound impact upon rural communities and farming. The products of the biogenetic revolution suggest important impacts for the future of farming. The ability to genetically alter plants and animals promises even greater adjustment for farming. For example, the transformation of annual crops to perennials eliminating the need for preparing seedbeds may greatly reduce soil erosion. Nitrogen fixation in grasses, such as corn, may require major adjustments in the fertilizer industry. Adapting the genetic code of plants to withstand harsh environments may shift traditional crop belts such as the Corn Belt or the Wheat Belt. The recently released bovine growth hormone, a product of the biogenetic revolution, which is predicted to raise milk production by 40 percent, will have major consequences for dairy farmers.

Consequences of Technology Change

These technological revolutions have brought major changes for the structure of agriculture and rural communities. Since it is not possible to provide a detailed listing of the adjustments, the focus will be limited to the broad areas of the structure of agriculture and rural communities.

Perhaps the most noted change in agriculture has been the precipitous decline in farm numbers. Since 1900 the number of farms in the United States has declined from about 6 million to 2.2 million in 1982, with an increase in farm size from almost two hundred acres to nearly six hundred acres. In Iowa during this same period, the number of farms has decreased from 225 thousand to 113 thousand. Essentially, the number of Iowa farms has declined by one-half in the last eighty years.

However, the decline in farm number has not occurred uniformly across all farm sizes. For example, farms of less than fifty acres have declined by about 21 percent between 1925 and 1982, while farms with acreages between fifty and five hundred acres have declined by 58 percent. The large farms—those with five hundred acres or more—have increased by 848 percent for this same period (Fig. 10.1 and Table 10.1). The changes are even more pronounced if one examines the recent past.

From 1964 to 1982, the number of small farms in Iowa (less than fifty acres) increased by 31 percent; middle-sized farms (fifty to five hundred acres) declined by 41 percent; and large farms (five hundred or more acres) increased by 118 percent. This trend of more small and large farms with fewer middle-sized farms is referred to as a "bimodal" or dual agricultural system. The pattern is the same at the national level.

Technology has also resulted in part-time farming. In Iowa between 1930 and 1982, part-time farming increased from 18 percent to 47 percent (Fig. 10.2). Modern technology has greatly reduced the need for labor, resulting in fewer people living on farms; and among those who remain in farming many do so on a part-time basis.

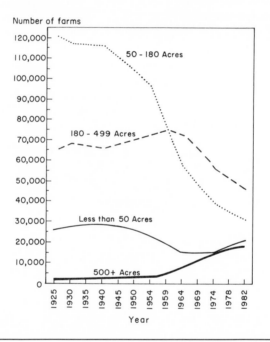

Fig. 10.1. Number of Iowa farms by size categories: 1925–1982. Source: U.S. Department of Commerce (1982).

Table 10.1. Change in Iowa farm numbers by size categories

Acres	Number of farms		Number change	Percent change
	1925	1982		
Less than 50	25,718	20,232	− 5,486	− 21.3
50–500	185,779	77,218	− 108,561	− 58.4
500 or more	1,893	17,963	+ 16,070	+ 848.9
Total	213,390	115,413	− 97,977	

Source: U.S. Department of Commerce (1925, 1982).

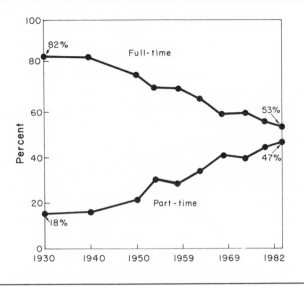

Fig. 10.2. Percentage of part-time and full-time farmers in Iowa: 1930–1982.
Source: U.S. Department of Commerce (1982).

Nationally the farm population declined from about 30 percent in 1920 to less than 5 percent in 1982. In Iowa the farm population declined from 40 percent in 1920 to 13 percent in 1980. For the nation as a whole, only about five million people live on farms (Fig. 10.3).

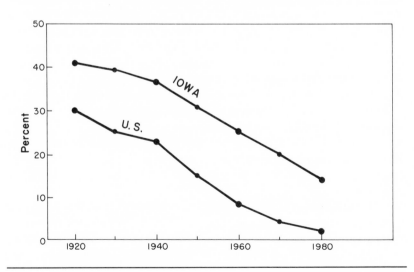

Fig. 10.3. Decline in United States and Iowa farm population. Source: U.S. Bureau of the Census (1980).

The consequences of this dramatic shift in farm population on rural communities remains evident. Throughout the rural landscape are artifacts of a past agricultural system: abandoned farmsteads stand in testimony of what used to be. Abandoned rural schools and churches are vivid reminders of past rural communities. In several Iowa counties the total population is only half of what it was one hundred years ago. This rural-to-urban migration, as Kirkendall has pointed out (see Chap. 9), is a product of the agricultural revolution.

Events in the 1970s

The events of the 1970s are crucial for understanding today's crisis. Several events came together to set the stage for the problems of the 1980s. In the early 1970s there was much concern about world hunger and impending food shortages. In 1972 world food production declined by thirty-three million tons, the first decline in twenty years. The United Nation's World Food Conference met in Rome in 1974 to discuss the impending food crisis. The conference projected that world production of cereal grain had to increase by about twenty-five million tons each year to meet the growing world food demand.

As world food production declined in 1972, the United States and Canada were heavily involved in supply management programs to control surpluses. During the same time, there was unprecedented global economic growth, especially in the oil-rich countries. Coinciding with these global events, the Soviet Union made a set of policy choices to upgrade the nutritional standards of its people. Beginning in 1972, the Soviet Union and the United States became important trading partners.

The conditions of the early 1970s triggered a period of great expansion in U.S. agriculture. Federal government policies were designed to increase agricultural output. Exports became increasingly important to trade, as our trade deficits mounted due to expensive energy imports. As new markets for U.S. farm products were opened, farmers were encouraged to increase food production by planting, in the now famous phrase of Earl Butz, fencerow to fencerow. As the energy crisis deepened, agriculture expansion was encouraged; the possibility of using crops in alternative fuels, especially alcohol fuels, was attractive indeed to farmers.

To facilitate agricultural expansion, enormous amounts of capital were invested in new equipment and machinery. The farm debt increased from about $80 billion in 1975 to about $210 billion in 1984 (Fig. 10.4). Farm income increased from about $25 billion in 1967 to about $60 billion in 1983 due mainly to off-farm earnings (Fig. 10.5). The growth

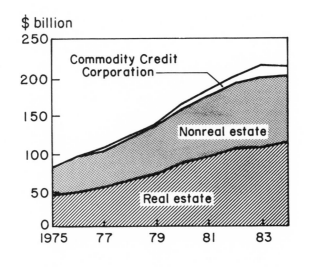

Fig. 10.4. Farm debt. Source: USDA (1984, 7).

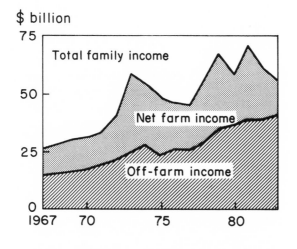

Fig. 10.5. Income of farm operator families. Source: USDA (1984, 5).

and expansion of the general economy contributed to high levels of inflation. As a result, inflation became accepted as a permanent feature of our economy and was incorporated into investors' expectations. It made sense, in economic terms, to be in debt as long as the inflation rate exceeded the nominal interest rates. It made sense to borrow heavily and repay those debts with cheaper dollars. Real estate, especially land, was viewed as a good hedge against inflation, which contributed to rapidly rising land values. Between 1969 and 1982 the average value of Iowa farmland increased from about $400 to $1,700 per acre. Average farm value in Iowa increased from $94 thousand to $471 thousand, a fourfold increase (Lasley and Goudy 1982). This newfound source of wealth provided for farm expansion through equity financing for new machinery and land purchases.

The 1970s brought other changes. Population growth in rural areas exceeded that of metropolitan areas. It was fashionable to talk about the urban to rural migration, an American rural renaissance. But while many rural communities enjoyed growth in the seventies, many did not. Between 1970 and 1980, 32 percent of the 839 rural Iowa communities with less than twenty-five hundred people lost population. Twenty-two percent of the ninety-five small Iowa cities with populations between twenty-five hundred and twenty-five thousand people also saw their populations decline.

The prosperity in agriculture during the seventies did not halt the historic decline in farm numbers. Nationally the number of farmers declined from 2.4 to 2.2 million. In Iowa the number of farms declined from 140 thousand in 1969 to 115 thousand in 1982, a decline of 27 percent (Lasley and Goudy 1984).

Late in the seventies inflation was called "public enemy number one." In response, the Federal Reserve Board adopted new anti-inflation policies in October of 1979. By raising the discount rate, the board forced interest rates higher; the effect was to curb inflation. But high interest rates also had the unintended effect of contributing to the farm crisis. Higher interest rates meant higher costs of production for farmers. U.S. farm exports began to decline (because of the value of the dollar relative to foreign currencies), and real estate values also began to fall. The consequences of the Federal Reserve Board action were most vivid in the case of land values. Since 1981 Iowa land values have declined by 56 percent ($2,147 in 1981 to $948 in 1985) (Olsen and Jolly 1985). Recent reports on national farmland values show declines in thirty-seven states and increases in eleven (Farmline 1986). The greatest decline has occurred in the Midwest where, for example, Iowa land values declined by $39 billion between 1982 and 1985. This translates into a loss of nearly $14 thousand for every resident of the state.

Current Conditions

Survey data from Iowa farmers illustrate the seriousness of the debt problem. In 1985, 11 percent of Iowa farmers had debt-to-asset ratios greater than 70 percent. A debt-to-asset ratio is found by dividing liabilities by assets. Those 11 percent of Iowa farm families with debt/asset ratios greater than 70 percent hold about 9 percent of the assets and 25 percent of the debt (Table 10.2). They are in imminent danger of financial insolvency. Another one-fifth of Iowa farmers have debt/asset ratios of 41 to 70 percent. Although this group of farmers own about 28 percent of the total farm assets, they own about 48 percent of the debt. This group is viewed as being at substantial risk. Taken together, these two highly leveraged groups represent about one-third of all farms in the state. But they hold three-fourths of the total farm debt (see Chap. 11.). Another one-third of Iowa farmers have debt/asset ratios of 11 to 40 percent. These farmers may have to tighten their belts; but most analysts believe that they will be able to pay off their debts. In addition, about one-third of Iowa farms have debt/asset ratios of less than 11 percent. They are in strong financial positions, because they have little debt to service.

Thus the financial status of farmers breaks into nearly equal groups: one-third without debt, one-third with manageable debt, and one-third heavily indebted. The national numbers are similar to those found in Iowa.

The Seriousness of the Farm Crisis

Since 1982 Iowa State University Agriculture and Home Economics Experiment Station and the Cooperative Extension Service have sur-

Table 10.2.   Debt-to-asset ratios among Iowa farmers

Debt-to-asset ratio (%)	Percentage of operators	Percentage of assets	Percentage of debts
0–10	35	29	2
11–40	32	34	25
41–70	21	28	48
71 +	11	9	25

Source: Jolly and Barkema (1985).

veyed Iowa farm families on current problems. The Iowa Farm and Rural Life Poll has been conducted every six months to monitor agriculture and rural development issues.

In the first statewide survey of approximately nineteen hundred randomly selected farmers in the fall of 1982, it was evident that farmers were apprehensive about the future of the farm economy (Lasley 1982). Fifty-four percent indicated that they felt the overall economic prospects for Iowa farmers would either become somewhat or much worse in the next five years. Forty-two percent indicated they felt the quality of life of farm families would deteriorate in the next five years. When asked about the principal reason for the financial problems of farmers, one-fourth said poor management, 15 percent said poor financial advice, and 61 percent said factors beyond farmers' control.

In the spring 1983 poll, nearly half of the 2,293 farmers in the survey (49 percent) felt that a national moratorium on farm foreclosures was needed, 38 percent said it was not, and the remaining 13 percent were undecided (Lasley 1983). The survey also revealed that farmers were not planning expansion of their farming operations. Only 4 percent planned to buy a new tractor; 9 percent planned to make a major farm equipment purchase; 9 percent planned to build new livestock facilities; 6 percent planned to build additional grain storage. Of the ten farm purchases asked about in the survey, less than 14 percent of the respondents were planning for these expenditures. Analysis of these data reveals that farmers were in a maintenance mode and were reducing expenditures. This, we may assume, will have a ripple effect on local agribusiness suppliers and main-street businesses.

In the spring 1984 survey of nearly two thousand randomly selected farmers, 94 percent felt it was very or somewhat likely that the number of farms would continue to decline (Lasley 1984a). Ninety-two percent felt that low prices for farm products would put many farmers out of business. Seventy-two percent agreed with the statement that "government agricultural policies were the primary cause of the present price problems."

The survey repeated some of the questions from the survey conducted in 1982 on perceptions about the future economic prospects for farmers. Fifty-two percent said the overall economic prospects for Iowa farmers would become worse in the next five years. Forty percent indicated their quality of life would become worse in the next five years.

When the survey conducted in the fall of 1984 asked about the objectives of the 1985 farm bill, developing new markets and lowering interest rates received the highest rankings (Lasley 1984b). In this survey, 67 percent indicated that Iowa farmers faced a very serious financial problem. An additional 25 percent indicated that farmers faced a moderate financial problem. Thirty-nine percent indicated that agribusiness faced a very serious problem, and another 40 percent felt that agribusiness faced a moderate financial problem. When asked, "How concerned are you about your farm's financial condition?," 42 percent indicated that they were very concerned, and an additional 24 percent said they were moderately concerned.

The spring 1985 survey asked farm families what adjustments they had made in their family financial expenditures in the past twelve months (Lasley 1985a). Responses to the sixteen expenditure categories provided further evidence of the seriousness of the farm crisis. For example, 72 percent had postponed a major farm purchase; 65 percent had delayed a household purchase; 72 percent had reduced social and entertainment expenses, and 55 percent had reduced charitable contributions. Fifty-nine percent of the farm families reported they had dipped into their savings to meet expenses; 28 percent had sold possessions or cashed in insurance to raise cash; and 27 percent had postponed medical care to save money. These drastic measures of financial management indicated the seriousness of the farm problem. When asked how they would describe their financial situation compared with a year ago, 34 percent said that they were somewhat worse off, and an additional 25 percent said that they were much worse off.

In the fall 1985 survey, farm families were asked to rate twelve state and national issues by checking a seven-point scale that ranged from "1" not concerned to "7" very concerned. The number one issue identified by farmers was "prices for farm products," followed by "closing of local main-street business" and "interest rates to borrowers" (Lasley 1985b).

Taken as a composite, the data portray an overwhelming crisis among Iowa farmers. Across the seven statewide surveys, the data are consistent. It is somewhat surprising that farmers were providing similar responses, regardless of their financial conditions. Previous analysis of these data have found consistent patterning based upon debt/asset ratios. However, given the uniformity in responses to the questions, it appears that there is a widely held "depression mentality" among Iowa farmers.

The data also reveal farmers' concern about the future of their rural communities. As farmers have made financial adjustments and reduced their expenditures for inputs, local main-street businesses and agribusinesses have been adversely affected. The data suggest that farmers are concerned about their own families' quality of life and financial conditions as well as the future of family farming and rural community viability.

Some Tough Ethical Questions

The farm crisis in the 1980s stems from an overreaction to the profitable and overly optimistic 1970s. Today's crisis, like the crises of Shay's Rebellion, the Farmers' Holiday movement, and the NFO's actions in the 1960s, calls attention to the contradiction in the phrase "profitable family farms." Whenever profits accrue to family farms, more people take up farming, existing farms expand, the price of land and other inputs are bid up, and before long, surpluses begin to mount. The result? Commodity prices drop. This is a contradiction inherent in the free-market system. Harold Breimyer (1977, 67–78) characterizes the problem by speaking of the farmers' noninstinct for self-preservation. Typical rural values—independence, self-reliance, and rugged individualism—do not contribute to a sense of group consciousness. These values, Breimyer notes, contribute little toward developing a sense of what is required for farmers to survive in an increasingly service-oriented society. As a result, farmers respond to economic signals according to how those relate directly to them. They remain oblivious to how these values might affect them in the long run, or how they might affect others.

Breimyer raises the ethical problem that is called the "tragedy of the commons." Everyone seeking out their own self-interest will, in time, contribute to the demise of the entire group. One farmer or even a group of farmers expanding production in the 1970s would have been inconsequential to the market. But when every farmer attempts to expand, the consequences are disastrous.

Is there a moral obligation to save the family farm? I believe this question raises the ethical issue of equity and fairness. Inconsistent government policies are immoral. It is wrong for us to tell farmers to plant fencerow to fencerow, only to follow that command with embargoes and PIK programs.

It is also wrong for us to adopt policies, without advance warning, that shift us from a period of high inflation and low interest rates to a period of high interest rates and low inflation. Government has the responsibility to devise and maintain consistent policy directions. Efficient business planning requires a stable environment. This is essential for family farm survival.

Government also has the responsibility to ensure that everyone plays on a level field. Tax policy is a good example where government can do more to ensure fairness and equity across various farm sizes (see Chap. 18).

Agriculture is currently caught up in a debate; should we let the market decide who survives or should we institute federal policies to save people and community? The question is often raised, "Is farming a way of life or is it a business?" Unfortunately, the debate is not quite so simple. How one answers the question posed in the title of this book rests upon one's values. How do you reconcile increasing farmers' income because farming is "a way of life," if it means higher food prices for the urban poor? Is the moral question of saving the family farm any different from the question of saving the mom-and-pop grocery store or the village blacksmith? To these questions, I do not have the answer.

I am sympathetic to the moral dimensions of what may be called the pauperization of family farms; my research provides ample evidence of the heavy toll the crisis has taken on human lives. There is little disagreement about the need to raise income for many farm families. The conflict is over how to do it. Some argue that if we could recreate the commodity price structure of the 1970s, the farm crisis would melt away. Others prefer the price ratio of 1909 to 1914, "The Golden Age of Agriculture." They call their plan "parity." Any plan designed to raise commodity prices, which limits production (e.g., set aside plans, marketing orders, production quotas) must prevent the profits from being capitalized into the land or marketing certificate. Any method to reduce production (such as marketing certificates, progressive taxes on inputs, or limits on farm size) must address the moral questions of how this affects the welfare of *all* members of our society. I am not arguing against production controls or against plans like the Harkin bill; but I believe that if these types of policies are going to be adopted, we must be able to articulate how society as a whole will be strengthened.

As more and more Americans lack ties to farm life, it becomes more important for us to learn how to show them that they have an interest in the future of family farming. Why should the inner-city black teenager, the senior citizen living on a subsistence pension, or the urban housewife, pay higher food prices so that family farmers can make a decent living?

The historical evidence is convincing; agrarian values are not omnipotent in the policy arena. Even when 90 percent of the U.S. population were farmers, Daniel Shay found little support for stopping farm foreclosures. Milo Reno, founder of the Farmers' Holiday movement, was not able to stem the tide at a time when one-third of Iowa's population lived on farms.

I suggest that we will not gather much nonfarm support for saving the family farm by arguing from a unique set of values and beliefs. As Breimyer points out, many of the values farmers hold have contributed to their plight.

Perhaps the farm crisis calls into question the values and beliefs that agrarians have treasured. Farming as a way of life, being one's boss, the ability to work outdoors, individual freedom, and the rest of the values embedded in the Jeffersonian ideology are not very convincing to those who do not share them. I believe we need to foster a new ideology, a new system of values and beliefs. While I am not prepared to specify all of its dimensions, a new ideology ought to embrace the values held by the farm *and* nonfarm population. This new ideology should include the values of community, family, environmental quality, efficiency, sustainability, cooperation, opportunity, equality, and adaptability. If it can be demonstrated that saving family farms contributes to these widely shared values, then public support to save family farms should be easier to achieve.

My fear is that as the farm crisis wears on, we may become insensitive to the human suffering. Indifference and callousness of this sort is immoral. As others have noted, one should judge a society based upon how it responds to those in need. The tradition of rural America has been one of self-help and voluntarism, whether combating disease or drought, or building a town hall. If there has ever been a time that we need to encourage such cooperation in rural America, it surely must be now.

References

Breimyer, Harold F. *Farm Policy: 13 Essays.* Ames: Iowa State University Press, 1977.

Farmline. *Farmland Values: Where's the Bottom?* Vol. 7, No.5. Washington, D.C.: Economic Research Service, 1986.

Jolly, Robert, and Alan Barkema. *1985 Iowa Farm Finance Survey.* Ames: Iowa State University Cooperative Extension Service Assist 8, May 1985.

Lasley, Paul. *Summary: Iowa Farm and Rural Life Poll.* Pm-1073. Ames: Iowa State University Cooperative Extension Service, 1982.

_____. *Summary: Iowa Farm and Rural Life Poll.* Pm-1100. Ames: Iowa State University Cooperative Extension Service, 1983.

_____. *Summary: Iowa Farm and Rural Life Poll.* Pm-1158. Ames: Iowa State University Cooperative Extension Service, 1984a.

_____. *Summary: Iowa Farm and Rural Life Poll.* Pm-1178. Ames: Iowa State University Cooperative Extension Service, 1984b.

_____. *Summary: Iowa Farm and Rural Life Poll.* Pm-1209. Ames: Iowa State University Cooperative Extension Service, 1985a.

_____. *Summary: Iowa Farm and Rural Life Poll.* Pm-1237. Ames: Iowa State University Cooperative Extension Service, 1985b.

_____, and Willis Goudy. *Changes in Iowa Agriculture, 1969–1978.* Pm-1152. Ames: Iowa State University Cooperative Extension Service, 1982.

_____. *Changes in Iowa Agriculture, 1978–1982*. Pm-1152. Ames: Iowa State University Cooperative Extension Service, 1984.

Olsen, Douglas, and Robert Jolly. *1985 Iowa Land Value Survey*. FM-1821. Ames: Iowa State University Cooperative Extension Service, 1985.

U.S. Bureau of the Census. *Census of Population*. Washington, D.C.: Government Printing Office, 1980.

U.S. Department of Agriculture. *1984 Handbook of Agriculture Charts*. USDA Handbook No. 637. Washington, D.C.: Government Printing Office, 1984.

U.S. Department of Commerce, Bureau of the Census. *Iowa Census of Agriculture,* Vol. 1, Part 15. Washington, D.C.: Government Printing Office, 1925 and 1982.

 NEIL E. HARL

The Financial Crisis
in the United States

The General Setting

Rapid economic and social change in agriculture is not a new phenomenon. Since the beginning of recorded history, agriculture has been adjusting to conditions of greater efficiency. As a consequence, the percentage of the population and the percentage of the capital stock needed to produce food and fiber products has declined steadily. The decline has been especially marked since the 1930s as developments in plant and animal breeding and machinery and chemical usage, and improvements in the level of management ability of farmers have combined to cause an acceleration in the movement of labor out of the sector. Agriculture has truly been a development sector as the industry has downsized itself in relative terms, freeing labor and capital for use in the nonfarm economy. The development occurring in agriculture has been enormously beneficial to the general economy, permitting the allocation of resources to a burgeoning service sector including space exploration and medical and scientific research, to mention only the more obvious growth sectors of the nonfarm economy, and to high-technology manufacturing and product development. Had agriculture been frozen by the implementation of highly protective policies in the condition it was in as of the early 1920s, at the beginning of two decades of severe economic trauma for agriculture, society could have been denied the resources needed to support the enormous development effort of the past half century.

However, what is now occurring in agriculture in terms of firms failing because equity is exhausted or operating credit is denied, has little to do with efficiency and does not represent a continuation of the long-

NEIL E. HARL is Charles F. Curtiss Distinguished Professor in Agriculture and professor of economics at Iowa State University.

term trend toward greater efficiency in agriculture. In fact, the firms now at risk are some of the most efficient in the industry and are operating at or near the minimum point on the long-term average total cost curve except for one factor: the amount of debt held is excessive as measured by the economic environment of the 1980s. Those who survive are not necessarily the most efficient and in fact tend to be the older, more cautious farmers with smaller operations and little or no debt. Thus, the phenomenon cuts across farm and ranch firms in a highly arbitrary manner.

The data are making it increasingly clear that agriculture is going through the most wrenching financial adjustment in a half century. Not since the 1930s have issues of debtor distress gripped rural America as they have in the 1980s. One need only look to our farms and rural communities for proof.

• In several agriculture states, land values have dropped by one-half or more since 1981, cutting enormous amounts of collateral value and wealth from balance sheets.

• The numbers of farm foreclosures, forfeitures of land contracts, and defaults on notes have reached levels not seen since the days of the Great Depression.

• The level of emotional trauma being suffered by indebted farmers and small businesspersons is a tragedy of awesome proportions.

The scope of the problem is much broader than farms. Although economic stress gained a foothold among the more heavily indebted farmers, the phenomenon has escalated rapidly so that today it threatens to engulf the entire rural community. In fact, few will escape unscathed. Many lenders are struggling to survive. Suppliers have taken and will continue to take enormous hits as unsecured creditors. Main-street businesses have felt the ravages of this cancer that gnaws at the very structure of rural communities.

The data make it clear that the problem is almost national in scope. The severity varies from area to area, and the upper Midwest has suffered the most from the ravages of this economic downturn, but the blight of agricultural stress virtually blankets the country. In many ways, it's been like a war against an invisible enemy. And that enemy is the cost of servicing a huge debt load with interest rates at unprecedented levels in real terms.

WHY THE PROBLEM EXISTS. It would be an unwise use of time to focus a great deal of attention on who is responsible for the plight of rural communities. Finger pointing and accusations of culpability will do little to remedy the situation. But in choosing remedial policy instruments, it is important to recognize the roots of the problem. Two principal categories of forces are responsible for many of the economic woes of agriculture: (1) three major federal policies that created an economic environment highly unfavorable for agriculture and other sectors that are both capital intensive and export sensitive and (2) forces operating at the farm or ranch level that moved some firms into a "window of vulnerability." Once within the window of vulnerability, the unfavorable economic environment was sufficient to move the firms inexorably toward insolvency.

Federal Policies. As noted, three federal policies operating over nearly two decades created an economic environment that has been highly unfavorable for agriculture in the 1980s. Although agriculture is not alone in being impacted adversely, the characteristics of a relatively low cash rate of return for many farm assets, a high level of capital intensity for U.S. agriculture, and sensitivity to change in export supply and demand conditions in international farm commodity markets have magnified the impacts upon farm firms.

The first federal policy contributing to the unfavorable economic environment for agriculture was the set of policies over five different federal administrations that came to treat inflation as an expected part of economic life. The relatively high rate of inflation from the budget strains of the Vietnam conflict was compounded by the effects of rapid increases in energy costs after 1972. By the late 1970s, the persistence of inflation in the economy had led to widespread efforts at accommodation. The most common strategy for accommodating inflation was to index one's economic fortunes to the rate of inflation. Thus, social security benefits and taxes were indexed, federal civil service compensation levels were indexed, and many labor union contracts were indexed as to basic compensation levels. Beginning in 1985, the entire income tax system was indexed.

Farmers, unable to index with the same degree of effectiveness, in some instances accelerated the purchase of capital assets in the face of consistent increases in the cost of machinery and equipment and in the price of land. The differential effect of the two responses to inflation became painfully clear in the early 1980s. Indexing is a benign strategy in an era of declining rates of inflation. Anticipating the purchase of capital assets is not benign and leaves the purchaser with financial commitments to be met.

The experience of the inflationary era of the 1960s and 1970s makes it clear that an enormous price is paid when expectations about conditions that should be viewed as aberrational in nature harden into a belief that the condition is permanent.

The second important factor was the decision by the Federal Reserve Board in October of 1979 to wring inflation out of the U.S. economy. The action to limit the supply of credit led almost immediately to high nominal rates of interest, which eventually served to dampen the level of economic activity. In the first half of the 1980s, inflation dropped from the 13 to 15 percent range to 3 to 4 percent. Thus, the gains from inflation, which were substantial during the 1970s, were dramatically reduced in the 1980s, leaving farm debt to be serviced largely from current income.

The third significant factor contributing to an unfavorable economic environment for agriculture in the 1980s appears to have been enactment of the Economic Recovery Tax Act of 1981 that cut federal revenues so sharply as to assure massive budget deficits. The 1981 legislation was enacted with the realization that an estimated $872 billion in revenue would be cut from the federal tax system through fiscal year 1986. Cuts of that magnitude assured that the outcome would be massive federal budget deficits.

The result of these policies has been an economic environment of low inflation and record-setting real interest rates as tight credit and strong private sector demand for capital have boosted interest rates. For agriculture, the result has been (1) a strong dollar that has set records against other currencies and that has cost U.S. agriculture dearly in terms of exports of farm commodities, (2) high interest rates that have boosted the cost of production for indebted farmers to high levels, and (3) falling land values as potential investors have been confronted with the reality of 8 to 12 percent real interest rates and the reassessment of land as an alternative investment.

FACTORS CONTRIBUTING TO FARMER VULNERABILITY. In the economic environment of the last four to five years, any factor that made a farmer vulnerable by increasing the debt load was sufficient to assure economic difficulty. It was the resulting "window of vulnerability" that set the stage for financial stress.

Beginning farmers are almost always vulnerable the first several years of operation. Part of the uniqueness of family farms is that families accumulate most of the equity capital for the firm from earnings. The result is economic vulnerability during the first several years of life of farm firms. That has certainly been the case in the 1980s. This factor alone assures that we are in danger of losing much of a generation of young farmers.

Adverse weather conditions in some areas with consequent loss of part or all of a crop have been costly to farmers affected. For many areas, agriculture has experienced an unusual sequence of adverse weather conditions beginning in 1980, both too wet and too dry.

Losses in cattle feeding in the 1970s and even losses in hog production in more recent times have increased debt loads and, thus, vulnerability. For about half of the months over the last five years, hog production has been at a loss. That is unprecedented in this country. Losses in cow-calf enterprises in recent years have been perhaps less visible but no less devastating.

Expansion to bring a family member into the operation has increased debt loads. The economics of farming in recent years has encouraged the continuation of family operations with ownership and management transferred to the next generation.

Major purchases of land, machinery, or livestock facilities in the late 1970s and early 1980s were factors increasing economic vulnerability.

Any event or series of events that placed a farmer in the window of vulnerability has proved to be economically devastating. Once in the window, high real interest rates moved the firm toward insolvency at a breathtaking pace.

AMOUNT AND DISTRIBUTION OF DEBT. The amount of debt in U.S. agriculture has increased dramatically since 1950 as shown in Fig. 11.1. Total farm debt outstanding in 1950 totalled $11.2 billion in 1950, rising to over $216 billion nationally in 1983, before declining in 1984 and 1985 to almost $212 billion as some debt has been paid off or discharged otherwise and as the economic environment has discouraged the contracting of new debt. Debt as a percentage of net farm income stood at 92 percent in 1950 but rose to 1350 percent of net farm income in 1983. The increase in personal, business, and federal government debt has been similar as shown in Fig. 11.2.

Nature and Severity of the Farm and Ranch Financial Problem

Never in the history of U.S. agriculture have problems of debtor distress occurred where there was greater heterogeneity in financial condition among farmers and ranchers.

EXTENT OF FINANCIAL STRESS. As of January 1985, approximately 22 percent of the farmers nationally had debt-to-asset ratios of greater than 40 percent and were responsible for nearly 62 percent of the farm debt.

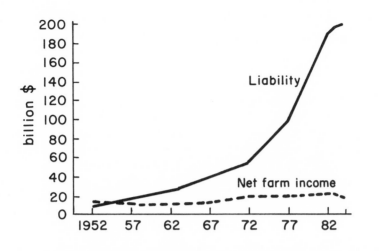

Fig. 11.1. Net farm income and liabilities. Source: *Financial Characteristics of U.S. Farms* (1985, 1 and Table 1).

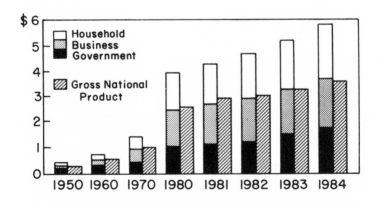

Fig. 11.2. National, personal, and business debt (in trillions of dollars). Sources: Federal Reserve Board and U.S. Department of Commerce.

In general, it has been thought that farmers with debt-to-asset ratios above 40 percent are unable to service their debt and pay other costs when due in a setting of real interest rates prevalent in the mid-1980s and the rates of return for agricultural assets common in the mid-1980s. (See Table 11.1)

Moreover, the problem in some regions is substantially more serious than the national data indicate. A January 1985 survey in North Dakota indicated that 36 percent of the farmers had debt-to-asset ratios over 40 percent, held 37 percent of the assets, and accounted for 74 percent of the debt. Table 11.2 shows the Iowa data as of January 1984.

Over one-third of the farmers in Iowa (averaging 59 years of age) had little or no debt as of January 1984 with debt-to-asset ratios of 10 percent or less. Approximately another third (37 percent) had debt-to-

Table 11.1. Percentage of farms and debt-to-asset ratio for each region and for the United States, January 1985

	41–70		71–100		Over 100	
	Farms	Debt	Farms	Debt	Farms	Debt
Northeast	8.7	28.4	5.3	23.5	3.0	N/A
Lake states	15.8	33.5	9.8	29.8	3.3	N/A
Corn Belt	14.6	35.5	10.1	31.3	3.9	N/A
Northern plains	16.3	37.0	10.5	25.8	4.0	N/A
Appalachian	7.2	34.4	2.3	14.7	1.0	N/A
Southeast	7.4	25.2	5.8	37.3	3.1	N/A
Delta states	10.9	38.0	6.9	27.8	4.1	N/A
Southern plains	5.9	24.5	5.5	33.4	2.1	N/A
Mountain states	15.5	32.9	6.6	22.6	2.5	N/A
Pacific states	10.8	28.9	6.0	36.9	3.2	N/A
United States	11.6	32.9	7.3	29.0	3.0	N/A

Source: *Financial Characteristics of U.S. Farms, January 1985* (1985).

Table 11.2. Financial condition of sample Iowa farmers by 1984 debt-to-asset ratio, January 1984

	Debt-to-asset ratio					
	0–10	11–40	41–70	71–100	Over 100	All farms
Operators (%)	38	37	19	4	1	
Assets (%)	31	42	24	3	1	
Debt (%)	4	39	47	8	2	
Age (average)	59	53	47	45	47	54
Assets per farm (average)	$503,000	$694,000	$745,000	$470,000	$217,000	$615,000
Debt per farm (average)	$11,000	$160,000	$383,000	$375,000	$262,000	$156,000
Equity per farm (average)	$492,000	$534,000	$362,000	$95,000	−$45,000	$459,000
Acres owned (average)	233	298	271	172	131	261
Acres rented (average)	121	189	306	382	198	193

Source: *1985 Iowa Farm Finance Survey* (1985).

asset ratios of 11 to 40 percent. In general, it has been thought that the 11–40 percent group would be able to stabilize its financial condition although the upper quarter of that group was encountering stress. Members of the remaining group, 24 percent of the operators, were severely impacted and were moving toward insolvency or were already insolvent. The more recent balance sheet data for Iowa (as of January 1985) are shown in Tables 11.3 and 11.4.

The data indicate that a movement has occurred of borrowers in the 41–70 percent category into the over 70 percent group. Moreover, a significant number from the 11–40 percent category have moved into the 41–70 percent group. On the average, farmers who were in the 71–100 debt-to-asset ratio category as of 1 January 1984, lost eighty-four thousand dollars (88.4 percent) of their equity during 1984. Thus, the rate of deterioration in financial condition has been great. Even those in the 0–10 percent debt-to-asset category on 1 January 1984, lost 20.1 percent of their equity in 1984 principally because of declines in asset values.

The U.S. Department of Agriculture estimates that 20 percent of all farms with annual sales of over forty thousand dollars had both negative cash flow and a debt-to-asset ratio of more than 40 percent. Just over 50 percent of all U.S. farms were experiencing a negative or zero cash flow as of January 1985. A total of 852,242 farms reported a negative or zero cash flow with the largest number (638,479) having debt-to-asset ratios of 40 percent or less.

Table 11.3. **Distribution of operators, assets, and debts of sample farmers, by 1985 debt-to-asset ratio, January 1985**

	Debt-to-asset ratio				
	0–10	11–40	41–70	71–100	Over 100
Operators (%)	35	32	21	7	4
Assets (%)	29	34	28	7	2
Debt (%)	2	25	48	17	8

Source: *1985 Iowa Farm Finance Survey* (1985).

Table 11.4. **Financial condition of sample Iowa farmers by 1984 debt-to-asset ratio, January 1985**

	Debt-to-asset ratio					
	0–10	11–40	41–70	71–100	Over 100	All farms
Assets per farm (average)	$411,000	$578,000	$625,000	$347,000	$171,000	$506,000
Debt per farm (average)	$18,000	$170,000	$388,000	$336,000	$244,000	$161,000
Equity per farm (average)	$393,000	$408,000	$237,000	$11,000	−$73,000	$345,000
Loss of equity in 1984 (average %)	−20.1	−23.6	−34.5	−88.4		

Source: *1985 Iowa Farm Finance Survey* (1985).

The financial condition of farm and ranch firms may also be evaluated on the basis of return to equity. As can be seen from Table 11.5, 32.6 percent of the operators have an estimated return to equity of less than -.05 percent. Those operators hold about 23.1 percent of the assets but are responsible for almost 42 percent of the debt. At the same time, more than 39 percent of the U.S. farm operators had a return to equity of greater than 5 percent. That group held about 36 percent of the assets and 36 percent of the farm debt.

The data make it abundantly clear that enough assets and debt are held by farmers who are unstable economically to assure that further weakness in land and machinery values (below 1985 levels) is a virtual certainty unless (1) farm incomes rise substantially, (2) real interest rates decline significantly, or (3) major public-sector intervention efforts are implemented to stabilize the agricultural sector. Lenders holding land as collateral, principally the Federal Land Bank and sellers under land contract, have reported sharply higher default rates in 1985. The willingness of short- and intermediate-term lenders to provide credit needed to keep land payments current appears to have diminished rapidly. Further increases in delinquency rates on land loans are anticipated.

Unless something dramatic is done or circumstances change, as many as one-third of the farmers nationally will move to insolvency, taking down their lenders, their suppliers, and other merchants and inflicting incalculable damage upon the fabric of rural communities. Discharged indebtedness goes ricocheting through local communities laying waste, with the unsecured creditors taking the greatest hit. However, with the weakness in land and machinery markets, even secured creditors are, in reality, only partially secured as collateral values have slipped below loan balances.

Table 11.5. Distribution of farm operators, debt, and assets by return to equity for the United States, 1 January 1985

	Insolvent farms	Less than −.20	−.20 to −.10	−.10 to −.05	−.05 to .05	.05 to .10	.10 to .20	Greater than .20	All farms
Operators (%)[a]	2.99	12.10	8.87	8.65	28.27	11.43	12.21	15.48	100.00
Debt (%)[b]	13.27	14.37	7.25	7.05	22.15	10.52	10.66	14.72	100.00
Assets (%)[c]	1.99	6.82	6.34	7.93	40.79	14.21	11.97	9.95	100.00

Source: *1984 Farm Costs and Returns Survey* (1985).
Note: Return to equity is net cash income from the farming operation plus nonfarm income minus estimated living allowance divided by operator farm equity.
[a]Percentage of U.S. farms.
[b]Percentage of U.S. operator debt.
[c]Percentage of U.S. operator assets.

RELATIONSHIP OF COMMODITY PRICES TO LAND VALUES. Lower farm commodity prices would be expected to lead to a reduction of the price at which land is economically supportable. That would be the case at least, if potential investors had a permanent expectation of lower commodity prices.

Iowa State University projections under an assumption of a 140-bushel corn yield produce the figures shown in Table 11.6. Thus, with an expected corn price of $3.00 and a capitalization rate of 8 percent, land would be economically supportable at $1988.00 per acre. If the expected price for corn were to decline to $2.25, based on income capitalization and under the same assumptions, the economically supportable price would be about $675.00 per acre. It is indeed clear that land prices are linked to expected levels of commodity prices.

POSSIBLE SCENARIOS. Undoubtedly the most crucial question in framing solutions to problems of farm debtor distress is what can be expected over the next two to five years with respect to (1) interest rates, (2) farm income, and (3) strength of the general economy both domestically and worldwide. Substantial uncertainty surrounds each of those variables. For purposes of discussion, four scenarios are identified.

1. Continued high real interest rates with stable or slightly lower farm commodity prices. High interest rates choke off economic activity in the general economy with a recession as the usual outcome.

2. The value of U.S. dollar relative to other currencies, presently high by historical standards although down from late February 1985 could decline sharply because of the effects of the record-setting trade deficit (which could climb to $160 billion in the 1985–86 fiscal year) and a decline in interest rates domestically. The result presumably would be increased exports with a positive effect on farm income. A decline in the value of the dollar would have inflationary effects, also.

3. The Federal Reserve, concerned about economic pressure on

Table 11.6. **Estimated land values based on income capitalization for high-grade land assuming continuous corn**

Corn price ($)	Net income ($)	Capitalization rate and land values			
		6%	8%	10%	12%
2.25	54.05	900.83	675.63	540.50	450.42
2.50	89.05	1484.17	1113.13	890.50	742.08
2.75	124.05	2067.50	1550.63	1240.50	1033.75
3.00	159.05	2650.83	1988.13	1590.50	1325.42

Source: Barkema (1985).

Third World debtor nations (over $900 billion owed, much of the total to U.S. financial institutions) and pressure on some sectors of the U.S. economy might relax credit controls with an increase in the money supply and resulting higher rates of inflation. After some lag, farmland values would likely be affected. However, it is unclear in a world of deregulated financial markets what the impact would be on real interest rates.

4. If high and rising interest rates cause Third World nations to default on their debt obligations, an international liquidity crisis of major proportions could occur. The effects would be highly destabilizing within and without the United States. Obviously, every effort will be made to avoid such a financial catastrophe. The probability of such a default would seem to be quite low.

GLOBAL IMPLICATIONS. Efforts to make U.S. agriculture more competitive on international commodity markets should be evaluated also in terms of likely impacts on other countries producing agricultural products, particularly Third World countries, and on importing nations and consumers. An aggressive program to move larger quantities of U.S. agricultural products into international trade channels could be profoundly destabilizing for some Third World exporters of agricultural commodities. Such a move could, for example, exacerbate the problems of Third World countries in meeting commitments to service their large and growing debt burden. Quite clearly, the analysis of the effects of changes in U.S. farm policy should be global in scope and comprehensive in nature with emphasis on general equilibrium outcomes as well as on the U.S. economy. The trade deficit is a hidden form of foreign aid. Unfortunately, only about 15 percent of the U.S. trade deficit was with Third World debtor nations in 1984. In 1980, the U.S. ran a small surplus ($293 million) with the same debtor nations.

Development and Evaluation of Intervention Efforts

There is an apparent necessity for large-scale federal intervention in agriculture if economic disaster of major proportions is to be averted. The financial situation in agriculture is of sufficient scale and severity to suggest that consideration of public intervention is justified. In general, if the benefits from intervention (on a present value basis) exceed the costs of intervening, it is appropriate to consider intervention.

PRINCIPLES OF INTERVENTION. Any intervention should be governed by agreed-upon principles. The following have been suggested for the United States:

1. Intervention should be as broad as the problem giving rise to the intervention effort. Thus, intervention should not be just for the Farm Credit System (comprised of thirty-seven banks organized into twelve Farm Credit Districts with a total loan portfolio of more than $70 billion), which is currently the driving force behind public intervention in the United States. If intervention were to be undertaken at the level of lenders, the program of intervention should reasonably extend to all lenders. Otherwise farmers with identical farm operations and debt loads would be treated differently depending upon who their lender was. Then, competitive equilibrium would likely be disturbed, perhaps irrevocably.

2. Although the Farm Credit System is in grave financial condition, *many* commercial banks involved substantially in lending to farmers face similar problems. As shown in Table 11.7, commercial banks hold 25.5 percent of the farm debt in Iowa but have 49.9 percent of the debt held by those with debt-to-asset ratios above 100 percent. Production credit associations with 8 percent of the farm debt, hold only 4.8 percent of the debt of insolvent farmers. The Federal Land Bank with 28.4 percent of the farm debt has 8 percent of the debt of insolvent farmers.

3. Intervention should preferably be directed at stabilizing farmers as borrowers. If farmers are not stable, lenders are unlikely to be or become stable. If farmers as borrowers are made substantially stable, then others—lenders, suppliers, and rural communities generally— should also become stable. It would be an extremely costly venture to attempt to stabilize lenders if farmers are not substantially stable.

4. Intervening on behalf of lenders could be justified on the grounds of expediency in avoiding collapse of the lending system by keeping lenders in a viable state. The result, after an initial period of adjustment, could be reduced cost of credit to all borrowers, not just those in financial difficulty. This poses the question of whether intervention should be targeted.

5. To limit the cost of intervention and to avoid perceptions of

Table 11.7. Distribution of debt in Iowa within debt-to-asset classes by lender

Debt-to-asset class	Banks	PCA	FLB	FmHA	Other	Individual	Total
%	%	%	%	%	%	%	%
0–10	40.3	5.5	11.1	0.5	25.8	16.8	100.0
11–40	23.1	5.3	27.0	2.9	15.7	26.0	100.0
41–70	23.7	10.7	32.6	3.1	14.2	15.8	100.0
71–100	23.7	6.0	28.4	10.8	12.6	18.5	100.0
100 +	49.9	4.8	8.0	7.2	12.1	17.9	100.0
Lenders' share of total debt	25.5	8.0	28.4	4.6	14.4	19.1	100.0

Source: *1985 Iowa Farm Finance Survey* (1985).

unfair treatment of farmers over nonfarmers (many of whom are also in financial trouble), targeting of benefits from intervention is necessary. Widespread public acceptance of realistic, hardheaded, equitable intervention efforts can reasonably be expected. But little public acceptance is likely if benefits flow heavily (even though not exclusively) to farmers not in financial difficulty. It is acknowledged that targeting of benefits from intervention poses fairness problems of a different sort, as farmers who are not under serious financial stress may resent benefits flowing to those in financial difficulty.

6. Programs of intervention should be flexible in nature such that if economic circumstances change, the program could be altered or terminated. This argues against heavy up-front expenditures and in favor of annual maintenance expenditures of a program of intervention.

7. Public intervention should not interfere unreasonably with adjustment and economic efficiency and should be governed by realistic long-term expectations as to demand-supply, price and profitability relationships. A major benefit of intervention is avoidance of overshooting of equilibrium conditions. The probabilities of serious overshooting in land price, for example, seem high under present conditions.

8. It is not unreasonable to request assistance from the general public, but the public's investment in intervention should be protected if economic circumstances were to change and agriculture were to return soon to profitability, or land values were to increase substantially for other reasons.

9. Agriculture is not the only sector of the economy experiencing serious economic difficulty. Any sector or subsector that is both capital-intensive and export sensitive is suffering from the effects of high interest rates, a strong dollar, and weak demand in countries pressed to keep their debt obligations serviced. Most of the sectors or subsectors experiencing stress, other than agriculture, can more easily respond to financial pressure by reducing output to obtain relief as to price. Because no single producer in agriculture is sufficiently large to influence price, reduction of output is less likely without intervention. In general, society benefits from this feature of agriculture in the form of greater output and lower product prices than would be the case otherwise. However, occasionally agriculture needs help in adjusting if serious economic damage is to be avoided from overproduction.

EVALUATING PROGRAMS OF INTERVENTION. An almost infinite array of public-sector interventions is possible for most policy problems. This is certainly the case with the current financial crisis in agriculture. Although evaluations are difficult to make inasmuch as proposals are understandably diverse in basic features and characteristics, it is essential to an objective review and appraisal that proposals be evaluated on the basis of an agreed-upon set of criteria. For the farm financial crisis, it is suggested that the set of criteria explain, for both intervention and nonintervention, (1) the direct and indirect costs to taxpayers and consumers, (2) who receives the benefits from intervention, (3) whether the proposal is likely to stabilize the farming sector and whether reasonable stability is likely to be extended to lenders and suppliers, (4) who bears the risks of further declines in asset values, (5) who bears the risks of future changes in interest rates and other costs of production, (6) who receives the benefits of future increases in asset values, (7) whether the proposal encourages necessary resource adjustment and promotes economic efficiency, and (8) the administrative costs expected to be associated with the implementation and operation of the specific proposal.

In the case of the farm financial crisis, intervention is viewed as a means to facilitate the adjustment process, minimize the costs of adjustment, and avoid the consequences of overadjustment or the overshooting of what should be equilibrium conditions. If present economic conditions continue, resource adjustment at the firm level will be needed under any reasonable scenario of intervention. In the event that the economic environment were to return to more favorable economic conditions for agriculture, the amount of adjustment needed would be proportionately less.

Possible Programs of Intervention

NO PUBLIC INTERVENTION. A policy of no intervention could be followed with the burden of adjustment left to borrowers, lenders, and others to pursue available remedies. Lenders would be expected to foreclose on real estate mortgages, proceed with remedies under the Uniform Commercial Code in the event of default on obligations with personal property as collateral, forfeit the rights of defaulting buyers under installment land contracts and work out repayment arrangements under informal compositions with borrowers. Among the latter are voluntary, privately arranged restructuring efforts as principal balances are written down or interest rates are reduced or both. Heavily indebted farmers at or approaching insolvency would be expected to file for bankruptcy under U.S. Bankruptcy Code Chapter 7 (liquidation) or Chapters 11 or 13 (reorganization) options, voluntarily turn over assets to creditors in satisfaction of debt obligations, or sell assets and apply the proceeds of sale on amounts owed.

In some areas of the United States, a policy of nonintervention would likely not create unacceptable levels of economic trauma. However, available data indicate that in some areas substantial economic costs would be incurred in terms of loss of wealth, failure of financial institutions, insolvency by suppliers, and shrinkage of the economic and social base of rural communities.

DEBT RESTRUCTURING WITH LOAN GUARANTEES. The debt restructuring program announced by President Reagan on 18 September 1984 was an effort in meeting the debt problems of commercial agriculture in the United States. If a farmer could show cash flow equal to 110 percent of costs and debt service on a projected basis, and the lender were to write down at least 10 percent of the principal value of the loan, a guarantee of up to 90 percent of the remaining principal balance could be guaranteed by the Farmers Home Administration. The program was referred to as the Debt Adjustment Program, or DAP, and was intended for loans classified as substandard by the lender's supervising agency. The rules specified that, if necessary, the lender would have to write down more than the initial 10 percent of principal to meet the cash-flow requirements. Loans with adequate security generally do not require a write down by the lender to obtain a guaranty under the regular loan guaranty program. Announcements on 6 and 22 February 1985 of modifications in the program reduced the cash-flow requirement for eligible participants from 110 percent to 100 percent of projected cash flow and permitted lenders to take the required prinicipal write-down in the form of interest rate reductions to borrowers spread over several years. Moreover, assurances were given that additional loan guaranty authority would be made available if needed. However, it became apparent in March 1985 that loan guaranty authority was not available to restructure real estate loans. Loan guaranty authority is available to restructure loans over seven years with the possibility of a balloon payment. Final regulations were published on 15 February 1985, so the program has the clear advantage of being in place.

If available for real estate loans and with adequate amounts of loan guaranty authority, debt restructuring through federal loan guarantees would provide buoyancy to land and machinery markets to help the asset restructuring that must take place to occur on a rational basis. Loan guarantees only minimally interrupt and distort economic relationships and represent a good solution in many ways. The farmer is encouraged to remain with debt obligations on a deferred payment basis rather than to file for bankruptcy or use other remedies.

An "upside" eligibility test is imposed by requiring a significant write-down of interest or principal or both by lenders. Borrowers who are likely to be able to service outstanding debt and stabilize their financial condition would not be admitted to the program. The "downside" eligibility test, rendering those ineligible for the program who have no reasonable likelihood of surviving financially, is administered in the form of the cash-flow requirement.

One of the most difficult features of the DAP was dealing with outstanding unsecured debt. The rules specifically required that the loans remaining after the debt restructuring must be adequately secured. Moreover, the rules required that the plan submitted deal with all debt, secured as well as unsecured. The secured and unsecured creditors are expected to negotiate for write-offs and repayment terms that may be different by the security position of creditors.

Every effort should be exerted to make the federal loan guaranty program work. It is the only general program available to ease the debt problems in the immediate future.

NEWLY CREATED FEDERAL ENTITY TO FACILITATE LAND HOLDING AND FINANCING. Because of the importance of interest rates in any effort to stabilize farm and ranch firms, one approach would be to channel state and federal funds directly into interest rate reductions for farm loans. At the same time, there is a need for assets, particularly farmland held by those so heavily indebted that retention of the assets is infeasible, to be insulated from the market.

It is believed that the two functions, interest rate reductions and a holding tank for farm assets, should be joined in one entity if possible. Unless the economic environment changes dramatically very soon, major adjustments in organization of farm and ranch firms must take place to reflect the realities of the 1980s. Farmers should be encouraged to develop realistic cash-flow/reorganization plans that will, if possible, stabilize the firm. Some interest rate reductions (on the order of 3 to 5 percentage points) should be available to assist in making the cash-flow/reorganization plans feasible. If the firm cannot be stabilized under those conditions, changes in enterprises, management approaches, and asset ownership may be necessary.

The proposal for formation of an agricultural financial corporation (AFC) would have two major components. One component would provide the supplemental financing for "buying down" interest rates on farm loans for feasible cash-flow/reorganization plans on a targeted basis but with an expectation that interest subsidies would eventually be repaid with some interest on amounts advanced. Another component would provide a mechanism for acquiring the assets, notably farmland, given up by farmers who are unable to develop a feasible cash-flow/

reorganization plan short of asset liquidation. This entity could acquire land (1) subject to foreclosure or bankruptcy, (2) from lenders holding land in inventory, or (3) from farmers who are unable to service the real estate debt. The land would be rented back to the farmer at a reasonable rental, and the farmer would be encouraged to repurchase the land as soon as possible.

Although various possible designs of entities would appear to be feasible, a federally chartered corporation, similar in some respects to the U.S. Commodity Credit Corporation, would be the basic vehicle. It is anticipated that the corporation, which might be referred to as the Agricultural Financing Corporation, would have a governing board that would be broadly representative of production agriculture, public and private sector lending, and agribusiness firms, with significant consumer and taxpayer representation.

References

Barkema, Alan. Unpublished Research Manuscript. Iowa State University, 1985.

Financial Characteristics of U.S. Farms, January 1985. Agric. Info. Bull. No. 495. Econ. Res. Service, U.S. Dept. of Agric., July 1985.

1985 Iowa Farm Finance Survey. Iowa Dept. of Agriculture, Iowa State University and Iowa Crop and Livestock Reporting Service. Ames, Ia., 1985.

1984 Farm Costs and Returns Survey. U.S. Dept. of Agriculture. Washington, D.C., 1985.

SUGGESTED READINGS

Doyle, Jack. *Altered Harvest*. New York: Viking Penguin, 1985.

Fite, Gilbert C. *American Farmers: The New Minority*. Bloomington: Indiana University Press, 1981.

Harl, Neil E., "The Architecture of Public Policy: The Crisis in Agriculture," *Kansas Law Review* 34(1986): 425–456.

Jolly, R. W., and D. G. Doye. "Farm Income and the Financial Condition of United States Agriculture." Staff Report #8-85. Ames, Iowa: Food and Agricultural Policy Research Institute, 1985.

Jolly, Robert W., and Damona G. Doye. "Farm Finance: Farm Debt, Government Payments, and Options to Relieve Financial Stress." A General Accounting Briefing Report to the Honorable Bill Bradley, United States Senate, 1986.

Saloutos, Theodore. *The American Farmer and the New Deal*. Ames: Iowa State University Press, 1982.

Shover, John. *First Majority—Last Minority: The Transformation of Rural Life in America*. DeKalb: Northern Illinois University Press, 1976.

The University and the Family Farm

IV

ACCORDING TO Neil Harl, we are facing a financial problem of critical proportion in the farm belt. "Unless something dramatic is done," he writes, "more than one-third of the farmers nationally will move to insolvency, taking down their lenders, their suppliers, and other merchants" (Chap. 11). If these are the facts—and we have little reason to doubt Harl's judgment—then we must ask ourselves whether we want this to happen.

When we ask what sort of future we desire, we begin to articulate values and hopes. In doing this we cannot help but enter what Aiken (1962) calls the ethical level of moral discourse, for the concern here is with fundamental convictions and principles. There are many ways in which human values bear on agricultural policy, but which values are most important with respect to the family farm? Kirkendall puts the matter succinctly: with little more than 1 percent of all Americans now living on family farms, why *should* we try to save them? Haven't we long since passed the point at which we could have done so? What ethical principle could possibly justify the expense and effort that is now required to save family farms? And who are we to trust for answers to these questions?

We often look to the university for help with difficult matters. Universities serve by educating: professors disseminate knowledge to their students. Perhaps colleges can lead during the farm crisis.

It would be a relief if we could simply turn these questions over to academics. But the relationship between the university and the people is not as simple or unproblematic as the previous paragraph makes it seem. It is not clear that agricultural economists have all the answers, or that sociologists or moral philosophers (or even theologians) do. Moreover, it has been alleged that the university is not an impartial observer of society, and its social scientists are reputed to influence and shape culture even more than they record and analyze it.

The essays in Part Four investigate the relationship between higher education and agriculture. The land grant universities in this country were established in 1862 "to teach . . . agriculture and the mechanic arts . . . in order to promote the liberal and practical education of the industrial classes." Few doubt that these institutions helped to spread education in rural areas. What is more arguable is that these state-funded schools have always put the interests of the industrial classes first. And it is here that principles of justice and democracy need to be discussed.

In a now classic essay called "Hard Tomatoes, Hard Times," Jim Hightower and Susan DeMarco lament the intimate relationship between agribusiness corporations and colleges of agriculture. Schools have put time, energy, resources, and money into the pockets of big business, they say, rather than into the overalls of family farmers. Hightower and DeMarco are by no means alone in making this claim; Wendell Berry repeats it in Chapter 27, and has stated elsewhere that the tax monies originally provided to serve "the liberal and practical education" of farmers were used instead as an educational subsidy for agriculture specialists and agribusinesspeople—the farmers' competitors (Berry 1977, 158).

What should agricultural economists at universities like Oklahoma State and Michigan State do? What sorts of roles ought they play? Glenn Johnson argues that there are many tasks to perform, and simply "analyzing data" is by no means the extent of it. In "Roles for Social Scientists in Agricultural Policy," he argues that economists should not only produce value-free knowledge of a scientific sort, but should also participate in generating "value knowledge," knowledge about whether certain situations are good or bad. This is a role for agricultural economists that is not always acknowledged by those in Johnson's profession. But he does not stop there; he goes on to say that social scientists should produce "prescriptive knowledge," actual solutions to practical problems. Again, as he notes, "not all academicians agree" that researchers should be engaged in this, but Johnson argues that, one way or another, they do it whether they want to or not. It is best to be honest about it, and then be as objective as one can.

Johnson's essay was not composed as a response to Hightower. We must read Johnson's article on its own terms: as the expression of an economist genuinely concerned about the wider social ramifications of his professional work.

The third essay, on the other hand, was written in part as a response to Johnson. Tony Smith comes from a different corner of the university and from a different conceptual scheme altogether. Smith is a political philosopher, not a social scientist, and is more concerned with the power relations between different classes in society than with the specific tasks of the university economist. Johnson and Smith use language that seems miles apart, but they are in fact talking about the same thing. Whereas Johnson wishes to expand his profession's focus beyond narrow technical questions, Smith is more radical. He questions the very bases and motivations of agriculture research at institutions like his own, Iowa State. He sharply criticizes land grant institutions, agribusiness corporations, and agricultural economists.

The responses of Johnson and Smith to each other conclude the section.

References

Aiken, H. D. *Reason and Conduct: New Bearings in Moral Philosophy.* New York: Knopf, 1962.

Berry, Wendell. *The Unsettling of America: Culture and Agriculture.* New York: Avon, 1977.

**JIM HIGHTOWER
and SUSAN DeMARCO**

Hard Tomatoes, Hard Times:
The Failure of the Land
Grant College Complex

Introduction

Corporate agriculture's preoccupation with scientific and business efficiency has produced a radical restructuring of rural America that has been carried into urban America. There has been more than a "green revolution" [the transfer of modern agricultural technologies to Third World countries occurring over the past several decades] out there—in the last thirty years there literally has been a social and economic upheaval in the American countryside. It is a protracted, violent revolution, and it continues today.

The land grant college complex has been the scientific and intellectual parent of the revolution. This public complex—composed of colleges of agriculture, agricultural experiment stations, and state extension services—has put its tax dollars, its facilities, its staff, its energies, and its thoughts almost solely into efforts that have worked to the advantage and profit of large corporations involved in agriculture.

The consumer is hailed as the greatest beneficiary of the land grant college effort, but in fact, consumer interests are considered secondarily, if at all, and in many cases, the complex works directly against the consumer. Rural people, including the vast majority of farmers, farmworkers, small-town businesspeople and residents, and the rural poor either are ignored or directly abused by the land grant effort. Each year about a million of these people pour out of rural America into the cities. They are the waste products of an agricultural revolution designed within the land grant complex. Today's urban crisis is a consequence of the failure in rural America. The land grant complex cannot shoulder all

JIM HIGHTOWER is commissioner of agriculture in Texas. SUSAN DeMARCO was codirector of the Agribusiness Accountability Project that undertook the study upon which this paper is based. The essay is reprinted with permission of the authors from *Radical Agriculture,* ed. Richard Merrill (New York: Harper, 1976), pp. 87–107.

the blame for that failure, but no single institution – private or public – has played a more crucial role.

The complex has been eager to work with farm machinery manufacturers and well-capitalized farming operations to mechanize all agricultural labor, but it has accepted no responsibility for the farm laborer who is put out of work by the machine. It has worked hand in hand with seed companies to develop high-yield seed strains, but it has not noticed that rural America is yielding up practically all of its young people. It has been available day and night to help nonfarming corporations develop schemes of vertical integration while offering independent family farmers little more comfort than "adapt or die." It has devoted hours to create adequate water systems for fruit and vegetable processors and canners, but thirty thousand rural communities still have no central water systems. It has tampered with the gene structure of tomatoes, strawberries, asparagus, and other foods to prepare them for the steel grasp of the mechanical harvesters, but it has sat still while the American food supply has been liberally laced with carcinogenic substances.

The land grant complex, as it is known today, has wandered a long way from its origins, abandoning its historic mission to serve rural people and American consumers.

This chapter independently examines America's land grant college-agricultural complex. Its message is that the tax-paid land grant complex has come to serve an elite of private, corporate interests in rural America while ignoring those who have the most urgent needs and the most legitimate claims for assistance.

It is the objective of the land grant college task force to provoke a public response that will help realign the land grant complex with the public interest. In a speech reordering agricultural research priorities, the director of science and education at the U.S. Department of Agriculture (USDA) said that "the first giant steps are open discussion and full recognition of the need." This chapter is dedicated to that spirit.

The Land Grant College Complex

As used throughout the report, "land grant college complex" denotes three interrelated units, all attached to the land grant college campus:

1. Colleges of agriculture – created in 1862 and 1890 by two separate Morrill acts.
2. State agricultural experiment stations – created in 1887 by the Hatch Act for the purpose of conducting agricultural and rural research in cooperation with the colleges of agriculture.

3. Extension service—created in 1914 by the Smith-Lever Act for the purpose of disseminating the fruits of teaching and research to the people in the countryside.

Reaching into all fifty states, the complex is huge, intricate, and expensive. It can be estimated that the total complex is approaching an expenditure of three quarters of a billion tax dollars appropriated each year from federal, state, and county governments. The public's total investment in this complex, including assets, comes to several billion dollars in any given year, paying for everything from test tubes to experimental farms, from chalk to carpeting in the dean's office.

The Research Effort

There is no doubt that American agriculture is enormously productive and that agriculture's surge in productivity is largely the result of mechanical, chemical, genetic, and managerial research conducted through the land grant college complex.

But the question is whether the achievements outweigh the failures, whether benefits are overwhelmed by costs. It is the finding of the task force that land grant college research is not the bargain that has been advertised.

The focus of agricultural research is warped by the land grant community's fascination with technology, integrated food processes, and the like. Strict economic efficiency is the goal, not people. The distorted research priorities are striking:

• 1,129 scientific man-years (smy) on improving the biological efficiency of crops and only 18 smy on improving rural income
• 842 smy on control of insects, diseases, and weeds in crops and 95 smy to ensure food products free from toxic residues from agricultural sources
• 200 smy on ornamentals, turf, and trees for natural beauty and a sad 7 smy on rural housing
• 88 smy on improving management systems for livestock and poultry production and 45 smy for improving rural institutions
• 68 smy on marketing firm and system efficiency and 17 smy on causes and remedies of poverty among rural people.

In fiscal year 1969, a total of nearly six thousand scientific man-years were devoted to research on all projects at all state agricultural experiment stations. Based on USDA's research classifications, only 289 of those scientific man-years were expended specifically on "people-oriented" research. That is an allocation to rural people of less than 5 percent of the total research effort at the state agricultural experiment station. (See Table 12.1.)

An analysis of these latter research projects reveals that the commitment to the needs of people in rural America is even less than appears on the surface. In rural housing, the major share of research has been directed not to those who live in them but to those who profit from the construction and maintenance of houses — architects, builders, lumber companies, and service industries.

Again and again, the point is made that industry needs help because it cannot do its own research and because it is affected by external factors. People, however, are responsible for their own condition. For industry, public research assistance is considered an investment; for people, that assistance is treated as welfare.

Mechanization Research

The primary beneficiaries of land grant research are agribusiness corporations. These interests envision rural America solely as a factory that will produce food, fiber, and profits on a corporate assembly line extending from the fields through the supermarket checkout counters. It is through mechanization research that the land grant colleges are coming closest to this agribusiness ideal.

Table 12.1. Scientific man-years of "people-oriented" research conducted at state agricultural experiment stations — 1966 and 1969

Research problem areas	1966 SMY at SAES	1969 SMY at SAES
Food choices, habits, and consumption	8	11.5
Home and commercial preparation of food	14	12.4
Human nutritional well-being	103	93.5
Selection and care of clothing and household textiles	18	15.0
Housing needs of rural families	11	6.5
Family decision making and financial management	20	16.0
Causes and remedies of poverty among rural people	11	17.1
Improvement of economic opportunities for rural people	42	27.7
Communication processes in rural life	17	18.3
Individual and family adjustment to change	28	25.6
Improvement of rural community institutions and services	29	45.3
Total	301	288.9[a]

Sources: USDA Science and Education Staff (1970, 247–78); also, USDA, USDA-NASULGC (1968, 5, 28, and 29).

[a]This allocation of scientific man-years indicates how meager the commitment to "people-oriented" research really is in comparison with the land grant community's rhetoric of concern. The experiment stations actually were doing *less* people-oriented research in 1969 than they were in 1966. The 289 smy allocated to people in 1969 represents only 4.8 percent of the total of 5,956 smy expended that year at state agricultural experiment stations.

Mechanization means more than machinery for planting, thinning, weeding, and harvesting. It also means improving on nature's design, that is, breeding new food varieties that are better adapted to mechanical harvesting. Having built machines, the land grant research teams found it necessary to develop a tomato that is hard enough to survive the grip of mechanical "fingers," to redesign the grape so that all the fruit has the good sense to ripen at the same time, and to restructure the apple tree so that it grows shorter, leaving the apples less distance to fall to its mechanical catcher. Michigan State University, in a proud report on "tailor-made" vegetables, notes that their scientists are at work on broccoli, tomatoes, cauliflower, cucumbers, snapbeans, lima beans, carrots, and asparagus.

If it cannot be done by manipulating genes, land grant scientists reach into their chemical cabinet. Louisiana State University has experimented with the chemical "Ethrel" to cause hot peppers to ripen at the same time for "once-over" mechanical harvesting; scientists at Michigan State University are using chemicals to reduce the cherry's resistance to the tug of the mechanical picker; and a combination of ferric ammonia citrate and erythorbic acid is being used at Texas A & M to loosen fruit before machine harvesting.

Once harvested, food products must be sorted for size and ripeness. Again, land grant college engineers have produced a mechanical answer. North Carolina State University, for example, has designed and developed an automatic machine that "dynamically examines blueberries according to maturity."

Genetically redesigned, mechanically planted, thinned, and weeded, chemically readied, and mechanically harvested and sorted, food products move out of the field and into the processing and marketing stages — untouched by human hands.

Who is helped and who is hurt by this research and development?

It is agribusiness that is helped. In particular, the largest-scale growers, the farm machinery and chemical input companies, and the processors are the primary beneficiaries. Big business interests are called upon by land grant staffs to participate directly in the planning, research, and development stages of mechanization projects. The interests of agribusiness literally are designed into the product. No one else is consulted.

Obviously, farm machinery and chemical companies are also direct beneficiaries of this research because they can expect to market products that are developed. Machinery companies such as John Deere, International Harvester, Massey-Ferguson, Allis-Chalmer, and J. I. Case almost continually engage in cooperative research efforts at land grant colleges.

These corporations contribute money and some of their own research personnel to help land grant scientists develop machinery; in return, they are able to incorporate technological advances in their own products. In some cases, they actually receive exclusive license to manufacture and sell the product of tax-paid research.

Mechanization of fruits and vegetables has focused first on crops used by the processing industries. Brand name processors (such as Del Monte, Heinz, Hunt, Stokely Van Camp, Campbell's, and Green Giant) are direct beneficiaries of mechanization research. Many of these corporations have been directly involved in the development of mechanization projects. In addition to the food-breeding aspects of mechanization, processors and canners also have benefited insofar as mechanization has been able to lower costs of production and insofar as that savings has been passed on to them. Of course, many food processors also are growers—either growing directly on their own land or growing indirectly, controlling the production of others through contractual arrangements.

Large-scale farming operations, many of them major corporate farms, also are directly in line to receive the rewards of mechanization research. In the first place, it is these farms that hire the overwhelming percentage of farm labor, thus having an economic incentive to mechanize. Second, these are the massive farms, spreading over thousands of acres. This scale of operation warrants an investment in machinery. Third, these are heavily capitalized producers including processing corporations, vertically integrated input and output industries, and conglomerate enterprises. Such farming ventures are financially able and managerially inclined to mechanize the food system.

Then there are the victims of mechanization—those who are directly hurt by research that does not consider their needs. If mechanization research has been a boon to agribusiness interests, it has been a bane to millions of rural Americans. The cost has been staggering.

Farmworkers have been the earliest victims. Again and again, the message is hammered home: machines are now or are on their way to replacing farm labor. There were 4.3 million hired farmworkers in 1950. Twenty years later, that number had fallen to 3.5 million. As a group, those laborers averaged $1,083 for doing farm work in 1970, making them among the very poorest of America's employed poor. The great majority of these workers were hired by the largest farms, which are the same farms moving as swiftly as possible to mechanize their operation.

Farmworkers have not been compensated for jobs lost to mechanization research. They were not consulted when that research was designed, and their needs were not a part of the research package that resulted. They simply were left to fend for themselves—no retraining, no effort to find new jobs for them, no research to help them adjust to the changes that came out of the land grant colleges. Corporate agribusiness received a machine with the taxpayer's help, but the workers who were replaced were not even entitled to unemployment compensation.

Independent family farmers—at least those who have sales under twenty thousand dollars a year (which [in 1972] includes 87 percent of all U.S. farms)—also have been victimized by the pressure of mechanization, and their needs also have been largely ignored by the land grant colleges.

Mechanization has been a key element in the cycle of bigness: enough capital can buy machinery, which can handle more acreage, which will produce greater volume, which can mean more profits, which will buy more machinery. Mechanization has not been pressed by the land grant complex as an alternative but as an imperative.

Mechanization research by land grant colleges either is irrelevant or only incidentally adaptable to the needs of some 87 to 99 percent of America's farmers. The public subsidy for mechanization actually has weakened the competitive position of the family farmer. Taxpayers, through the land grant college complex, have given corporate producers a technological arsenal specifically suited to their scale of operation and designed to increase their efficiency and profits. The independent family farmer is left to strain private resources to the breaking point in a desperate effort to clamber aboard the technological treadmill.

Like the farmworker, the average farmer is not invited into the land grant laboratories to design research. If the farmer were, the research "package" would include machines useful on smaller acreages, assistance to develop cooperative ownership systems, efforts to develop low-cost and simpler machinery, a heavy emphasis on new credit schemes, and special extension to spread knowledge about the purchase, operation, and maintenance of machinery. In short, there would be a deliberate and major effort to extend mechanization benefits to all with an emphasis on at least maintaining the competitive position of the family farm in relation to agribusiness corporations. These efforts do not exist or exist only in a token way. Mechanization research has left the great majority of farmers to "get big" on their own or to get out of farming altogether.

Mechanization also has a serious impact on the consumer, and that impact puts America's "bargain" food prices in serious question. Land grant researchers are not eager to confront the issue of quality impact on mechanization, choosing instead to dwell on the benefits that food engineering offers agribusiness.

The University of Florida, for example, recently has developed a new fresh-market tomato (the MH-1) for machine harvesting. In describing the characteristics that make this tomato so desirable for machine harvest, the university points to "thick walls, firm flesh, and freedom from cracks." It may be a little tough for the consumer, but agricultural research can't please everyone. The MH-1, which will eliminate the jobs of thousands of Florida farmworkers who now handpick tomatoes for the fresh market, is designed to be harvested green and to be "ripened" in storage by application of ethylene gas.

Agribusiness Versus Consumers

The colleges also are engaged in "selling" the consumer on products he or she neither wants nor needs, and they are using tax money for food research and development that should be privately financed. At Virginia Polytechnic Institute, for example, eight separate studies have been conducted to determine if people would like a blend of apple and grapefruit juice.

Another aspect of selling the consumer is "knowing" the consumer. There are many projects that analyze consumer behavior. Typically, these involve consumer surveys to determine what influences the shopper's decision making. If this research is useful to anyone, it is to food marketers and advertisers, and reports on this research make clear that those firms are the primary recipients of the results. The corporations that benefit from this research should pay for it and conduct it themselves.

The consumer is not just studied and "sold" by land grant research; the consumer is also fooled. These public laboratories have researched and developed food cosmetics in an effort to confirm the consumer's preconceptions about food appearances thus causing the consumer to think that the food is "good." Chickens have been fed the plant compound xanthophyll to give their skin "a pleasing yellow tinge," and several projects have been undertaken to develop spray-on coatings to enhance the appearance of apples, peaches, citrus, and tomatoes. Following are some other cosmetic research projects that were under way at land grant colleges in the early 1970s:

• Iowa State University was conducting packaging studies, which indicated that color stays bright longer when bacon is vacuum-packed or sealed in a package containing carbon dioxide in place of air thus contributing to "more consumer appeal."
• Because of mechanical harvesting, greater numbers of green tomatoes are being picked; scientists at South Carolina's agricultural experiment

station have shown that red fluorescent light treatment can increase the red color in the fruit and can cause its texture and taste to be "similar to vine-ripened tomatoes."

• Kansas State University Extension Service, noting that apples sell on the basis of appearance rather than nutrition, urged growers to have a beautiful product. To make the produce more appealing, mirrors and lights in supermarket produce cases were cited as effective selling techniques.

Sold, studied, and fooled by tax-supported researchers, there finally is evidence that the consumer actually is harmed by food engineering at land grant colleges. Ethylene gas, used to speed up the growth of produce has been shown, when used on tomatoes, to provide lower quality with less vitamin A and C and inferior taste, color, and firmness. There is strong evidence that DES, a growth hormone fed to cattle, causes cancer in humans. Yet DES has added some $2.9 million to the treasury of Iowa State University, where the use of the drug was discovered, developed, patented, and promoted—all with tax dollars. Eli Lilly & Company, which was exclusively licensed by Iowa State to manufacture and sell the drug, has enjoyed profits on some $60 million in DES sales to date.

More and more, chemicals are playing a role in the processing phase. Ohio State University reports that "chemical peeling of tomatoes with wetting agents and caustic soda reduces labor by 75 percent and increases product recovery." One wonders if the consumer will recover. Lovers of catfish might be distressed to learn that this tasty meat now is being skinned chemically for commercial packaging.

Three assumptions are made by the task force. First, if there is to be research for firms that surround the farmer, benefits of that research should flow back to the farmer. Second, no public money should be expended on research that principally serves the financial interests of agricultural input and output corporations; they may be a part of modern agriculture, but they also are very big business and capable of doing their own profit-motivated research. Finally, anything that is good for agribusiness is not necessarily good for agriculture, farmers, rural America, or the consumer.

Failure of Land Grant College Research

Except for agribusiness, land grant college research has been no bargain. Hard tomatoes and hard times is too much to pay. That does not mean a return to the hand plow. Rather, it means that land grant college researchers must get out of the comfortable chairs of corporate board rooms and get back to serving the independent producer and the common man of rural America. It means returning to the historic mission of taking the technological revolution to all who need it, rather than smugly assuming that they will be unable to keep pace. Instead of adopting the morally bankrupt posture that millions of people must "inevitably" be squeezed out of agriculture and out of rural America, land grant colleges must turn their thoughts, energies, and resources to the task of keeping people on the farm, in the small towns, and out of the cities. It means turning from the erroneous assumption that big is good, that what serves Ralston Purina serves rural America. It means research for the consumer rather than for the processor. In short, it means putting the research focus on people first — not as a trickled-down afterthought.

The greatest failing of land grant college research is its total abdication of leadership. At a time when rural America desperately needs leadership, the land grant community has ducked behind the corporate skirt, mumbling apologetic words like "progress," "efficiency," and "inevitability." Overall, it is a pedantic and cowardly research system, and America is the less for it.

A change in the focus of land grant research will not happen simply because it should happen. Change will come only if those interests now being abused by the research began to make organized demands on the complex. If independent family farmers, consumers, small-town businesspeople, farmworkers, environmentalists, farmer cooperatives, small-town mayors, taxpayers' organizations, labor unions, big-city mayors, rural poverty organizations, and other "outsiders" go to the colleges and to the legislatures, changes can occur. These interests need not go hand in hand, but they all must go if land grant college research ever is to serve anyone other than the corporate elite.

Making Research Policy

The short-range research policy of the land grant system is the product of the annual budgeting process, and the substance of that research budget is determined by the Agricultural Research Policy Advisory Committee (ARPAC), which reports directly to the secretary of agriculture. Its members are taken from USDA and the land grant community; in fact, they are the agricultural research establishment.

The National Association of State Universities and Land Grant Colleges (NASULGC) is the home of the land grant establishment. Its particular corner in the association is under the title of Division of Agriculture, composed of all deans of agriculture, all heads of state experiment stations, and all deans of extension. With eight members on the twenty-four person ARPAC board, NASULGC's agricultural division plays a major role in the determination of research priorities and budgets. The division also represents the land grant college complex before Congress on budget matters.

The top rung on the advisory ladder is USDA's National Agricultural Research Advisory Committee. In the early 1970s, this eleven-member structure included representatives from the Del Monte Corporation, the Crown Zellerbach Corporation, AGWAY, Peavey Company Flour Mills, the industry-sponsored Nutrition Foundation, and the American Farm Bureau Federation.

Most national advisory structures are dominated by land grant scientists and officials, but whenever an "outsider" is selected, chances are overwhelming that the person will come from industry. A series of national task forces, formed from 1965 to 1969 to prepare a national program of agricultural research, were classic examples of this pattern. Out of thirty-two task forces, seventeen listed advisory committees containing non-USDA, non–land grant people. All but one of the outside slots on those seventeen committees were filled with representatives of industry, including General Foods on the rice committee, U.S. Sugar on sugar, Quaker Oats on wheat, Pioneer Corn on corn, Liggett & Myers on tobacco, Procter & Gamble on soybeans, and Ralston Purina on dairy. Only on the "soil and land use" task force was there an advisor representing an interest other than industrial, but even there, the National Wildlife Federation was carefully balanced by an advisor from International Minerals and Chemical Corporation.

There are also state and local advisory structures to the land grant complex. Commenting on such groups and their impact on the allocation of research resources, USDA's Roland Robinson wrote (1964): "Many of the advisory groups, similar to those of the Department of Agriculture, are established along commodity and industry lines. Consequently they are oriented toward traditional research needs. The rural nonfarmer, the small farmer, the leaders of rural communities and the consumer are not usually represented on experiment-station advisory committees."

Land grant policy is the product of a closed community. The administrators, academics, and scientists, along with USDA officials and corporate executives, have locked themselves into an inbred and even incestuous complex and are incapable of thinking beyond their self-interest and traditional concepts of agricultural research.

Congress holds hearings each year on the appropriations requests for agricultural research. It is here that the public might expect some serious questioning of research focus and some assertion of other than private interests. It does not happen.

Hearings on agricultural research budgets are left pretty much to the land grant community, buttressed by its agribusiness colleagues. The appropriations process falls far short of being a careful, substantive scrutiny; in fact, it is little more than a chance for special interests to press for particular research projects or facilities.

Public witnesses appearing before the agricultural subcommittees overwhelmingly represent agribusiness interests. Technically, anyone can testify, but it is industry that has the resources to maintain Washington representatives and to fly witnesses in and out the capital for a day of testimony. There are dozens of agribusiness lobbyists in Washington, ranging from the full-scale operation of the American Farm Bureau Federation to coveys of Washington "lawyers" retained to look out for the special interests of practically every corporate name in agriculture.

The few Washington organizations representing the interests of farmers, sharecroppers, small businesses, the poor, minorities, con-sumers, or environmentalists either do not have the resources and staff to deal effectively with the agricultural research budget or have failed to perceive their self-interest in that budget. Tax-exempt public interest groups are prohibited by law from lobbying and cannot appear to testify on appropriations unless invited to do so by the committee.

There are hundreds of pages of testimony on the land grant complex each year but no tough questioning of how those resources are being used. With two thousand farm families leaving the land each week, with some eight hundred thousand people a year being forced out of rural America, and with all other stark evidence of rural failure, it seems that some representative of the people would probe a bit into the nature and impact of the land grant complex.

Congress has relinquished its responsibility and authority to nar-rowly focused officials at USDA and within the land grant community. Like spokespersons of the military-industrial complex, these officials and their allies come to the capital at appropriations time to assure a docile Congress that its investment in agricultural hardware is buying "progress" and that rural pacification is proceeding nicely.

Agribusiness Links to Land Grant Campuses

In dozens of ways, agribusiness gets into the land grant college complex. It is welcomed there by administrators, academics, scientists, and researchers who share the agribusinessperson's vision of integrated, automated agriculture. Corporate executives sit on boards of trustees, purchase research from experiment stations and colleges, hire land grant academics as private consultants, advise and are advised by land grant officials, go to Washington to help a college or an experiment station get more public money for its work, publish and distribute the writings of academics, provide scholarships and other educational support, invite land grant participation in their industrial conferences, and sponsor foundations that extend both grants and recognition to the land grant community.

Money is the web of the tight relationship between agribusiness interests and their friends at the land grant colleges. It is not that a huge sum of money is given — industry gave only twelve million dollars directly to state agricultural experiment stations for research in 1969. Rather, it is that enough money is given to influence research done with public funds.

But to a larger extent, agribusiness was welcomed into the community because its attitudes and objectives were shared by the land grant communities. Agribusiness corporations wanted help with their new chemical, with their hybrid seed, with their processing facility, or with their scheme for vertical integration. The scientists, engineers, and economists of the land grant community had both the tools and the inclination to deal with those needs.

Industry money goes to meet industry needs and whims, and these needs and whims largely determine the research program of land grant colleges. A small grant for specific research is just good business. In the first place, the grant is tax deductible either as an education contribution or, if the research is directly related to the work of the corporation, as a necessary business expense. Second, the grant will draw more scientific attention than its value warrants. One scientist will consult with another, and graduate assistants and other personnel will chip in some time. If the project is at all interesting, it will be picked up and carried on by someone working under a public budget or assigned to someone working on a Ph.D. Finally, not only is the product wrapped and delivered to the corporation, but with it comes the college's stamp of legitimacy and maybe even an endorsement by the scientist who conducted the research. If it is a new product, the corporation can expect to be licensed, perhaps exclusively, as its producer and marketer. Everything considered, it amounts to a hefty return on a meager investment.

There is a long list of satisfied corporate customers. As would be expected, half of industry's research contributions to state agricultural experiment stations in fiscal year 1969 went to just four categories: insect control, weed control, plant and animal biology, and biological efficiency.

Prime contributors are chemical, drug, and oil corporations. Again and again, the same names appear—American Cyanamid, Chevron, Dow Chemical, Exxon, Eli Lilly, Geigy, FMC-Niagara, IMC Corporation, Shell, Stauffer, Union Carbide, and the Upjohn Company are just a few of the giants that gave research grants to the University of Florida, North Carolina State University, and Purdue University. Chemical, drug, and oil companies invested $227,158 in research at Florida's Institute of Food and Agricultural Science, for example, accounting for 54 percent of research sponsored there by private industry in 1970.

Where does the corporation end and the land grant college begin? It is difficult to find the public interest in the tangle. These ties to industry raise the most serious questions about the subversion of scientific integrity and the selling of the public trust. If grants buy corporate research, do they also buy research scientists and agricultural experiment stations?

Land Grant Research Foundations

At least twenty-three land grant colleges have established foundations to handle grants and contracts coming into their institutions for research. These quasi-public foundations are curious mechanisms, handling large sums of money from a wide array of private and public donors but under practically no burden of public disclosure.

A funding source can give money to a private research foundation, which then funnels money to a public university to conduct research. By this shell game, industry-financed research can be undertaken without obligation to make public the terms of the agreement. The foundation need not report to anyone the names of corporations that are making research grants, the amounts of those grants, the purpose of those grants, or the terms under which the grants are made.

These foundations also handle patents for the colleges. When a corporation invests in research through a foundation, it is done normally with the understanding that the corporation will have first shot at a license on any patented process or product resulting from the research. On research patents that do not result from corporate grants, the procedure for licensing is just as cozy. At Purdue University, for example, a list is drawn of "responsible" companies that might have an interest in the process or product developed, and the companies are approached one by one until there is a taker.

Extension Service

The Extension Service (ES) is the outreach arm of the land grant college complex. Its mandate is to go among the people of rural America to help them "identify and solve their farm, home, and community problems through use of research findings of the Department of Agriculture and the State Land Grant Colleges."

Three hundred thirty-one million dollars were available to the Extension Service in 1971. Like the other parts of the land grant complex, extension has been preoccupied with efficiency and production — a focus that has contributed much to the largest producers but that has slighted the pressing needs of the vast majority of America's farmers and ignored the great majority of other rural people.

Extension Service has not lived up to its mandate for service to rural people. The rural poor, in particular, are badly served by the service, receiving a pitiful percentage of the time of extension "professionals," while drawing temporary assistance from the highly visible nutrition aides program and irrelevant attention from the 4-H program. In 1955, a Special Needs Section was added to extension legislation, setting aside a sum of money to assist disadvantaged areas. Extension has failed to make use of this section.

The civil rights record of ES comes close to being the worst in government. Policy-making within ES fails to involve most rural people, and USDA has failed utterly to exercise its power to redirect priorities and programs.

The Extension Service's historical and current affiliation with the American Farm Bureau Federation casts a deep shadow over its claim that it can ever be part of the solution of the problems of rural America.

Black Land Grant Colleges

In 1862 at the time of the first Morrill Act, 90 percent of America's black population was in slavery. The land grant colleges that developed were white bastions, and even after the Civil War, blacks were barred from admission both by custom and by law. When the second Morrill Act was passed in 1890, primarily to obtain more operating money for the colleges, Congress added a "separate but equal" provision authorizing the establishment of colleges for blacks. Seventeen southern and border states took advantage of the act, creating institutions that are still referred to euphemistically as "colleges of 1890."

The black colleges have been less than full partners in the land grant experience. It is a form of institutional racism that the land grant community has not been anxious to discuss. From USDA, resource allocations to these colleges are absurdly discriminatory. In 1971, of the $76.8 million in U.S. Department of Agriculture funds allocated to those sixteen states with both white and black land grant colleges, 99.5 percent went to the white colleges, leaving only .5 percent for the black colleges.

Less than 1 percent of the research money distributed by the Cooperative State Research Service went to black land grant colleges. This disparity is not by accident but by law.

Public Disclosure

It is difficult to discover what the land grant complex is and what it is doing. For example, most agricultural experiment stations offer an annual report in compliance with the Hatch Act disclosure provisions, but these reports are less than enlightening:

• Some do not list all research projects but merely highlights.
• Some list research projects but only by title without even a brief description.
• Most do not include money figures with the individual projects, and very few reveal the source of the money.
• None contains any element of project continuity to show the total tax investment over the years in a particular investigation.
• Most contain only a very general financial breakdown, listing state, federal, and "other" funds received and expended.
• Few offer any breakdown of industry contributions, naming the industry, the contribution, and the project funded.

These are the basic facts. There is no listing of more esoteric items, such as patents developed by the station and held by the college, or advisory structures surrounding the stations.

Data is not supplied uniformly, collected in a central location, reported in a form that can be easily obtained or understood. Even more significant is the fact that many fundamental questions go unasked and many fundamental facts go unreported.

Millions of tax dollars are being spent annually by an agricultural complex that effectively operates in the dark. It is not that the land grant community deliberately hides from the public. The farmer, the consumer, the rural poor, and others with a direct interest in the work of the land grant complex can get no adequate picture of its work. Congress is no help; it does not take the time to probe the system, to understand it in detail, and to direct its work in the public interest.

The land grant college complex has been able to get by with a minimum of public disclosure, and that has meant that the community has been able to operate with a minimum of public accountability.

Conclusion

There is nothing inevitable about the growth of agribusiness in rural America. While this country enjoys an abundance of relatively cheap food, it is not more food, not cheaper food, and certainly not better food than that which can be produced by a system of family agriculture. And more than food rolls off the agribusiness assembly line—rural refugees, boarded-up businesses, deserted churches, abandoned towns, broiling urban ghettos, and dozens of other tragic social and cultural costs also are products of agribusiness.

Had the land grant community chosen to put its time, its money, its expertise, and its technology into the family farm rather than into corporate pockets, then rural America today would be a place where millions could live and work with dignity.

The colleges have mistaken corporate need for national need. This is proving to be a fatal mistake—not fatal for the corporations or for the colleges but for the people of America. It is time to correct that mistake, to reorient the colleges so that they will begin to act in the public interest.

Recommendations

The Task Force on the Land Grant College Complex does not presume to prescribe an agenda for the land grant college complex. That is the proper role of constituencies with a direct interest in the work of the complex.

Rather, the task force seeks, through its recommendations, to open the closed world of the land grant complex to public view and to participation by constituencies that today are locked out.

Generally, these recommendations call for full-scale public inquiry, both in the Congress and state legislatures, regarding the nature, extent, and national impact of the land grant complex. There should be a General Accounting Office audit of the land grant complex. An immediate reopening of the hearings on the 1972 to 1973 agricultural research budget by the House and Senate is also necessary. Also, the secretary of agriculture should act immediately to restructure the national advisory and policy-making apparatus so that there is a broadened input from "outside" constituencies for research planning.

The task force calls for an immediate end to racial discrimination within the land grant complex and the withholding of federal money from any state that does not place its black institutions on an equal footing with the white colleges.

Legislation is also needed that would prohibit land grant officials and other personnel from receiving remuneration in conflict of interests, prevent corporations from earmarking contributions to the land grant complex for specific research that is propitious in nature, and insure that land grant patenting practices do not allow private gain from public expenditures.

Laws requiring full public disclosure from the land grant complex are of crucial importance. Detailed, complete, and uniform reports from each college should be filed annually with the secretary of agriculture, who should compile them and make them easily available to the public.

The land grant colleges must get out of the corporate board rooms, they must get the corporate interests out of their labs, and they must draw back and reassess their preoccupation with mechanical, genetic, and chemical gadgetry. The complex must again become the people's university — it must be redirected to focus the preponderence of its resources on the full development of the rural potential.

References

Robinson, Roland. *Money and Capital Markets.* New York: McGraw-Hill, 1964.

U.S. Department of Agriculture, Science and Education Staff. *Inventory of Agricultural Research, FY 1969 and 1970.* Washington, D.C.: Government Printing Office, 1970.

U.S. Department of Agriculture, USDA-NASULGC. *A National Program of Research for Rural Development and Family Living.* Washington, D.C.: Government Printing Office, 1968.

GLENN L. JOHNSON

Roles for Social Scientists in Agricultural Policy

THIS ESSAY ORIGINATED when I was asked to participate in a debate with a Marxist philosopher on the topic "The social scientists' role in policy decisions is to analyze facts, not protect farming interests." I was to argue that the proposition was true. I declined to participate in such a debate; agricultural policy is too important to be confused by an ideological exchange not particularly relevant for U.S. agriculture in 1986. In addition, the affirmative position implies that social scientists should analyze facts but not values, a position with which I fundamentally disagree.

However, the role of the social scientist in agricultural policy is a topic on which I have some thoughts. The main topics I will address are these: (1) how market-controlled, agricultural economies operate through time to create crises of the type now being experienced; (2) how family farms survive, in part, as a consequence of how agriculture operates; (3) the historical record of U.S. agricultural production since 1880 and especially since 1918; (4) the kinds of policies and programs needed to control overproduction and avoid crises for agriculture; (5) how solutions to the overproduction problem may eliminate family farms; and finally (6) the roles social scientists can reasonably be expected to play in studying two areas: the issue of agricultural production control, price support, surplus storage policies and programs and the issue of what should be done about the family farm. These roles include participation in efforts to produce prescriptive knowledge, generation of value knowledge, and generation of value-free knowledge.

GLENN L. JOHNSON is professor of agricultural economics at Michigan State University, a fellow of the American Agricultural Economics Association, and past president of the International Association of Agricultural Economists.

How Market-Controlled, Agricultural Economies Operate

In all but about eight of the sixty-six years since the close of World War I, U.S. farmers have faced either adverse price pressure or have experienced production controls, price supports, and government-held surplus. The shortages, favorable prices, and an absence of government-held stocks in the middle and late 1970s was distinctly atypical. The present situation should be regarded as normal for a market-controlled agricultural sector.

We need to understand why this is so. Farmers typically make long-term, major investments in durables and in the production of such intermediate inputs as feed grains, roughages, livestock herds, feeder animals, and the like. Most of these durables and intermediate inputs have acquisition prices substantially higher than their liquidation values. These prices rise and fall in concert with the prices of the products produced, land being the most important of these. (See Johnson and Quance 1972.)

Another part of the explanation is that farmers are not and cannot be very well informed about the highly dynamic complex situations in which they invest in durables, buy their inputs, produce intermediate inputs, and produce and market their products. As a consequence of being imperfectly informed, farmers make a large number of mistakes in committing resources to production. They can easily correct their under-commitment mistakes as it will still pay to buy more inputs. On the other hand, mistakes of overcommitment or overproduction are very difficult to correct. When a farmer buys too much of something that can be sold only at a price substantially below acquisition price, the farmer incurs liquidation losses if he or she disinvests in it. With disposal prices below acquisition costs, farmers typically find it better to keep overcommitted resources in production than to liquidate them. This is because the value of the earnings of the overcommitted resources in production exceed the low price they would command if liquidated. Their overcommitted resources include themselves. When a forty-five-year-old farmer goes broke, that farmer does not hold union seniority or even possess skills and attitudes for nonfarm work. Too often he or she has to seek employment in low-paying jobs in local government and service industries (Chennareddy and Johnson 1968).

Thus, even if farmers make under- as well as overinvestment mistakes, there is a tendency toward overproduction because farmers can easily correct their underproduction mistakes but not their overproduction mistakes. The overall effects of overproduction mistakes by thousands of farmers is low earnings on farm investments and, consequently, capital losses on those assets. This leads to pressure for government programs and price supports followed by storage programs. Then as subsidy and storage costs increase, a need for production controls emerges in order to hold down those costs. To the extent that low earnings and capital losses are permitted to occur as is currently being permitted, cash-flow problems and bankruptcy soon materialize for leveraged farmers. These conclusions describe the situation for midwestern farmers today.

It may be of interest to know that in 1926 my leveraged father went broke in Mower County, Minnesota. His main difficulty was that he was in an age group whose members leveraged themselves during World War I in order to get started in farming. After the shock and trauma of going broke, my father moved to western Illinois where he worked for John Deere and farmed part-time. By 1928, he had saved enough money to start up again as a highly leveraged rental farmer. In 1929 the Great Depression came along. After the crash, my father's debt so far exceeded the liquidation value of his mortgaged assets that no creditor did him the favor of closing him out. He would have been better off had he gone bankrupt because he would not have had to pay off substantial parts of his debt after World War II. Going bankrupt would have been less traumatic for him during the Great Depression than in 1926; one had so much company in the 1930s. In my western Illinois community, almost every moderately leveraged farm family went broke. Those who managed to ride through the Depression were those who were not leveraged and those who had not deposited much money in the banks.

It is important to note that the losses associated with overproduction and overinvestment in agriculture are social as well as private (Johnson and Quance 1972; Boyne 1964). When a farmer invests sixty thousand dollars in a tractor he cannot recover in the length of time it takes to wear it out, both he and society lose. They both give up a sixty thousand dollar tractor, which only produces over its lifetime, say forty thousand dollars, thereby losing twenty thousand dollars.

The differences between acquisition costs and salvage values of identical agricultural resources are due to real costs. These real costs include the costs of transporting, assembling, disassembling, arranging credit, commissions, and the like. If these costs are not recovered, both social and private losses occur. It is also important to note that both leveraged and nonleveraged farmers incur such losses. Society still loses whether or not a farmer is leveraged and goes broke. However, both society and the leveraged farmer lose still more when liquidation is forced.

My unleveraged wife has a small farm in western Illinois. The

farmer who rents her land also farms land owned by another unleveraged landlady. Still further, he owns land of his own. If leveraged at
all, the renter-owner is only moderately leveraged. Thus, this farm unit
is in no danger of going broke or being driven into bankruptcy. However, it would not be difficult to find a leveraged farmer in the same
community going broke on a farm of about the same size using about
the same methods and quantities of inputs to produce about the same
acreages and yields of corn and soybeans.

Some of my colleagues in general economics tell me that it is efficient for society to drive leveraged farmers into bankruptcy by putting
them through the financial wringer. Many political and agricultural
leaders also believe this to be true. I, for one, fail to see that our farm in
western Illinois is any more efficient than an otherwise similar but
leveraged farm whose operator is going bankrupt. I see no more reason,
as far as efficiency is concerned, for driving the leveraged but not the
unleveraged farmer into bankruptcy. I do, however, see that both the
leveraged farmer and society can be damaged still further by driving him
or her into bankruptcy.

One of the reasons that a typical threatened farmer does not want to
go bankrupt is that the farmer's assets (including the farmer's own labor)
earn more on the farm than anyone is willing to pay for them when the
property is auctioned off at a liquidation sale and he or she seeks alternative employment. The same thing was true a few short years ago for
Chrysler Motor Company. Chrysler officials and stock owners did not
want the firm to be liquidated because the resources in the firm were
capable of generating income more than equivalent to their liquidation
value. Neither did its employees. It is my opinion that society benefited
by the government programs that bailed Chrysler out of its cash-flow
problem. The benefits society is receiving from the bailout originate in
the higher earning power of Chrysler's assets and its laborers in place
than they would have had had they been liquidated. I also believe that
the Farm Credit Administration bailout of leveraged farmers in the
Great Depression of the thirties yielded similar net private and social
benefits.

The mistakes that farmers make in investing originate in their imperfect knowledge of future changes. Some of the important things that
change are foreign demand, the influence of wars and peace on markets,
technology, the value of the dollar, weather, export embargoes, diseases,
pests, governmental programs, and the demands and preferences of consumers. Of these sources of uncertainty, technical change is not very
important. The creation of new technology does not automatically produce surpluses, as new technologies do not produce anything until

farmers buy the inputs that carry them and put those inputs to use. Farmers create surpluses by making overinvestments; technology alone does not create surpluses.

If one examines the history of U.S. agriculture from World War I to 1933, one finds a period of overproduction, depressed prices, lost markets, and financial stress for farmers similar to the present situation (except that there were no government programs then to be blamed for the overproduction). In longer sweeps of history, one finds a bewildering array of changes. As highly educated economists, military strategists, political scientists, natural scientists, and professors are incapable of predicting the changes, it should not be surprising that farmers have failed to do so. As a case in point, remember that less than ten years ago political leaders, agricultural natural scientists, humanists, activists, and church leaders feared continuing food shortages and widespread starvation. Hence, they advocated the investments and production expansions that now plague farmers. Farmers were urged by all of these groups to make the production expansion mistakes for which they and society are now paying the costs.

Consequences for Family Farms

This very brief description of how market-controlled, agricultural economies operate is recognizable by all who have historical knowledge of the United States and other market-controlled, agricultural economies. There are several important consequences of this analysis for the survival of family farms (Johnson 1969, 1970).

Small unleveraged farmers tend to survive periods of stress. This was true in the Great Depression and is being observed now. It is not subsistence farmers and small part-time farmers who are going broke and being driven into bankruptcy. Instead, their numbers are currently increasing. Large, leveraged farmers, on the other hand, are being quickly driven into bankruptcy or liquidation. Medium-sized and larger unleveraged farms tend to survive, particularly if they are family farms and not highly dependent on hired labor; however, such farmers are now discouraged from further expansion because labor earnings are not high enough to pay hired wage rates, capital earnings are not high enough to pay current interest rates, and land earnings are not high enough to pay interest on present land values. Incorporated farms (other than closely held family farm corporations) are by their nature leveraged and must pay acquisition prices for labor and service their debts to stockholders. Such corporate farms now face severe competition from family farmers who contribute batches of labor and capital to produce agricultural products to consumers at prices that yield substandard rates of return on both labor and capital. Historically, the overall consequence is that family farms have survived despite the apparent success of large-scale leveraged farms in such abnormal periods as those from the early 1970s to the early 1980s. Ironically, the survival of family farms seems to depend on the malfunctioning of market-controlled, agricultural economies.

The Historical Record and Prospects for the Next Fifty Years

Before considering policy changes needed to control overproduction, we should examine the record of U.S. agriculture (Johnson and Quance 1972; Johnson and Wittwer 1984).

In all but about eight of the years since 1918, much of U.S. agriculture has experienced *either* adverse prices and earnings *or* price supports, surplus storage programs, and production controls. From 1920 to 1929, adverse prices and incomes were experienced by farmers *in the absence of government programs.* Also, much of our present excessive investments in agriculture were made in the middle 1970s when market prices exceeded price-support levels. In those periods, the market-controlled U.S. agricultural economy outproduced effective demand and, in the absence of price supports and storage programs, experienced or set the stage for the kinds of adverse price pressures that put my father out of farming in 1926. Anyone who blames the present farm crisis solely on government programs should carefully study the 1920s and the mid-1970s. The current crisis is very similar to that of the twenties, the main difference being that the 1980 manifestations of overproduction now appear in the form of government surpluses and deficiency payments rather than in still more adverse price relationships and incomes than those farmers experience today.

For the benefit of those who decry the small number of farms remaining, it is important to note that if it had not been for the farm consolidation and off-farm migration that has reduced the number of farms in the last eighty years, present per capita farm and per family farm incomes would be so unbelievably low that we would need anti-off-farm migration laws to keep "true peasants" "down on the farm." We now have less than 2.4 million farms as compared to 6.4 million in 1920. One can hardly wish that we had managed our affairs so that the farm income generatable in the United States from the present effective demand of U.S. and world consumers were to be divided up among four million additional farms.

I now turn to a long-term historical view of the hundred-year period starting in 1880. Since 1880, we have increased U.S. agricultural production approximately sevenfold. (See Fig. 13.1.) This great long-term increase in production was needed and has been achieved despite programs since 1933 to control short-term overproduction. Somehow we did succeed in exercising short-term control over production and prices while at the same time making the public and private investments necessary to improve agricultural technologies, institutions, and human skills to attain the needed long-term growth. As a nation we have benefited greatly from this sevenfold increase. Thinking persons find ample reason to prefer our problems of overproduction and surpluses or adverse price pressures to those of underproduction, food deficits, and food rationing or high food prices so characteristic of centrally controlled economies whether under leftist or rightist governments.

Despite our current concerns about the high cost of price and production control, storage, and subsidy programs, this country benefited greatly from the agricultural surpluses it had at the beginning of World War II and the larger production plant it had in agriculture to meet war needs in 1942. When Secretary of Agriculture Earl Butz succeeded in disposing of burdensome surpluses, many soon regretted the resultant increase in food prices and the absence of grain reserves and heaped at least partially undeserved hot coals on his head. In many instances we have also used our food surpluses for humanitarian purposes both at home and abroad. Thus, some of our expenditures on price and production control programs and subsidy and storage schemes have been offset by these good uses of our surpluses and by the food security we and much of the rest of the world have enjoyed. However, having noted this,

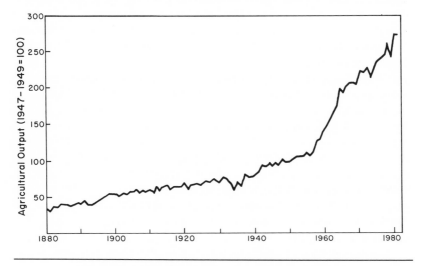

Fig. 13.1. Agricultural output, 1880–1982. Source: Johnson and Wittwer (1984).

we must also raise questions as to whether or not superior new policies and institutions can be created to control overproduction in the short run at lower costs than we have incurred while still getting the needed long-term growth in productive capacity. When we try to answer such questions, we also have to ask questions about what should be done concerning the family farm since the policy and institutional changes needed to solve the overproduction problems of market-controlled agriculture may also eliminate the family farm's reason for surviving.

In a 1984 study, Sylvan Wittwer and I concluded that the United States should plan on doubling its capacity to produce agricultural products in the next fifty years (Johnson and Wittwer 1984). Having capacity to double output, however, does not mean that we should double output. What we should do in the next fifty years about actual output will be revealed by conditions that cannot be reliably foreseen. Doubling production in the next fifty years would probably involve increasing average crop yields 70 percent, the acreage of land farmed by 16 percent, and the intensity of cropping mixes by 125 percent. In addition, substantially improved private and public infrastructure, public institutions, and human skills would be required. We will also have to invest substantial amounts in our land resources in order to make them more productive and to protect the additional more fragile soils we well may need to crop. Mere conservation of the land resources we now have will be inadequate.

In the next section I will look at some policy and institutional changes that should be studied as part of our effort to control short-term overproduction. In the section after that, I will look at what such policies and programs may do to the family farm.

Policies, Programs, and Institutions Needed to Control Overproduction

I believe we should subjectively investigate policies that would cause the federal government to play a smaller, less expensive role than it now plays in agricultural production, price control, and storage programs. I also believe different policies could advantageously shift much of the cost of controlling overproduction and supporting prices from taxpayers to consumers. The added costs to consumers can be substantially reduced with the greater efficiency that would come from reducing the wastes of overusing resources in agricultural production. Among the changes that should be studied are:

1. Improvement of research on when investments and resources committed to agricultural production are in danger of outproducing effective demand.

2. Improvement of dissemination.of such research in order to discourage overinvestments by farmers.

3. Enabling legislation to permit associations of producers to regulate investments and resource use in agriculture. (The National Farmers Organization has long advocated something like this while the Farm Bureaus' activities include price bargaining for crops, particularly specialized crops. Our present marketing orders are steps in the direction I have in mind. The objectives of the government's current "dairy herd buyout scheme" under the 1985 farm bill could also have been attained by an organization of dairy farmers as well as by the federal government had appropriate enabling legislation been passed.)

4. A shift of the U.S. government's role in agriculture away from direct control towards regulation of the producer organizations recommended above. (Such organizations would have a neutral tendency to become monopolistic and exploitive of consumers and/or farmers. We should not wait several years before establishing controls of such organizations as we did when we passed enabling legislation (1) to give labor unions greater control over the delivery of labor services and wage rates and (2) to permit the formation of industrial corporations large enough to exercise monopolistic and monopsonistic powers.)

5. Research on a residual continued role for direct government price and production controls and storage programs. (I am not hopeful that better research, better extension, enabling legislation for farm organizations to control production and prices would be adequate for some of the large, important, widely dispersed commodities such as corn, wheat, soybeans, cotton, milk, pork, beef, and the like.) (Johnson and Quance 1972, 180–83)

Creative research work is needed from social scientists to investigate such institutional, policy, and programmatic changes. Such research could result in a substantial improvement over the policies, programs, and institutions we have had and over the programs and institutions we will have under the 1985 farm bill.

What Would Policies and Programs That Successfully Control Agriculture's Overproduction and Overinvestments Do to the Family Farm?

The family farm has long survived partly because farm families have been trapped into contributing the services of batches of land, labor, and capital to the production of agricultural products at substandard rates of return. The high output of these family farms has kept returns to labor and capital low enough in agriculture to keep adverse pressure on expanding leveraged farms. They have also kept pressure on those farms so large that they have to compete with nonfarm business in the national capital and labor markets.

If we were to succeed in materially improving our policies for controlling the overcommitment of resources, we would no longer be trapping farm families into supplying these batches of land, labor, and capital at substandard rates of return (Johnson 1969, 1970). At the same time we would be maintaining resource earnings at levels that would permit expanding leveraged farmers and corporate farms to expand by acquiring land and competing with industry for capital and hired labor in national markets. This would ease the adverse pressure of family farms on larger-than-family farms. Freed of such price pressure, we should expect individual farm entrepreneurships and full-fledged corporations to be able to take over the family farms much as the supermarkets took over mom-and-pop grocery stores in the mid-1930s and early 1940s.

Whether or not our agricultural policies and institutions continue to attain the kinds of results they have in the past sixty years, we must continue to research the family farm. If the family farm survives as the result of a malfunctioning agricultural economy, there are many problems associated with it. Farms that trap families into furnishing batches of land, labor, and capital at substandard returns and that waste resources are clearly deficient in several important respects. If, on the other hand, we successfully devise policies and institutions to control overproduction so the family farm will disappear, we should research the question of whether or not we want it to disappear. If the answer is that we want to keep family farms, then we need to research policies and institutional changes that will permit us to retain them as the dominant production unit in American agriculture while controlling short-term overproduction and attaining needed long-term increases in the productive capacity of U.S. agriculture.

Roles for Social Scientists

This paper has already indicated general roles for social scientists to play in (1) designing policy, program, and institutional changes for controlling the universal tendency of market-controlled agricultural economies to outproduce effective demand and (2) considering the fate, preservation, and improvement of family farms. Before being more specific, it will be helpful to outline the steps involved in making public decisions on such problems.

The essential steps in solving problems—in reaching decisions and putting such decisions into effect—apply to both public and private problem solving. (See Fig. 13.2.) There are six interdependent steps including definition of the problem, the accumulation of knowledge, the analysis of that knowledge, decision making, putting the decision into effect, and bearing responsibility for the consequences.

Fig. 13.2 also has two information banks, one containing value-free positivistic information and the other information about values. These two information banks are conceived to supply information to the process and to receive information generated by the process for storage and subsequent use. An overarching pragmatic loop is included for the benefit of pragmatists who hold that truth depends on consequences in a way that makes value-free positivistic information and value knowledge dependent upon each other (Runes 1961, 248; Parsons 1958).

The output of the decision step in Fig. 13.2 is a decision or prescription for solving the problem at hand. A simple equation indicates the relationship between decisions or prescriptive knowledge on one hand and knowledge of values and value-free knowledge on the other. The equation is:

Prescription = f(knowledge of values and value-free knowledge)

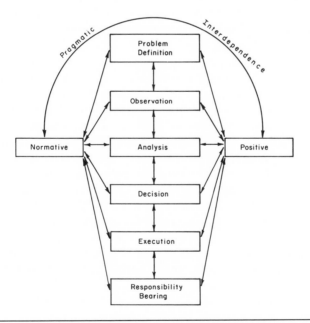

Fig. 13.2 Steps in problem solving. Source: Johnson (1976).

The "f" in this equation is a decision rule. This rule converts knowledge of values and value-free knowledge into a prescription to solve a problem. If value-free knowledge and knowledge of values are perfect, the decision rule is simply one of subtracting bad from good and maximizing the difference. However, when knowledge is imperfect (as it always is in practice because perfect knowledge is infinitely expensive for mere humans), available decision rules become more complicated and often have to include distributions of power. Such distributions of power are often legitimated by covenants. The power distributions sustained by such covenants enable us to resolve conflicts among different persons and groups involved in the decision process. An example of such a covenant includes acceptance of the democratic principle of one vote per person.

Prescriptive knowledge is knowledge about what "ought to be" or "ought not to be" done. It is right to do that which we ought to do and wrong to do that which we ought not to do. Knowledge about rightness and wrongness differs fundamentally from knowledge of values — knowledge of the goodness and badness of conditions, situations, and things. As the philosopher C. I. Lewis (1955) points out, rightness and wrongness are not the same as goodness and badness. It is not always right to do that which is good if something still better can be done. Conversely, it may be right to do that which is bad if it is the least bad that can be done in the circumstances.

Whatever decision rule is used, that rule defines which act or set of acts is the best, in some sense, to do. Maximization in some form or another is an essential part of deciding what it is right to do.

The problem-solving process outlined in Fig. 13.2 usually includes a conscious or unconscious attempt to establish the preconditions for finding whatever is best as defined by the decision rule. The preconditions include a common denominator among the values (concepts of goods and bads) involved. If decisions are to be considered that hurt some in order to benefit others, the common denominator must be interpersonally valid. Also, use of a decision rule to find "the best," however defined, requires reasonable assurance that a "best" exists. The final prerequisite is an agreed-on decision rule including covenants as to (1) what distributions of power can be used in resolving conflicts, and (2) whether one can resort to chance (in the absence of agreement) to reach a decision, as when draft numbers are determined by lottery.

Ethics has to do, among other things, with the evaluation of our prescriptive decisions. Decisions about what to prescribe to solve a problem can be faulty in a number of ways. They can be based upon inaccurate knowledge of values. The accuracy of both value-free and value knowledge can be judged on the basis of experience and logic. The test of logic is the test of coherence, while the test of experience is the test of correspondence. Most philosophers will agree that the test of correspondence can be applied to value-free knowledge. Though some will disagree, it seems distinctly possible to apply the test of correspondence or experience to the goodness or badness of many conditions, situations, and things (Johnson 1986b) important to the design of policy, program, and institutional changes to solve overproduction problems, and ensure the survival of the family farm. Some farmers and farm families are now experiencing acutely the badness of bankruptcy and financial insolvency. Many of us enjoy the goodness of being able to satisfy our nutritional needs with a very small proportion, by world standards, of our income. Others, both in this and other societies, experience the badness of malnutrition and starvation. Such experiences and logic can be used to generate and test the truth of value knowledge as descriptive of "real" values.

As there seems to be no counterpart in reality of a prescription to observe, it is difficult to use the test of correspondence to test the descriptive truth of prescriptive knowledge. About all we seem to be able to do is to act on the prescription and then test the value-free and value-laden consequences of our acts to see if they agree with expectations. The prescription itself is definitional rather than existential and hence is not testable as to correspondence with reality.

In reaching a decision to solve a practical problem, choices are made as to which decision rule to use. Such choices involve the establishment and, in some instances, the breaking of covenants concerning the use of power distributions in reaching prescriptive conclusions. In making such prescriptive choices, additional value-free and value knowledge is used and processed through still another decision rule involving further distributions of power. There seems to be an infinite regression of decisions about decision rules to use in making decisions about decision rules and so on *ad infinitum*. Objectivity in reaching prescriptive knowledge seems to consist of applying the tests of correspondence and coherence to the value-free and value knowledge used in solving the immediate practical problem and in making choices about which decision rules to use in solving the problem created by vested choices about decision rules.

All of this tells us quite a bit about the roles social scientists should play in setting agricultural policies and in establishing institutions to implement them. The roles of social scientists include those of generating three bodies of knowledge. One is to establish the value-free positive knowledge indicated in the right bank of Fig. 13.2. Another role is to generate the normative or value knowledge in the left bank. A third role is to generate prescriptions as to what kinds of policies, programs, and institutional changes will solve the overproduction problem, attain necessary long-term growth, and preserve the family farm, if we judge it worth saving. Each of these three roles is addressed in the following subsections. Because one of the kinds of knowledge is about values and another is prescriptive, ethics and the roles of social scientists are basically inseparable (a point that will become more apparent as the discussion unfolds). The roles of social scientists in generating prescriptive knowledge will be discussed first followed by discussion of their roles in generating value and value-free knowledge.

PRESCRIPTIVE KNOWLEDGE. Social scientists often participate as advisors, consultants, and problem-solving researchers in the generation of prescriptions (solutions) to practical problems. Not all academicians agree that social scientists, as researchers, should engage in problem-solving research. Despite this and in accordance with the tradition of land grant universities and in compliance with the needs and demands of society, social scientists do perform multidisciplinary research, consulting, and advising on prescriptive recommendations for solving problems. The ethical nature of such involvement requires that social scientists be honestly objective in doing such work.

Ethics requires objectivity. Objectivity is particularly vital to the improvement of agricultural policies because so many unobjective assertions circulate on these subjects. Free-market advocates, antimarket critics, antiagricultural establishment activists, self-serving disciplinarians from academia, governmental officials, bureaucrats, conservationists, environmentalists, religious leaders playing God, and many others play unobjective activist roles. In these roles, both value and supposedly value-free positivistic assertions are made that are not based objectively on experience and logic and, indeed, are maintained despite experience and logic to the contrary. Social scientists bear the special responsibility for testing such questionable assertions and for generating relevant knowledge that has been objectively evaluated on the basis of the tests of experience and logic.

Because prescriptions depend on value-free and value-laden knowledge, this section deals mainly with choices of decision rules and associated power covenants. (We saw above that these are inherent parts of such rules.)

Probably the most important conflict about decision rules to be used in changing agricultural policies, programs, and institutions has to do with the reliance on market versus governmental ways of making decisions. When resource allocation, production, storage, and consumption decisions are controlled mainly by market forces, the basic distribution of power that counts is of the power that arises as a consequence of the distribution of ownership of income producing rights and privileges. When, on the other hand, major reliance is placed on government instead of the market to control agricultural resource use, the relevant distributions of power have to do with political and administrative power. In a real sense the basic conflict is one between those who own or "hold" market power and those who hold or "own" political power. Differences in points of view here also depend upon the amounts and kinds of power individuals and groups expect to hold in the future. While market versus government control appear to be polar opposites, they in fact depend on each other. Private ownership of rights and privileges is enforced and maintained with police power based on political and administrative power. In turn, political and administrative power is conditioned by the distribution and ownership of income-producing rights and privileges.

For many decades, agriculture and farmers possessed sufficient political and administrative power in government to shape policies and institutions to their advantage (Paarlberg 1985). Our earlier examination of history indicates that the policy and institutional changes made by government when farmers were dominant in political and administrative processes were reasonably good, viewed from a long-term social perspective. The distribution of political and administrative power vis-à-vis agricultural policies and institutions has shifted substantially from farmers to urbanites, activists, intellectuals, and members of administrative bureaucracies, while the alliances and allegiances that formerly maintained the political power of the farm bloc have now virtually disappeared.

In a related manner, James Bonnen (1983) argued that as an agricultural society becomes more industrialized and specialized, its political power becomes more fragmented. He contends that U.S. farmers no longer think politically about the totality of agriculture but think instead about policies, programs, and institutions for the special commodities they produce and the special inputs they use. Oftentimes such fragmented specialized interests can be favored with political actions agreed upon in a political logrolling process in which various specialized interests, both agrarian and nonagrarian, trade favors.

Though the market power of farmers and agribusinesses has not disappeared, there is increasing reason to doubt that farmers and agribusinesses will continue to have enough political power to maintain expensive governmental programs to control overproduction, get needed long-term growth, and preserve the family farm. For this reason, farmers and agribusiness people will probably find it advantageous to rely proportionally more on market than on political power and governmental expenditures. However, it should not be forgotton that the market and political power depend on each other. Further, it should not be forgotten that police and military power also interrelate with market and political power. The latter two kinds of power maintain and enforce the first two and are, in turn, maintained and enforced by them.

There has always been a substantial farm leadership with faith in market power, in letting free markets and prices make allocative decisions. As new holders of power in government have become disenchanted with farmers, they have formed something of an alliance with the "free-market and free-price advocates" that favor the use of the market mechanism as a decision rule or procedure.

Some consumers simply do not own enough income-producing rights and privileges to generate enough income to convert their nutritional needs into effective market demands. Shift toward use of the market mechanism as a decision rule means that governmental programs that increase the market power of the poor and disadvantaged tend to be neglected. In the subsection to follow on knowledge of values, the values of equality and equity will be discussed. At this point, we note that equality in the distribution of decision-making power is typically attained with the loss of decisiveness and capacity to resolve conflicts (Johnson 1981). As agriculture has lost the concentration of political power it formerly possessed, political processes for changing agricultural policies, programs, and institutions have slowed down and become chaotic and indecisive. As a result, there now seem to be reduced opportunities to legislate for the general good of society and even for the general good of agriculture; instead, what decisiveness there is seems to concentrate on the welfare of special interests within agriculture. (This claim is found in the argument by Bonnen noted earlier.)

The certainty with which power is held is related to honesty and objectivity. The powerful need not rely on dishonesty to protect their interests. Those who are uncertain of their power or who are trying to establish power find it advantageous to mislead in order to "legitimatize" their power. Currently the weakened political power of farmers seems to have raised the hopes of other potential power holders vis-à-vis agriculture including various activist groups, biological and physical scientists outside the USDA/land grant system, consumer groups, conservationists, religious groups, and the like. Often these groups have been opposed by farmers and their "USDA/land grant agricultural establishment" whose power is now less certain than formerly. Thus, farmers, agribusinesses, and the USDA/land grant establishment as well as their critics are motivated to be unobjective as they jockey to establish or maintain their political and regulatory (police) power positions.

Activists, religious leaders, some humanists, and others often regard power distribution as somewhat unethical. To these persons there are two worlds — a world of ethics and a materialistic world of politics, business, and science. This division of the world somewhat parallels the distinction made by some religious leaders between God's world and the material world, between the sacred and the profane. Both divisions seem constraining. The first deprives ethics of the value-free knowledge so essential in deciding what is right and wrong, while the second limits God to the sacred world. Bonhoeffer (1965) objects strenuously to the constraint on God and does it in a way that also rejects the constraint on ethics.

Power distributions are tested, changed, and established with power. True certain power is power and is not changeable. Uncertain power is changeable. Tests of power are often costly and destructive whether the power being tested is political, market, police, social, military, or religious. As a form of power, knowledge has greater strength in the long run than in the short run. As academic social scientists, we must recognize the short-term weakness of the power of knowledge but still carry out our responsibility to produce, organize, store, and transmit it from one generation to another. Prescriptive knowledge is also useful in predicting the behavior of resource owners, producers, agribusiness people, and consumers. Such predictions are required in choosing among alternative policies, programs, and institutions. For the most part, economists use theories of market behavior in making such predictions. Behavior is predicted by assuming that resource owners, producers, and consumers will do (within the freedom permitted by institutional constraints) what some maximizing theory (decision rule) prescribes as optimal for them to do. Such predictions are quite accurate (as is soon found out by those who think you can use price supports but no production controls and still not create unused surpluses).

The broad roles of social scientists in generating prescriptive knowledge are (1) to participate as consultants, advisors, members, and leaders of research teams in solving farm problems. Such problems include the short-term control of overproduction, the attainment of needed long-term growth, the expansion of the market power of the nutritionally needy, and the possible preservation of the family farm; and (2) to study changes in and make recommendations about changes in decision rules and power covenants important in making decisions on farm issues.

VALUE KNOWLEDGE. This section is based on the premises that the "real" goodnesses and badnesses of at least some conditions, situations, and things can be experienced and that we can reason about them to attain objective knowledge of real goodnesses and badnesses (Johnson 1981,

1986a, 1986b). Descriptive knowledge of values so generated is, like our value-free knowledge, unprovable and culture dependent, but nonetheless can be sufficiently reliable to be agreed upon for purposes at hand. Such knowledge of values, like the value-free knowledge of biological and physical sciences, orginates in part with insight and inspiration and is not generally accepted in a community until there is empathetic agreement on the experiential and logical bases for it. This view of value knowledge and of ethics is consistent with Bonhoeffer's rejection of the division of the world into the profane and sacred mentioned earlier.

Both monetary and nonmonetary values are of concern to ethicists and to policy analysts designing changes in agricultural policies and institutions. Both are normative. Both economics and ethics are concerned with nonmonetary and monetary values. Of the three major subdivisions of economics (production economics, consumption economics, and welfare economics), two are concerned almost exclusively with nonmonetary values. Ethics and economics have too much in common to be regarded as dichotomous.

Economists find it relatively easy to work quantitatively and objectively with monetary values. They measure prices of products and inputs, quantities produced and used, and compute gross income, gross expenses, net income, and many indexes of output and input in which quantities are added together on the basis of their prices or monetary values. Knowledge of monetary values is extremely important in making decisions to change agricultural policies and institutions. Capital losses due to declining monetary values of fixed assets in agriculture are of similar crucial importance.

Generally speaking, our knowledge of agricultural prices, production, yields, resource use, expenditures, and incomes are now of poorer quality than they were a few years back (Bonnen 1977). Professional and clerical personnel for producing such data now command relatively higher wages than they received a few decades ago. While computers make it possible to process data at less cost than in previous years, the cost of acquiring primary data is still determined largely by the increasing salaries paid to professional and semiprofessional workers.

One role of social scientists in improving our agricultural policies and institutions is to improve our data and information systems for agriculture. Of particular significance are the data needed on the capital losses incurred by farmers when there is widespread overproduction (Johnson and Quance 1972; Boyne 1964). Also needed are better data on the disposal and liquidation values of farm assets, their values in use, and their acquisition prices. In analyzing such data, greater objectivity is needed, particularly by general economists, who commonly use sophomoric theories that ignore real-world differentials between acquisition costs and liquidation prices. Such unrealistic theories disregard capital losses such as those now occurring in agriculture. They also lead to the unwarranted conclusions that leveraged farms should be run through the wringer of the market because the resultant adjustments will increase efficiency and will benefit society.

Nonmonetary values are more difficult for social scientists to work with than monetary ones and are even more crucial in policy and institutional choices. The nonmonetary goodnesses of capital gains often conferred without being earned and the badnesses of capital losses imposed (often unjustly) are important in setting production and price controls and storage policies. They are important for farmers, the credit systems that finance farmers, the taxpayers who pay subsidies, and the consumers who pay the higher prices resulting from production limitations and price supports. Associated questions of justice or equity arise (Johnson 1983). So do questions of equality and inequality. Policies and programs that relieve well-to-do farmers of unjustly imposed capital losses are often regarded as wrong by those who believe equality is good and do not distinguish between equality and equity.

We also need to know more about the goodness or the justness of having farm incomes equal to those received by other members of society exerting similar efforts, exercising similar managerial skills and discipline, and contributing the services of equal investments. This goodness is often neglected by critics of farming and by family farm advocates. Some critics are concerned about programs that help farmers whose incomes are higher than others in society. Also, some family farm advocates would like to insure the preservation of family farms (whose resources are not adequate to earn acceptable standards of living) by paying subsidies such as those extended to such farms in Europe and Japan.

There is much work to be done by social scientists on nonmonetary values. Agricultural policymakers have long been concerned with questions of equality and equity. The concept of parity prices in agricultural policy history is related to the values of both equality and equity. The dictionary defines equity and equality very differently. An equitable distribution of power or payments or incomes is a *justified but not necessarily equal distribution*. When we seek justice in a court action, we do not seek equality. We seek to attain or keep possession of that which is rightfully ours. There is much confusion in agricultural policy discussions of equality with equity that needs to be straightened out by objective social scientists.

Food security has nonmonetary value but there are costs in attaining it, some of which are monetary and some nonmonetary. These values need more investigation.

Efficiency is a difficult value to measure and analyze. There are more or less efficient ways of attaining any value, monetary or nonmonetary. Both Frank Knight (1951) and Kenneth Boulding (1981, 152–55) point out that questions of efficiency arise in trying to attain any good or avoid any bad. The distinction between economic and noneconomic values and between economic and other kinds of efficiency, therefore, evaporates and becomes meaningless. It is relatively easy to answer questions about efficiency if the distribution of the ownership of rights and privileges is fixed. Efficiency loses its meaning when such distributions are changed unless we are able to measure nonmonetary values conferred on some in terms comparable to our measures of the values of what is taken away from others. This is a fundamental measurement problem faced by social scientists. One of the roles of social scientists is to improve their ability to measure gains and losses in welfare that result from changing the distribution of various kinds of power among people.

More and better information is needed about the goodness and badness of the consequences of market adjustments and the demise or preservation of family farms. Family farms are complex social institutions with many characteristics, some of which are bad and some good. Rosy mythical stereotypes of the family farms of the "good old days" do not provide adequate value knowledge for making decisions to preserve the family farm. There seems to be little doubt about the family farm's ability to outproduce effective demand; this is a mixed blessing to both the family farmer and society. What about the badness of the social, educational, and medical isolation of family farm communities that caused large numbers of farm boys to run away from drudgery and autocratic parents in the "good old days"? In the future, can communities of family farms each large enough to earn levels of living comparable to those earned elsewhere sustain adequate social services? Experience and logic can be used by social scientists to arrive at better knowledge than we now have of the values involved in answering these and similar questions.

It is also important in our thinking about nonmonetary values to distinguish between intrinsic and extrinsic values or at least between values – some of which are more intrinsic than others. Life, for instance, has a value that is more intrinsic than the value of food, which is an instrument for maintaining life. Among the extrinsic values, we should probably distinguish between values that are instrumental and the exchange values that arise out of exchanges in markets. Instrumental values arise because something is a means of attaining a more ultimate or more intrinsic value. Exchange values arise when people in a society trade the attainment of one value off against the attainment of another value. In the latter case, the two values traded off against each other may be more or less intrinsic. In any case, market trade-offs establish exchange ratios between additional or reduced quantities of the two items being exchanged. In one sense, all exchange values are instrumental values, because we obtain one value by giving up some of another.

VALUE-FREE POSITIVISTIC KNOWLEDGE. Value-free positivistic knowledge is typically regarded as much more objective and less culture dependent than it really is. Such philosophers as Popper (1959) question the idea of there being any provable empirical knowledge. But there is much value-free knowledge that is adequate for purposes of revising our production and price control, storage, and family farm policies, institutions, and programs. Unfortunately, much of such information that should be provided by social scientists is also lacking.

This essay has presented considerable value-free knowledge about how a market-controlled economy operates. More such knowledge is needed from social scientists. We need to understand better the decisions of young people to commit their lives to farming and of farmers to vary the rate at which they extract productive services from their durable assets (including themselves and their family members).

Smaller family farms (particularly the small, old-fashioned, diversified ones) have often been alleged to be more regenerative than and ecologically superior to large modern farms. They are also alleged to have been more soil conserving. This does not square entirely with my experience (starting in the 1920s when the consequences of another earlier thirty years were readily observable in the landscapes). I believe one role of social scientists is to supply us with more and better knowledge about the ecological impacts of family versus other kinds of farms. Much more and better value-free positivistic knowledge is needed about how and whether different kinds of farming pollute the environment and contaminate the food chain. We need more and better value-free positivistic knowledge about erosion and the contamination of ground and surface waters. Too often available information on such subjects is inaccurate and the result of biased self-serving investigations by institutions, academic disciplines, and activist groups generating what passes as "data and knowledge." Such practices and "information" undermine the integrity of the institutions that generate our policy, program, and institutional changes and technological advances, and educate the personnel who operate our farms, agribusinesses, and agricultural institutions.

References

Bonhoeffer, Dietrich. *Ethics.* New York: Macmillan, 1965.

Bonnen, James T. "Agriculture's System of Developmental Institutions: Reflections on the U.S. Experience." In *L'agro-alimentaire Quebecois et son developpement dans l'environnement economique des annees 1980,* 43–64. Quebec, Canada: Agricultural Economics Department, University of Laval, 1983.

_____. "Assessment of the Current Agricultural Data Base: An Information System Approach." In *A Survey of Agricultural Economics Literature,* Vol. 2, edited by Lee R. Martin, 386–407. Minneapolis: University of Minnesota Press, 1977.

Boulding, Kenneth. *Evolutionary Economics.* Beverly Hills, Calif.: Sage, 1981.

Boyne, David H. *Changes in the Real Wealth Position of Farm Operators, 1940–1960.* Technical Bulletin 294. East Lansing: Michigan State University Agricultural Experiment Station, 1964.

Chennareddy, Venkareddy, and Glenn L. Johnson. "Projections of Age Distribution of Farm Operators in the U.S. Based Upon Estimates of the Present Value of Incomes." *American Journal of Agricultural Economics* 50, no. 3 (1968): 606–20.

Johnson, Glenn L. "Economics and Ethics." Twenty-Fourth Annual Centennial Review Lecture, 9 April 1985. *Centennial Review* 30, no. 1 (winter 1986a): 69–108.

_____. "Effects on Farm Management Decisions of the Institutional Environment." *Policies, Planning and Management for Agricultural Development.* Proceedings of the Fourteenth International Conference of Agricultural Economists, Minsk, USSR. Oxford: Express Litho Service, 1970.

_____. "Ethical Issues and Energy Policies." *Increasing Understanding of Public Problems and Policies 1980.* Proceedings of 30th National Public Policy Education Conference. Oak Brook, Ill.: Farm Foundation, 1981.

_____. "The Modern Family Farm and Its Problems." In *Economic Problems of Agriculture in Industrial Societies,* edited by U. Papi and C. Nunn. New York: Macmillan, 1969.

_____. "Philosophic Foundations: Problems, Knowledge and solutions." *European Review of Agricultural Economics* 3, no. 2/3(1976): pt. II, 226.

_____. *Research Methodology for Economists: Philosophy and Practice.* New York: Macmillan, 1986b.

_____. "Synoptic View." In *Growth and Equity in Agricultural Development,* edited by A. Maunder and K. Ohkawa, 592–608. Oxford: Gower Publishing Co., 1983.

Johnson, Glenn L., and C. L. Quance, eds. *The Overproduction Trap in U.S. Agriculture.* Baltimore, Md.: Johns Hopkins University Press, 1972.

Johnson, Glenn L., and Sylvan H. Wittwer. "Agricultural Technology Until 2030: Prospects, Priorities, & Policies." Special Report 12. East Lansing: Michigan State University, Agricultural Experiment Station, 1984.

Knight, Frank. *The Economic Organization.* New York: Augustus M. Kelley, Inc., 1951.

Lewis, C. I. *The Ground and Nature of the Right.* New York: Columbia University Press, 1955.

Paarlberg, Robert. *Food Trade and Foreign Policy: India, the Soviet Union and the United States.* Ithaca, N.Y.: Cornell University Press, 1985.

Parsons, Kenneth. "The Value Problem in Agricultural Policy." In *Agricultural Adjustment Problems in a Growing Economy,* edited by E. Heady et al. Ames: Iowa State College Press, 1958.

Popper, Karl. *The Logic of Scientific Discovery.* New York: Harper and Row, 1959.

Runes, D. D. *Dictionary of Philosophy.* Patterson, N.J.: Littlefield Adams & Co., 1961.

Social Scientists Are Not Neutral
Onlookers to Agricultural Policy

IN AGRICULTURAL SCIENCE, as elsewhere, academic experts confidently claim objectivity and impartiality, and unhesitatingly assume that they are fully capable of distinguishing scientific assertions from value judgments. I do not believe that this self-confidence is warranted. While I share Johnson's hope that the university might be a place where the problems facing the family farmer can be attacked in a way that serves society as a whole, I am very pessimistic regarding the chances of this hope becoming a reality. I am afraid that it is much more likely that university experts — intentionally or not — will serve the very corporate interests that in my view are the root cause of the difficulties. In social theory, "ideology" is a technical term used to describe assertions that claim to be objective, impartial, and scientific, but that actually serve the interests of ruling groups. In this paper I shall explore some of the ways agricultural science is ideological.

The paper has two parts. In part one I discuss some examples of ideological thinking taken from agricultural science in general. In part two I discuss the family farm specifically. The examples discussed in part one are taken from the Agricultural Science/Liberal Arts Faculty Workshop I attended in the summer of 1985 funded by the Kellogg Foundation.

For two weeks a dozen faculty members from liberal arts departments at Iowa State University listened to presentations from a variety of academic experts connected with agriculture. Again and again these men (in the two weeks only one woman spoke to us) shifted back and forth from relatively straightforward empirical assertions to more or less crass apologetics for the corporate system of agriculture. These shifts took place within the course of a page, a paragraph, a single sentence, or even a single clause. It became evident to many of us attending the workshop

TONY SMITH is associate professor of philosophy at Iowa State University.

that the aura of scientific objectivity often masked a style of thinking where ideological considerations so interpenetrated scientific considerations that one could not say with confidence where one began and the other ended.

This sort of speech is not innate. It must be learned. But neither is it, as far as I could tell, something consciously taken on. The vast majority of the speakers, at least, sincerely felt that their talks met strict canons of objectivity and would be offended at any accusation of partiality. And yet partiality and objectivity were somehow fused. In the course of academic training (if not before), scientists are somehow socialized in this strange style of speech, and they come to accept it as second nature. This discourse has its own rules, its own patterns, its own logic. I would now like to list some of the discourse mechanisms that were used at the workshop, mechanisms that ensured that the presentations reinforced the powers that be in agriculture, although this was not usually the conscious intention of the speakers. Many of these mechanisms are rather obvious, and no claim for the completeness of this list is being made. Nonetheless, it is hoped that this catalog, provisional as it is, can illuminate some of the forms of ideological speech within agriculture. The ideological mechanisms I shall discuss are: the exclusion of relevant questions, the omission of known facts, the retreat to abstract models, and the failure to consider relevant power relations. In the second part, I argue that the same mechanisms regularly operate when agricultural experts from the university discuss the family farm.

Ideological Mechanisms in Agricultural Science

EXCLUSION OF RELEVANT QUESTIONS. An agricultural scientist involved with a consortium that coordinates the transfer of agricultural technology from land grant colleges to the Third World gave a lecture on the economics of world hunger. The central thesis of the lecture was that however important a factor the unequal distribution of income and power may be, the most significant issue in the hunger problem is the lack of advanced technology in the Third World. As the crucial evidence for this thesis, the speaker referred to the fact that even if one totaled all the food presently grown in Africa with Africa's entire food imports and then divided this sum equally, there still would not be enough food to provide the African population with sufficient nutrition. This may or may not be the case; there is certainly reason to be suspicious of this claim (Dinham and Hines 1984, 187).

But even if we accept it, before we conclude that the problem is primarily a technical one, we should ask what sorts of crops are being produced in Africa. Specifically, suppose that an ever-greater proportion of African cropland is devoted to crops with low nutritional content. Then it could very well be that the problem lies first and foremost with the social relations of economic power that leads to this sort of production, and only secondarily with technical matters.

In fact, over the last twenty years in Africa, coffee production has increased by 400 percent, tea production increased sixfold, sugar production went up by 300 percent, cocoa and cotton production doubled, and so on (Dinham and Hines 1984, 187). This occurred because the native African population, lacking disposable income, offers a poor market for agricultural products. So production controlled by giant transnationals shifts from the production of foodstuffs for local nutritional needs to luxury items of low nutritional level for the export market.

The speaker was an expert on the topic of world hunger in general and of hunger in Africa specifically. His talk made use of a vast array of statistical information. And yet, not only did he not incorporate this shift to export production in his talk, he also did not seem to be familiar with how massive this shift has been.

Of course no one can know everything. But that a specialist in this area would not know basic relevant factors reveals more than a simple gap in one isolated individual's knowledge. It reveals, I believe, how ideological factors have permeated agricultural science. Somehow, something like a filtering device is operating here that tends to filter out from consideration obviously relevant issues when these issues threaten the status quo. Such a filtering device operates as the scientist is doing science, so that the very act of doing science is at one and the same time ideological.

OMISSION OF KNOWN FACTS. A specialist in agricultural economics spoke at the workshop on the factors that contribute to the price of food. He especially stressed what a large component of final prices was made up by labor costs. It soon became clear that what he termed "labor costs" included the salaries of agribusiness executives as well as that of workers in the food industry, the $339 thousand paid to the head of Hormel alongside the Hormel workers who had a wage cut of 23 percent despite the firm's profitability. When pressed, the speaker did not know how to divide up these "labor costs" in a more accurate fashion. Here we have another case of a filtering device at work, filtering out a clearly relevant question. But another omission, even more striking, points to a different sort of ideological mechanism.

In the course of an hour presentation on food prices, not once did the speaker mention oligopolistic concentration as a factor in the price of food. When pressed, he admitted that economists do accept a correlation between concentration in a sector and higher prices. And, when pressed, he agreed that in every food group (bakery, dairy, tinned fruits and vegetables, processed meats, sugar, etc.) four firms or fewer control over half the market. And, when pressed, he admitted that the USDA did a study that asserts that as a result of these quasi-monopoly conditions, consumers suffer overcharges amounting to 10 percent of their total food bill (in the breakfast industry alone these overcharges are estimated at more than two hundred million dollars a year). And yet he did not see fit to mention any of this in what was otherwise a quite exhaustive discussion of the factors affecting food prices. Subsequent informal discussion with him convinced me that he had sincerely tried to present the relevant facts regarding food prices as he saw them. How then could he have omitted factors that he himself knew and that he himself later admitted to be relevant?

Once again we must go back to the idea that somehow in the course of being socialized into an academic discipline, one gets more than scientific information and procedures. One is also simultaneously socialized into a world view determined as much by what it excludes as by what it includes. Besides filtering devices that screen out relevant questions that could challenge the established state of affairs, apprentice scientists somehow also internalize ways to filter out known relevant facts that could have the same effect. As a result, each individual assertion made by a given scientist could be fully warranted by the strictest standards of scientific reasoning, and yet his or her reasoning could nonetheless be thoroughly ideological in character due to what was systematically omitted.

RETREAT TO ABSTRACT MODELS. When the shift to cash crops was introduced by participants as a possible factor creating hunger in the Third World, a number of agricultural economists introduced the concept of "comparative advantage." According to this notion initially formulated by David Ricardo, if each country specializes in what it can produce most efficiently and then trades with other nations that have comparative advantages in other areas, all nations will be better off than if each country attempted to produce all it required itself. According to this reasoning, it makes good economic sense for Third World countries to specialize in the export crops they can produce efficiently, sell these crops to developed countries, and then use the money they receive from these sales to import grain and industrial products more efficiently produced in the developed countries.

In the abstract this reasoning has some cogency. But it ignores the concrete context of world trade. Third World producers are in a tremendously competitive sector of the world economy. Yet when they purchase the inputs they require, they must turn to a handful of transnationals based in the developed countries. Likewise when they sell their output, they must turn to a handful of firms in developed countries. A competitive sector sandwiched between two oligopolistic sectors will inevitably experience disadvantageous terms of trade. (Keep this phrase in mind, as it will become important again when we turn to the family farm.) In fact, the prices of the crops sold predominately by industrial countries (for example, grains and soybeans) have risen much faster than the prices of the commodities exported by the underdeveloped countries. So have the prices of manufactured goods imported from the developed countries. In 1960 three tons of bananas could buy a tractor. In 1970 the same tractor cost the equivalent of eleven tons of bananas, and in 1980 the figure stood at twenty tons.

Third World countries receive back very little of the final price of their agricultural products. For every one dollar spent on an agricultural product from the Third World, only about fifteen cents goes back to the producing country (George 1984, 10). In the concrete world, a full year of Africa's exports can pay for only twenty-seven days worth of imports.

In the light of all this, to refer to the *abstract* model of comparative advantage when discussing the *concrete* impact in the Third World of export crops on local nutritional needs is not simply to be doing science, whatever the scientific rigor used in constructing that abstract model. It is a way of using science to provide an ideological legitimation for existing patterns of trade.

FAILURE TO CONSIDER RELEVANT POWER RELATIONS. Many within the academic community have come, somewhat belatedly, to the realization that the current use of agricultural chemicals cannot continue. (In the United States alone, for example, over 1.1 billion pounds of pesticides are used each year.) The financial burden this places on the farmer and the environmental burden placed on nature can no longer be ignored. Few agricultural extension scientists, however, are prepared to call for a major shift to organic farming. More are slowly beginning to advocate integrated pest management (IPM), where the selective use of chemicals is combined with biological defenses against predators and the development of environmentally safe forms of biotechnology.

This is a progressive development. Nonetheless, claims were made for IPM that cannot be justified. The claim was made by agricultural scientists at the workshop that with the move to IPM along with recent developments in biotechnology, the trend has been already set in motion away from heavy reliance on agricultural chemicals.

This grossly understates the power of agricultural chemical companies. In 1985 these companies enjoyed a more profitable year than ever before. This year, in 1986, they are spending twice as much on market development as they did six years ago. And today the agricultural chemical companies, who control most of the biotech research being done, are not giving priority to research developing plants that are resistent to pests. That would be ecologically beneficial but not especially profitable for companies that sell products to get rid of pests chemically. So instead, as Congress's Office of Technology Assessment has stated, "It should be kept in mind . . . that much of the agricultural research effort is being made by the agricultural chemical industry, and this industry may see the early opportunity of developing pesticide-resistant plants rather than undertaking the longer-term effort of developing pest-resistent plants" (*Nation* 1984, 13). Ciba-Geigy, CalGene, Monsanto, Du Pont, and other chemical companies are presently developing strains of plants able to withstand potent herbicides. The damage to the ecosystem that results is liable to be even more massive than the profits these developments will bring to these chemical companies.

The predictions made by agricultural specialists that we are well on our way to solving the problem of environmental damage from chemicals ignore the interests of the chemical companies in a future world of bioengineered crops that grow well only in combination with chemicals, and they ignore the power of these companies to bring about such a world. Power relations within the agricultural system are not taken into account, power relations that make it likely that the problem of agricultural chemicals will *worsen* despite the best efforts of a few extension scientists. Predictions that fail to take into account power relations are, in short, more ideological than scientific in character, whatever the private intentions of those making the predictions might be. They hamper our grasp of social reality rather than enhancing it.

On the Family Farm

In the remainder of this chapter I shall argue that the standard academic discussions of the family farm betray the same sorts of ideological mechanisms as those just discussed. Recall the discussion of the fate of Third World agricultural producers. Did that sound familiar? It should have. For the structural position of the family farmer is pretty much the same. Family farmers are also caught in a competitive sector sandwiched between sectors of extremely concentrated economic power.

Iowa family farmers are independent and dispersed producers. Because of this structural fact, they are incapable of influencing the prices they must pay for necessary agricultural inputs or the prices they receive for their agricultural output. On the input side of the equation, the family farmer confronts a tractor attachments industry where the four largest companies accounted for 80 percent of sales in 1977, a harvesting machinery industry where the top four firms had 79 percent of sales in 1977, an industry producing nitrogenous and phosphatic fertilizers where the eight largest firms had 64 percent of total agricultural chemical sales in 1977, and the pesticides industry where the top four firms had a 60 percent market share in 1976 (Wessel 1983, 116).

When dispersed producers confront concentrated suppliers, the terms of exchange always favor the latter. The Federal Trade Commission in 1972 found monopoly overcharges of 5.784 percent for farm machinery and 4.2 percent for the feed industry. If these percentages are applied to sales revenues in these industries, farmers were overcharged $827 million for farm machinery and $775 million for feeds in 1980. Taken together, the overcharges accounted for over 8 percent of net farm income that year. If similar rates of overcharge existed in the other major farm supply industries that year, the total loss to farmers amounted to at least $2.8 billion or one-seventh of their total net income. (See Scanlon 1972, Table 1.) Another manifestation of the structural weakness of the farmer in confronting this concentrated economic power is that between 1973 and 1980 the cost of production outpaced the price increases farmers received by over 40 percent. With terms of trade this unfavorable, it is no wonder that the family farmer fell into deep debt. In 1970 the interest on farm debt equaled about 24 percent of total net farm income; a decade later, 80 percent.

It is true that as this debt accumulated and farmers cut back their purchases, farm suppliers eventually became squeezed too and had to lower retail prices. But this is not likely to bring much relief to farmers over the medium to long term. Instead, it would not surprise me if, in the future, farmers will confront an even greater concentration in the farm supply industries. John Deere, for instance, is expected to increase its market share from 35 to 40 percent to around 50 or 60 percent over the next few years. This will allow farm suppliers to demand even more favorable terms of trade from farmers.

On the output side of the economic equation, the same situation holds. Dispersed farmers must sell their grain on a market where about 85 percent of the world's trade in grain is handled by just six companies. These companies, by the size of their transactions and their access to information the family farmer lacks, are able to influence the market to their advantage and at cost to the independent producer. Likewise regarding food processing and distributing: .25 percent of food companies now control two-thirds of the industry's assets. Here too there is a direct correlation between market power and higher prices. An internal FTC report noted that "if highly concentrated industries were deconcentrated to a point where the four largest firms control 40% or less of an industry's sales, prices would fall by 25% or more" (Weiss cited by Green 1972, 14).

The situation is serious indeed. Thirty-five years ago, the farmer received about half the consumer's food dollar. Today the farmer gets much less as a result of the structural features I have sketched. In 1979, for instance, about sixty-nine cents of every dollar spent on food went to firms that transport, process, or retail domestically grown food commodities. Of the remaining thirty-one cents, seventeen cents went to the farm supply industries, and ten cents went to pay farm expenses such as rent, interest, capital depreciation, and taxes. This left the farmer with about four cents to pocket. In 1980 even this small sliver of the domestic food dollar received by the farmer declined by nearly 40 percent to only 2.5 cents, and the situation has not significantly turned around since. This situation should come as no surprise given the extremely concentrated economic power that confronts the farmer with respect to both input and output factors.

This is the underlying structural logic of the farm crisis. And yet throughout the farm crisis, how often have we heard speakers from the agricultural scientific community mention it? It was not mentioned once at the Kellogg workshop. It was not mentioned once in any article by or about agricultural economists on the farm crisis published in the *Des Moines Register* these last years. It was not mentioned once by an agricultural economist at the conference at Iowa State in February of 1986. Can this be an accident? I think not. Instead this is because relevant questions have not been asked, because known facts have been omitted, because agricultural scientists think in terms of an abstract model of markets as neutral mechanisms for allocation and ignore the concrete reality of markets as places where economic power is exercised. For whatever the reason, whenever agricultural scientists ignore the structural power held by agribusiness corporations over the family farm, their discussion is more ideological than scientific, whatever their intentions might be.

And if the diagnosis is ideological, then the policy recommendations coming from the university will be no less so. Many proposals—for instance those advocating that credit be eased or that the government bail out agricultural bankers in danger of going under, leave the underlying structural factors that caused the problem in the first place completely untouched. They are Band-Aids for a gaping wound. Conservative "compete or die" policies of academic defenders of the Reagan administration—for example, vast reductions in grain price supports—would in the short term lead to massive windfall profits for food processors, who today continue to raise prices despite using ever-cheaper grain. In the short to medium term, these policies will eliminate vast numbers of farmers. The result will be a tremendous increase of huge corporate farms, which already account for more than 50 percent of agribusiness cash receipts, although they represent only 5 percent of all farms. In the long term, farming itself might well become almost as concentrated as the farm supply and food processing industries. At that point the few farmers that remained would not suffer from the disadvantageous terms of trade today's family farmer faces. But then *every* section of the food stream leading to the consumer would be controlled by oligopolies. The inevitable result: a tremendous increase in food prices for consumers.

I am afraid that, ironically, the liberal solution would lead to exactly the same results. Many liberal agricultural experts in the university support Tom Harkin's Save the Family Farm Act. This attempts to save family farms through mandating by law that they act as if they were oligopolies and undertake a drastic reduction of production in order to raise prices. But then food processors would simply pass on these higher costs to consumers. According to some studies, this should amount to no more than an extra one cent per loaf of bread. But given that food processors raise prices significantly even when they use cheaper grain, I am afraid that without strict price controls—which are not a feature of the Harkin bill—we must expect that they would use any rise in their grain prices as an excuse to gouge consumers and that they would then shift the blame onto farmers.

Who benefits when food prices are raised at a time when 21.5 million people in the United States are out of work? Who would benefit from the Harkin plan? A proposal like this simply plays off one group of troubled Americans against another, while leaving the interests of agribusiness untouched. And to talk of cutting the production of food while hunger remains a major social problem strikes me as morally questionable.

If the problem of the family farm is to be resolved in a manner that addresses the roots of the problem, then a new social movement must arise, one that is not afraid to attack the underlying structural problem, one that is not afraid to challenge the concentrated economic power of the Cargills and the Beatrices. The fact that you will not hear agricultural scientists talking this way has everything to do with ideology and very little to do with science.

Consider an argument actually presented to us at the Kellogg workshop by an economist: "Farmers trusted the bankers who lent them money. They have since discovered that bankers will put their own interests far above the interests of farmers. Therefore, farmers should accept the process presently reducing many of them to tenancy and should align with their prospective landlords." In itself, this is simply a bad argument, an extremely bad argument. From an assertion regarding the relationship between some Group A (farmers) and some Group B (bankers), one logically cannot conclude anything regarding a relationship between A and some Group C (landlords) different from B.

Yet, once again, more is going on here than a simple mistake by a given individual. This argument *would* work if we could presuppose that the only groups farmers can work with are bankers and landlords. Then if the farmer has given up on the former, it would indeed be valid to conclude that the only alternative is to cooperate with the latter. Precisely this assumption was silently presupposed not just by this speaker, but by almost all the agricultural experts who spoke to us. The very possibility of other forms of alliance was systematically repressed in a process of which few, if any, of the speakers were conscious.

But of course there are many other groups with whom family farmers can ally in a new social movement. For every farmer in the United States there are five farm laborers slaving for low pay under dangerous conditions, harassed by immigration officials, and outside the protection of national labor laws. There are tens of thousands of truck drivers getting the produce to market. There are tens of thousands who work in packing plants and supermarkets, most of whom have had huge wage cuts imposed upon them these past years. There are black farmers, who have been losing their land at a rate twice that of white farmers while the federal government has pursued a deliberate policy of driving them out of business. (Between 1979 and 1983, federal farm loans to black farmers fell by 71 percent.) There are tens of thousands of rural women getting ready to take on leadership roles. And there are, of course, the tens of millions of food consumers whose nutritional levels will be threatened by any rise in food prices resulting from concentrated economic power.

All of these groups, who together constitute a vast majority of our nation, have a common interest against corporate agriculture. If these groups could somehow be united within a common organization — say a labor/farmer/consumer party — then social processes that agricultural economists assure us are "inevitable" and "irreversible" would lose that appearance at once. Such a movement could demand strict interpretation and enforcement of existing antitrust laws and the legislation of new ones. Such a movement could demand that family farms receive parity for their products while simultaneously demanding strict price controls to prevent oligopolies in food processing industries from passing on those higher prices. Such a social movement could demand that any processor who went on an investment strike to protest its loss of profits from these price controls would have its assets seized and handed over to its workers. Such a social movement could assert that basic nutrition is a fundamental right of all humans and not something for only those with sufficient disposable income.

Today in this country our political imagination is so stunted that we cannot even conceive of such radical solutions. But the demise of the family farmer and the eradication of our rural communities will continue unless the underlying structural forces behind that demise and that eradication are removed, and this will require a massive social struggle. It is my hope that the university community will participate in that struggle. It is my fear that, on the whole, it is part of what must be struggled against.

References

Dinham, Barbara, and Colin Hines. *Agribusiness in Africa.* Trenton, N.J.: Africa World Press, 1984.

George, Susan. *Ill Fares the Land: Essays on Food, Hunger, and Power.* Washington, D.C.: Institute for Policy Studies, 1984.

Nation, 7–14 July 1984, 13.

Scanlon, Paul D. "FTC and Phase II: The McGovern Papers." *Antitrust Law and Economics Review* 5, no. 3(1972): 33.

Weiss, Leonard. "Economic Studies of Industrial Organization." Internal Report, Antitrust Division of Federal Trade Commission, cited in *The Closed Enterprise System* by Mark Green. New York: Grossman, 1972.

Wessel, James. *Trading the Future: Farm Exports and the Concentration of Economic Power in Our Food System.* San Francisco: Institute for Food and Development Policy, 1983.

GLENN L. JOHNSON

A Response to Smith

SMITH'S PRESENTATION at the 1986 Iowa State conference ("Is There A Moral Obligation to Save the Family Farm?") and the version printed here are too important to be passed over lightly. Because they provide a form of questionable intellectual support for the divisive positions taken at the conference by various participants—Denise O'Brien, Carol Hodne, David Ostendorf, Wendell Berry, Marty Strange, Bishop Dingman and others—Smith's ideas should not escape careful and critical scrutiny.

Smith divides the members of our society into two parts, one wearing black hats and the other white. He accuses the black hats of being unobjective and partial. Then, apparently using their alleged unobjectivity and partiality as an excuse for his own, he proceeds with an unobjective, biased attack. If successful, this attack would divide those concerned about agriculture into two classes. In my opinion, we need to evaluate his attack and carefully avoid the danger of being misled by it into advocating an unnecessary, destructive, class struggle.

In his speech, Smith saw in agriculture an ideological struggle between the black-hat agricultural establishment and the white-hat migrant farm workers, family farmers, rural women, consumers, environmental groups, and laborers in agricultural processing and distributing industries. He saw the agricultural establishment as consisting of "agri-corporations and their allies." Their allies include the farm organizations such as the Farm Bureau, the two political parties, the USDA, large corporate farms, Congress (which passes the laws the USDA executes), the land grant colleges of agriculture, and the scientists he feels have sold out to corporate agriculture. In the essay included in this volume, he fears that "university experts will serve the very corporate interests that . . . are the root cause of the difficulties." And he calls for "a new social movement . . . one that is not afraid to attack the underlying structural problem, one that is not afraid to challenge the concentrated economic power of the Cargills and Beatrices." And he goes on, "Our political imagination is so stunted that we cannot conceive of such radical solutions . . . this will require a massive social struggle. It is my hope that the university community will participate in that struggle. It is my fear that, on the whole, it is part of what must be struggled against."

Smith deplores at great length an alleged "lack of objectivity and impartiality" on the part of agricultural scientists (biological and physical as well as social) in the agricultural establishment. He describes agriculturalists attending a Kellogg Foundation workshop (his representative case study) as shifting "back and forth from relatively straightforward empirical assertions to more or less crass apologetics for the corporate system of agriculture." He adds that at the workshop there was an "aura of scientific objectivity [that] often masked a style of thinking where ideological considerations so interpenetrated scientific considerations that one could not say with confidence where one began and the other ended."

In what follows, I will demonstrate that Smith uses what he calls "methods of discourse" in the unobjective and biased manner he attributes to the black hats. In my last section, I offer a constructive plea for a greater effort on the part of both defenders and critics of the agricultural establishment to be more objective and impartial, more cooperative and constructive, and far less divisive than Smith is. The moral and ethical issues now facing agriculture are too important to be addressed in a divisive spirit.

Smith's Own Unobjective, Partial Use of Discourse Mechanisms

Smith enlightens his readers on how to spot the unobjectivity of what he regards as reactionary agricultural scientists. He catalogs the following "discourse mechanisms," which one should look for: (1) exclusion of relevant questions, (2) omission of known facts, (3) retreat to abstract models, and (4) failure to consider relevant power relations.

Smith correctly deplores use of these mechanisms. Such methods of discourse are not generally approved for use in academic circles. Also, their use is a breach of academic ethics and fundamentally offends Christian ethics by relying on the "bearing of false witness." From a practical standpoint, biased, inaccurate, unobjective knowledge should be avoided because it leads to faulty prescriptions for solving important agricultural problems.

It is instructive to examine Smith's own discourse. He has used all four of the mechanisms he listed.

SMITH HAS EXCLUDED RELEVANT QUESTIONS. Numerous relevant questions have to be excluded from short presentations and Smith should not be held responsible for all omissions. Some questions, however, are so relevant for his class struggle argument that their omission is important. Among such important excluded questions are:

1. What does the ruling class Smith deplores contribute to the exploited class as he views those classes? These contributions are important. They should not be ignored even by one promoting a class struggle.

2. What constructive roles are played by what Smith views as an exploiting class that have to be performed in all societies? And would another group better perform those roles? What agency would manage our exports if the Cargills were eliminated? Who would administer the "strict price controls" Smith advocates? No country has successfully managed its agribusiness sector with such price controls. The state trading agencies of the eastern bloc countries have not been very successful and parastatal trading agencies in West Africa have been exploitive of farmers and almost disastrous (Bauer 1981).

3. Have societies that have shifted performance of the functions described in (2) above to the state (either socialist or fascist) experienced more or less equality and exploitation than before? Think about Argentina, The People's Republic of China, fascist Italy, Poland, Hungary, nazi Germany, the Spanish Inquisition, Nyerere's Tanzania, Nkruma's Ghana, Cuba, Nigeria, and the Soviet Union (Bauer 1981; Johnson 1983).

4. Would a movement powerful enough to overcome the corporate powers Smith finds so deplorable concentrate political, police, economic power unto itself to leave ordinary farmers even more powerless? Smith does not consider this important question.

SMITH HAS OMITTED KNOWN FACTS. Again, the body of facts that must be omitted is very large. The question here is whether facts particularly relevant for his conclusion have been omitted. Smith has omitted these facts in reaching his conclusion about the need for a class struggle:

1. There is an extensive, long-standing literature on monopoly in agriculture. Many studies of monopoly in the food industry have been carried out in recent years. Smith omits facts that such studies exist while implying, with only one questionable example, that agricultural scientists suppress facts on monopoly in agribusiness. He fails to mention Bruce Marion's (1986) *The Organization and Performance of the U.S.*

Food System, which reports recent research by agricultural economists on monopolies in the food system. And he omits the classic studies of monopolistic competition started in the pre–World War II period at Iowa State by William Nicholls. Furthermore, agricultural economists have studied the damaging effects of state-run monopolies in both capitalistic and Marxist countries. Smith's conclusion that agriculturalists do not study monopolies and that they are consciously or unconsciously co-opted into supporting them cannot be substantiated without omitting well-known relevant facts.

2. Available historical facts concerning the consequences of policies similar to those advocated by Smith are omitted. See the above section on excluded relevant questions.

SMITH HAS RETREATED TO ABSTRACT MODELS. While abstract models are often extremely useful in studying practical problems, Smith attempts to illustrate instances where abstract models are used in the class struggle by a privileged class to the disadvantage of what he views as an exploited class.

1. He asserts that use of the law of comparative advantage requires an abstract (unrealistic) assumption of perfect competition that leads to unrealistic conclusions in analyzing monopolies and oligopolies. Of course, such an assumption would do this. Fortunately, the law of comparative advantage is capable of dealing with monopolies and oligopolies and, hence, is not necessarily unrealistic in this respect. Smith's unique version of the comparative advantage model is, itself, unrealistic and unnecessarily abstract.

2. More importantly, Smith uses an abstract historical Marxist class struggle model to form his conclusions. In doing or permitting this, he used biased terms, excluded relevant facts, and omitted relevant questions in order to maintain his model and to provide us with an abstract *unrealistic* picture of class struggle in U.S. agriculture in which agricultural scientists are regarded as partial, unobjective participants.

SMITH DID NOT CONSIDER IMPORTANT RELEVANT POWER RELATIONS. There is long-standing, widely available research on the holding of power with respect to U.S. agriculture (McFadyen Campbell 1962; Hardin 1946, 1947, 1950; Guither 1980; Talbot and Hadwiger 1968; Galbraith 1952; Schmid 1985a, 1985b, 1978; Bonnen 1985, 1983; Mueller 1986; Cotterill 1984; Shaffer 1975; Busch and Lacy 1983). Yet Smith asserts that power distributions are ignored in the scientific work of social scientists in the agricultural establishment. In doing this he uses the omission of known facts mechanism. In addition and perhaps more significantly, he fails to present the available facts on the well-known consequences of concentrations of various kinds of power vis-à-vis agriculture in the hands of Marxist states based on the ideological class struggle view he embraces. One should consider the following centralizations and decentralizations of military, police, social, religious, and economic power with respect to agriculture: (1) the centralization of power in the Red Guard period of The People's Republic of China and the subsequent modest decentralization, (2) the centralization of power and later modest decentralization in Hungary, (3) Poland's two centralizations and one short-term decentralization, (4) Cuba's centralization of power, and (5) Tanzania's centralization of several kinds of power and its current efforts to decentralize.

To be symmetrical, we should recognize that non-Marxist (some nazi or fascist) concentrations of power over agriculture have had similar consequences. Relevant cases include: (1) Nigeria's and Ghana's parastatal organization, which centralized economic power to exploit poor farmers (Bauer 1981; Johnson et al. 1969), (2) Peron's centralization of power in the Argentine, (3) power centralizations in Franco's Spain, and (4) Germany's nazi and Italy's fascist centralizations of power.

SUMMING UP. I conclude that Smith's criticism of the agricultural establishment and its agricultural scientists is seriously flawed by his own exclusion of relevant questions, omission of known facts, retreat to an abstract model, and failure to consider power relationships relevant to his own position.

Need for Objective, More Healing, Less Divisive Discourse on the Future of the Family Farm

I now ask what is involved in objectively investigating problems of justice and equality vis-à-vis agriculture and the family farm? In answering this question I refer to ideas in my chapter of this book. Though it is entitled "Roles for Social Scientists in Agricultural Policy," its contents could easily pertain to biological and physical agricultural scientists, humanists, Christian ministers and laypersons, activists, circuit speakers, political leaders, administrators, and others concerned with farming, rural communities, and related aspects of general welfare.

In the section "Roles for Social Scientists" I point out that value-free and value knowledge are used to derive prescriptive knowledge for solving practical problems. Furthermore, I state that "ethics requires objectivity." Objectivity is discussed as the honest application of the tests of experience (correspondence), logic (coherence), and clarity (lack of ambiguity). In addition, in many instances, the pragmatic test of workability can and should be applied. The use of such tests does not imply that the knowledge of mere humans is ever entirely provable and self-contained (Popper 1959)—instead, it is implied that we can honestly and adequately test our knowledge "for the purposes at hand" (Johnson 1986b). Furthermore, the use of these tests does not preclude important roles for insight, inspiration, and, indeed, religious revelation in perceiving the meaning of experiences and in improving the theories we use to interrelate those experiences. In order for our knowledge to be used, it has to be accepted "for purposes at hand" by communities of people.

Our knowledge of what really has value (as opposed to our knowledge of *who* assigns what *value* to *what*) is similar in many respects to our value-free knowledge (Johnson 1986a, 1986b). In connection with these same arguments, I and others have recognized that prescriptive choices are based on decision rules involving distributions of power. Power goes along with the ownership of rights and privileges. As societies create, maintain, and/or acquiesce to distributions of power, societies may also be able to change them. However, changing power distributions may involve costly conflicts. The costs that may be involved can be social, political, and physical. These costs are also personal and individual. Assessing the benefits and costs of changing a power distribution is an essential part of making decisions to change it. Such assessments require accurate knowledge of who holds how much of what kind of power.

Smith and others stress the unfairness of what is happening in agriculture. There is little doubt about the existence of unfairness. Many formerly well-off farmers are now financially destitute. Many agribusinesses have gone broke and their employees have lost employment. Our agricultural problems are numerous, large, and complex with many having only solutions that are redistributive in nature.

Clearly, redistributive agrarian reforms are important for agriculture, human nutrition, and the future of the family farm and must often be done (Johnson 1983, 1986b; Johnson and Quance 1972). I believe redistribution is generally better done quietly and with minimal class struggle. My general conviction is that objective investigation of family farm problems will help us avoid costly class struggle and turmoil. Many past class struggles have unnecessarily hurt people while concentrating the ownership of rights and privileges in the hands of the state and its controlling party elites to the disadvantage of those who were supposed to be helped. This has occurred on both the right and the left. As someone has said, "the extremes of fascism and Marxism meet in the murky ground of tyranny!"

Impartial investigation will require restraint in use of the methods of discourse deplored (but nonetheless used) by Smith and others. I am not so naive as to think the use of such weapons and methods can be entirely avoided. We are all too human to succeed in being entirely objective and impartial. However, academic and religious codes of conduct as well as common honesty require us to make the effort. My own experience is that we often do better than Smith has done and that members of the agricultural establishment typically do better than he portrays them to do.

I include among the members of the agricultural establishment not only the agricultural scientists Smith castigates in his essay but corporate agribusiness leaders, large-scale farmers, the leadership of farm organizations, USDA and college of agriculture administrators, the farm leadership of the major political parties, and those running larger than family farms. Many of these people know and understand family farm people and problems more objectively than Smith seems to think they do and, evidently, better than Smith does.

I take comfort from the essays by Richard Kirkendall, Paul Lasley, and Michael Boehlje in this collection. I believe they demonstrate what academic objectivity is all about. The even-tempered responses of Charles Lutz, Erwin Johnson, Luther Tweeten, and Merrill Oster also demonstrate considerable impartiality in a manner to renew my faith.

In fairness to Smith, it should be pointed out that his was not the only divisive inflammatory presentation at the conference. Others offended with the unobjective use of the "discourse mechanisms" Smith deplores. Both pro- and antiagricultural establishment positions were unobjectively made with, I believe, the preponderance of such presentations being on the antiestablishment side (Hodne, O'Brien, Ostendorf, and Berry). I perceived the proestablishment presentations by Erwin Johnson and Merrill Oster to be more objective than those of the antiestablishment speakers.

I have stressed objectivity and have deplored divisiveness. I now want to add a constructive plea for, in Christian terms, a healing ministry. When farmers and agribusiness persons including their employees are hurting as badly as they are now, we need to help them by bringing to bear all of the resources of the agricultural establishment. Almost without exception, all parts of that establishment make constructive contributions to society. Our need is more for unity and healing than for divisiveness and magnification of present hurts. Furthermore, our humanity includes our fallibility. This human fallibility, in turn, demands that we be humble about our "facts," analyses, opinions, and prescriptions. I close by recognizing the danger that my own humanity and fallibility may make my analysis of Smith's use of discourse mechanisms more divisive, less constructive, and more arrogant than it should be.

References

Bauer, P. T. *Equality, the Third World, and Economic Delusion.* Cambridge, Mass.: Harvard University Press, 1981.

Bonnen, James T. "Agriculture's System of Developmental Institutions: Reflections on the U.S. Experience." In *L'agro-alimentaire Quebecois et son developpement dans l'environnement economique des annees 1980,* 43–64. Quebec, Canada: University of Laval, 1983.

_____. "United States Agrarian Development: Transforming Human Capital and Institutions." *United States-Mexico Relations: Agriculture and Rural Development.* Stanford, Calif.: Stanford University Press, 1985.

Busch, L., and W. B. Lacy. *Science, Agriculture and the Politics of Research.* Boulder, Colo.: Westview Press, 1983.

Cotterill, Ronald W. *Modern Markets and Market Power: Evidence from the Vermont Retail Food Industry.* Working Paper #84, North Central Project 117. Madison: University of Wisconsin, 1984.

Galbraith, J. K. *American Capitalism: The Concept of Countervailing Power.* Boston, Mass: Houghton Mifflin Co., 1952.

Guither, Harold. *The Food Lobbyists.* Lexington, Mass.: Lexington Books, 1980.

Hardin, Charles. "The Bureau of Agricultural Economics under Fire." *Journal of Farm Economics* 29 (1947): 359–82.

_____. "The Politics of Agriculture." *Journal of Farm Economics* 32 (1950): 571–83.

_____. "Programmatic Research and Agricultural Policy." *Journal of Farm Economics* 29 (1947): 359–82.

Johnson, Glenn L. "Economics and Ethics." Twenty-Fourth Annual Centennial Review Lecture, 9 April 1985. *Centennial Review* 30, no. 1(Winter 1986a): 69–108.

_____. *Research Methodology for Economists: Philosophy and Practice.* New York: Macmillan, 1986b.

_____. "Synoptic View." In *Growth and Equity in Agricultural Development,* edited by A. Maunder and K. Ohkawa, 592–608. Aldershot, England: Gower Publishing Co., 1983.

Johnson, Glenn L., et al. *Strategies and Recommendations for Nigerian Rural Development 1969/1985.* Consortium for the Study of Nigerian Rural Development (CSNRD), Report No. 33. East Lansing: Department of Agricultural Economics, Michigan State University, 1969.

Johnson, Glenn L., and C. L. Quance, eds. *The Overproduction Trap in U.S. Agriculture.* Baltimore, Md.: Johns Hopkins University Press, 1972.

McFadyen Campbell, Christiana. *The Farm Bureau and the New Deal.* Urbana: University of Illinois Press, 1962.

Marion, Bruce W. *The Organization and Performance of the U.S. Food System.* Lexington, Mass.: D. C. Heath and Company, 1986.

Mueller, Willard F. "The New Attack on Antitrust." Paper presented at the Conference on Management and Public Policy Toward Business in honor of Robert F. Lanzillotti, University of Florida, 4 April 1986.

Popper, Karl. *The Logic of Scientific Discovery.* N.Y.: Harper and Row, 1959.

Schmid, A. Allan. "Biotechnology, Plant Variety Protection, and Changing Property Institutions in Agriculture." *North Central Journal of Agricultural Economics* 7, no. 2(1985a): 129–38.

––––––. "Property Rights in Seeds and Micro-Organisms." In *Public Policy and the Natural Environment,* edited by R. K. Godwin and H. Ingram. Greenwich, Conn.: Jai Press, 1985b.

––––––. *Property, Power and Public Choice.* New York: Praeger, 1978.

Shaffer, James D. "Power in the U.S. Political Economy — Issues and Alternatives." In *Farm Foundation, Increasing Understanding of Public Problems and Policies, 1975,* 14–30. Chicago: Farm Foundation, 1975.

Talbot, Ross, and Don Hadwiger. *The Policy Process in American Agriculture.* San Francisco, Calif.: Chandler Publishing Co., 1968.

TONY SMITH

A Response to Johnson

A REPLY TO A REPLY should be as brief as possible. I shall try to respond to the main objections made by Johnson more or less in order.

Johnson missed a key point of my paper. I was not arguing that inherently political matters can be depoliticized successfully by academic scientists, nor that agricultural scientists are to be criticized for failing to do what in principle can be done. My point was that inherently political issues *cannot* be successfully depoliticized, and that agricultural scientists are to be criticized for claiming to do what *cannot* in principle be done. And so when Johnson points out the obvious, that my own presentation itself has a political viewpoint behind it, I would plead guilty were it not the case that the concept of "guilt" usually implies that one could have chosen to act differently. The only choice here is whether political viewpoints are stated openly or are hidden. The fact that I chose the former option while the latter seems to be the custom does not necessarily imply that I am the guilty one here.

The real question has to do with whether the fate of the family farm specifically and the agricultural sector in general is an "inherently political" matter. To me the answer is obvious. If issues like what sort of community we shall have (dispersed economic power or concentrated economic power), what sort of community we shall leave to our children (safe drinking water or water ruined by chemical runoffs), what sort of standard of living we have a right to expect, what sorts of communities will flourish and what sorts will decline—if questions like these are not political questions, then what are? These are certainly not technical matters to be left to "the experts." They are public matters that ought to be decided by the public itself after an extended period of informed discussion. Whether they know it or not, agricultural scientists are contributing to public discussion, either through empowering the public by helping it to be informed of the various possibilities open to it, or through demoralizing it by suggesting that omnipotent economic forces are at work to which we must helplessly submit. No matter how abstract or technical an agricultural scientist's work is, it always has this public, political component.

Now, I would not claim that all issues here are political in nature. The number of farm animals raised in a particular place at a particular time, their average weight, the average cost of raising them, and so forth, are straightforward and easily quantifiable matters that it is possible to know in a fairly value-free manner. But if this were all that science were about, we would not need any scientists. All we would need are people who could count! Of course science is more than this. It involves a questioning of these data and the attempt to interpret them within a theoretical framework. It is here that things get interesting. For some questions rather than others will be asked, and some terms rather than others will be used in the theoretical framework. Yet the data themselves do not determine which questions will be selected or which terms will be employed. Scientists do this. And scientists are not disembodied intellects living timelessly outside of human society. They are flesh and blood men and women historically situated in a given social context.

If a cultural worldview is dominant in a given historical and social context — and can anyone deny that a procorporate worldview is dominant in the U.S. today? — would it not be exceedingly strange if scientists alone were unaffected by it? And would it not also be exceedingly strange if this did not show up now and then in the questions they asked or the terms they used? If one answers no, then one must think that scientists are either mere machines who simply record what is "out there," or godlike beings who have somehow transcended the influences to which the rest of us are subjected. In my view, scientists are neither less than human nor more than human. Is this controversial?

Johnson writes that I pit the "black-hat" agricultural establishment against the "white-hat" migrant farm workers, family farmers, rural women, consumers, environmental groups, and laborers. He adds that my article may mislead people into advocating "unnecessary, destructive, class struggle." This is one of a number of unfortunate ad hominems. I wish he would have avoided that sort of attack altogether.

ON EXCLUDED RELEVANT QUESTIONS. On the issue of the "contributions" and "constructive roles" played by the exploiting classes, let me refer the interested reader to the chapter entitled "Contribution, Sacrifice, Entitlement" in *Capitalism or Worker Control? An Ethical and Economic Appraisal* (Schweickart 1980). It takes David Schweickart thirty-five pages to work out what I take to be the correct answers to these questions. I will leave it to the reader to decide whether the role of ideology in agricultural science could be adequately discussed in a paper that does not also include these other issues; I do not agree with Johnson that my omission of them was deliberate. Schweickart's book is also relevant to the third question Johnson raises in his section on excluded questions. He assumes that because I am critical of our present oligopolistic capitalism I must be a defender of state socialism. I am not. I am in favor of a system where all exercises of public power are subject to public control, a form of democracy that goes beyond *both* oligopolistic capitalism and bureaucratic socialism.

ON ABSTRACT MODELS. Two points on abstract models: first, the unrealistic model of comparative advantage was presented to us as realistic by agricultural economists at the Kellogg workshop at least a half dozen times. I have seen it presented as realistic countless times in economics textbooks, on the op-ed pages, in the business press, and in business ethics textbooks. If Johnson has seen through this model as the sham that it is, more power to him. But he should not assume that there are not economists out there who discuss it as if it were accurate. There are. (One example is the widely used text *International Economics* by Kenen and Lubitz.)

Second, class struggle is no abstraction. It is a reality. It may be a reality that is difficult for tenured professors to attend to, as I myself know. But let's look at some of those objective value-free facts Johnson likes. In 1984 the wealthiest fifth of the population received 43 percent of the national income, while the poorest fifth received less than 5 percent. Between 1978 and 1984 the nation's impoverished population went up by 38 percent. Of the 7.3 million families living in poverty in 1984, more than half had at least one worker and more than 20 percent had two or more workers. Also, some 1.4 million able-bodied poor sought work but were unable to find it. Only one-third of poor families received public-assistance payments, and only 43 percent received food stamps. The combined value of aid to families with dependent children and food stamps has declined 20 percent in real terms over the past dozen years.

And while the poor got poorer, the rich got richer. The gap between rich and poor today is the widest since the government began collecting this information in 1947. While the impoverished population jumped by eleven million between 1978 and 1984, the share of families with incomes of thirty-five thousand dollars or more (in 1984 dollars) grew from 29 percent to 34 percent, and the biggest boom in history transferred tremendous wealth to the large investors (*Business Week* 1986, 16). Anyone who can consider facts such as these and still consider class struggle an abstraction is like, well, someone who can look at the present situation in the agricultural sector and not see a struggle between agribusiness and the family farmer. Or like someone who can look at a burning barn and not see the fire.

ON IMPORTANT POWER RELATIONS. Once again Johnson assumes that the only political choice we have is between ever-increasing corporate control of our lives and centralized state control. I reject this dichotomy as vehemently as I reject Johnson's implication that my position somehow commits me to advocating tyranny. In Johnson's view, history is over, there is no point to struggling for future social institutions different from those we see today.

This is a cynical perspective used in both oligopolistic capitalist countries and bureaucratic socialist societies to justify their respective status quo. But history is not over. Phenomena ranging from the most progressive aspects of the present farm movement in the United States to the Solidarity movement in Poland point to the fact that a third alternative is on the historical agenda. Rural communities struggling to survive against oligopolistic markets that condemn them do not want to substitute rule by a state elite. The workers in Poland do not want to substitute a capitalist elite for their party bosses. Despite all the differences between these two movements, they are both struggling for the same sort of economic democracy, for the institutionalization of the principle that all forms of public power should be subject to public control. This goes far beyond issues of redistribution, as important as those issues are. It goes far beyond the "rich diversity of political alternatives" presented by the tweedle-dumb and tweedle-dumber parties. It goes far beyond anything we have seen in history thus far. But it should not be beyond the limits of our political imagination.

References

Business Week, 5 May 1986, 16.

Kenen, Peter, and Raymond Lubitz. *International Economics.* Englewood Cliffs, N.J.: Prentice-Hall, 1971.

Schweickart, David. *Capitalism or Worker Control? An Ethical and Economic Appraisal.* New York: Praeger, 1980.

SUGGESTED READINGS

Connor, John M., et al. *The Food Manufacturing Industries: Structure, Strategies, Performance, and Policies.* Toronto: Lexington Books, 1985.

George, Susan. *Feeding the Few: Corporate Control of Food.* Washington, D.C.: Institute for Policy Studies, n.d.

Hightower, Jim. *Hard Tomatoes, Hard Times.* Cambridge: Schenkman, 1973.

Johnson, Glenn L. "Economics and Ethics." *Centennial Review* 30, no. 1(Winter 1986): 69–108.

LeVeen, Phillip E. *Towards a New Food Policy: A Dissenting Perspective.* Berkeley: Public Interest Economics West, 1981.

Merrill, Richard, ed. *Radical Agriculture.* New York: Harper and Row, 1976.

Ruttan, Vernon. "Moral Responsibility in Agricultural Research." *Southern Journal of Agricultural Economics,* July 1983, 73–80.

Justice and the Family Farm

HAVE AGRICULTURAL POLICIES benefited rich farms and agribusiness while putting poor farms and farm laborers out of work? Tony Smith has raised the question of justice in the preceding section; *who* benefits from university research and policy recommendations? In Part Five we plunge into the complexities of the question of economic justice and ask whether family farmers have been treated unfairly. We are now fully enmeshed in what Aiken calls moral and ethical discourse.

"When some farmers receive upwards of $100,000 each in tax money, farm subsidies offend everyone's sense of justice." So concludes James Trager's analysis of the largest commercial transaction in history, the 1972 grain sales to Russia. In *Amber Waves of Grain* (1973), Trager tells the story of "the secret Russian wheat sales that sent American food prices soaring." Trager's charge that federal farm subsidies have benefited large farmers at the expense of smaller ones is a claim that has not disappeared in the dozen or so years since the publication of Trager's book. Even as the Senate Finance Committee was passing a tax reform bill in May of 1986, the measure was being hailed as one that would close the loopholes benefiting large corporations who enter farming in order to find tax write-offs.

Have federal policies, taken as a whole, favored corporate farms over family farms? In essays that follow, some authors allege that the government has allowed the formation of an oligopoly—a shared monopoly—in the food industry. While Lerza and Jacobson admit that "big is not necessarily bad," they agree with Tony Smith and Jim Hightower that the consequences of corporate concentration are disastrous; family farmers are "squeezed out," trapped between powerful oligopolies in the farm supply and food manufacturing sectors. "When one processor or wholesaler dominates a regional or product market, farmers have little leverage in determining the sale price of their livestock or crops." A technical study by Russell C. Parker and John M. Connor lends strong credence to the claim that consumers, at least, have lost money due to the oligopolistic structure of the food manufacturing industries. One is tempted to infer that farmers have been similarly cheated.

Luther Tweeten, one of this country's best known agricultural economists, responds to some of these charges. In his first piece, "Has the Family Farm Been Treated Unjustly?" he considers the specific claim that federal policies have unjustly favored large farms over family farms. Tweeten finds that, all things considered, this assessment does not seem justified. Federal policies in the areas of investment tax credit laws and rapid depreciation allowances have favored farms with large amounts of capital. But in many other areas—credit, monetary regulations, property taxes—the same inequities cannot be found. He concludes that, on the whole, federal policies have not unjustly favored large farms over family farms.

In a second essay, he responds to the

disjunction found in Lerza and Jacobson's title. We should not think, he argues, that it is either food for people *or* profits. We can have both—food for people *and* profits.

Reference

Trager, James. *Amber Waves of Grain.* New York: Arthur Fields, 1973.

JIM HIGHTOWER

The Case for the Family Farm

CONGRESSWOMAN SHIRLEY CHISHOLM first arrived in Washington in January 1969 promising to defend the interests of the big-city constituency. The congresswoman's first act was an indignant and highly publicized protest about her assignment to the House Agriculture Committee. Her successful protest reveals a great deal about urban America's myopia about farm policy.

The relevance of the Agriculture Committee is now clearer to urban legislators. The high cost of agricultural subsidy programs is causing an urban-oriented Congress to finally realize that what happens down on the farm affects the city as well. Traditionally, farm policy has been left to Corn Belt senators, cotton congressmembers, and the slick lobbyists of the agribusiness interests. The result has been a steady industrialization of the American food supply.

Anyone who admires what Detroit's Big Three have done for automobiles will be pleased with the new agriculture, for the same quality, cost competition, and reliability will soon be built into the food we eat. The new agriculture, once properly labeled "Butzculture" after Agriculture Secretary Earl Butz, can now also be labeled "Reaganculture" after the president who brought the old vision of corporate farming to rural America.

Since World War II, corporate power and government programs have combined to eliminate three million small farms and accelerate big-business domination of farming. But until now, the family farmers have persisted, hunkered down, and become more efficient. Reaganculture promises to finally drive them to the suburbs and the factories.

JIM HIGHTOWER is commissioner of agriculture in Texas. A slightly different version of this essay appeared as "The Case for the Family Farmer," in *Food for People, Not for Profit,* eds. Catherine Lerza and Michael Jacobson (New York: Ballantine, 1975), pp. 35–44, reprinted with permission from Center for Science in the Public Interest, copyright 1975.

The loss will be more than sentimental, for by replacing small entrepreneurs with large corporate enterprises, Reaganculture will make farming less efficient. If this sounds odd, it is because Americans have a fatalistic assumption that bigness is the only course toward efficiency and modernization. Even those who lament the rise of conglomerates in other spheres often reason that there is simply no other way to produce the food we need. It is no surprise that Reaganculture is being marketed on the promise that larger, corporate farms will produce cheaper food than family-sized units.

It's just not so. The family farm in 1986 is no anachronism, no nostalgic remnant of American Gothic. In size, family farms can range from a five-acre tobacco allotment in North Carolina to a five thousand-acre wheat spread in Nebraska. These are not the simple yeomen of Jefferson's ideal. The family farm is defined not in terms of acreage or sales but in terms of independent entrepreneurship. It is small-scale capitalism, alive and performing well — if not flourshing — throughout the countryside.

For all the rhetoric about traditional American verities, the message that the current administration carries across the country is that farming is no longer a way of life — it is a business. That may have a nice ring on the banquet circuit, but it doesn't mean much on the farm. Farming really involves more art than science or business. Producing an abundant, high-quality crop — whether a field of tomatoes or a brood of chickens — simply cannot be reduced to impersonal factory techniques. Farming requires a degree of personal involvement that can't be found either in the corporate board room of General Electric or on a Ford assembly line.

The price, taste, and nutritional value of our food supply depends on the farm family's willingness to plow a bit of itself into the land. Confucius said it centuries ago: "The best fertilizer is the footsteps of the landowner." No amount of chemical spray can replace it.

Since 1952 agricultural efficiency (output per work-hour) on farms has increased by 330 percent, compared with a twenty-year increase of 160 percent in manufacturing industries. That record is unmatched in the rest of the economy. More importantly, the record has been set on average-sized farms, not on the ponderous landholdings of the conglomerates.

Apologists for agricultural gigantism are fond of trotting out the phrase "economies of scale," used by academics to express the relationship between size and efficiency. In agriculture the phrase expresses a very simple economic concept — for every crop there is an optimum size for efficient production. Obviously there are inefficiencies if acreage is too small, but it is often forgotten that there are also inefficiencies if acreage is too large.

"Farming offers no great operating efficiency from large-size units," Harold Breimyer and Wallace Barr concluded. "Scale economy in crop production is handicapped by space and distance—there are cost disadvantages in farming acreages located far from headquarters." Any gains in scale of operation or in integration of economic functions are offset by such cumbersome and inefficient factors as absentee decision making, added levels of management, employment of administrative staff, increased labor bills, and heavier capital investments. The individual farmer or family corporation can meet, and many times surpass, the efficiency of larger units. The family farm turns out to be the optimum size for most crops.

The surprising implication is that today's inefficiencies are more often a product of agricultural bigness than of smallness. For example, in materials submitted in the U.S. Senate migratory labor subcommittee in 1972, the USDA reported that the optimum size for a California vegetable farm was 440 acres. Yet, the average size of the corporate farms that dominate vegetable production in California is 3,206 acres— eight times larger than efficiency warrants. Seventy-three percent of California vegetables are produced on farms that are much larger than optimum size.

But if the family farmer is so efficient, why isn't he or she rich? The farmer's financial problems stem from being squeezed from either side by big business. On one side the pinch comes from "input" suppliers, the companies that sell such necessities as land, seed, fertilizers, pesticides, feed, and machinery. While the price farmers receive for their products has increased only 6 percent since 1952, overhead has risen by a whopping 122 percent.

On the other side the farmer feels the pinch from the "output" corporations that process, market, and retail products. This is a highly concentrated oligopolistic industry, involving such familiar names as Del Monte, General Mills, Ralston Purina, Minute Maid, Kraft, United Brands, Dole, A & P, and Holly Farms. These are the middlepeople, separating the farmer from the consumer and taking two-thirds of the food dollar along the way.

This concentrated power can come down hard on family farmers, who are numerous, overproductive, and unorganized. The farmer has no choice but to sell vegetables, hogs, feed grains, fruits, and other commodities to these corporate processors, marketers, and retailers. In effect, output corporations are able to control the farmer's access to market. This means that a relatively small number of firms are able to dictate what will be produced — as well as how much, what quality, and at what price.

For example, fruit and vegetable farmers technically could bargain with any of twelve hundred U.S. canners. In fact they must deal with just the four canners that control a quarter of the market and set industry practices. Just one of those corporations — Del Monte — boasts of controlling "roughly 16 percent of total U.S. sales" of all canned fruits and vegetables. Within selected commodities, market control is frequently much tighter. A Tenneco subsidiary controls 70 percent of the date industry, while Del Monte accounts for nearly half of all canned peaches. A farmer who does not want to produce peaches on Del Monte's terms can look forward to a long winter of peach eating.

Agribusiness corporations have put even greater pressure on the farmer through the trend toward "vertical integration." In simple terms, this means that a corporation that previously had sold to or bought from a farmer decides to become a farmer itself. In most cases, vertical integration is accomplished through contracts with farmers; the corporation does not become a farmer, it rents one.

Chronic low prices for their product, unavailability of credit, and the sheer market power of the corporate integrator all combine to force farmers into signing contracts that promise a corporate middleman a certain amount at a given price. Corporations can lock up an entire market, forcing farmers either to sign contracts or to go under.

Despite the farmers' resistance, vertical and contractual integration is increasing, affecting nearly a fourth of total farm output. Integration of certain food commodities has gone much further than the norm. Pressed from one side by rising input costs and from the other by output corporations that pay too low a price, and from top and bottom by corporations offering vertical integration contracts, many family farmers are now also confronted with the crushing weight of another corporate force — the conglomerate.

To gauge its impact, take a look at Sunday dinner. It just isn't what it used to be. The turkey is from Greyhound, and the ham is from ITT. The fresh vegetable salad is from Tenneco with lettuce from Dow Chemical. Potatoes by Boeing are placed alongside a roast from John Hancock Mutual Life. The strawberries are by Purex, and there are after-dinner almonds from Getty Oil.

It is raw economic power and favorable government treatment rather than increased efficiency that has given agricultural conglomerates their competitive advantage. The family farmer receives only a small share of the billions of dollars the government spends annually in his or her name. Income and price-support payments are based on volume and acreage, thus putting a premium on bigness and awarding the largest payments to those least in need. Tax-supported research and extension programs have developed a technological arsenal that is rarely adaptable to the family farmer's scale of operation. Tax laws have attracted the noneconomic competition of rich, urban investors looking for a rural tax shelter.

Take irrigation. Water from federal reclamation projects is supposed to go only to farm units of less than 160 acres, but this limitation has never been enforced, allowing some of the largest landholders in the country to prosper from subsidized water. The massive Russian wheat deal was made in the name of farmers, but it was kept secret until the producers had sold their crop cheaply to grain-exporting middlemen.

While tirelessly praising the family farmer over the past thirty years and basking in the farmer's productivity and efficiency, the government has been selling him or her out to corporate America. There has been a lot of handwringing over the elimination of three million farms in three decades, but agricultural economists and other corporate apologists have rushed forward to assure us that it was all "inevitable."

The reason agribusiness corporations are inevitably dominating agriculture is that Earl Butz and Ronald Reagan and all their colleagues have made it so.

Before the family farmer is discarded while still efficient, productive, and competitive, we ought to take a hard look at where we are being led. It is not just the family farm that is threatened by corporate agribusiness and the new agricultural policy, it is the price and quality of the American food supply.

The plan for an integrated, concentrated agriculture is not put forward in the name of corporations, which will profit, but in the name of consumers, who will pay. In May of 1973, Secretary Butz said that "a market-oriented agriculture lets production shift and adjust as consumer preferences change and as foreign demand grows." The bait there is the suggestion that sovereign consumers will have this concentrated economy firmly in hand. Consumer to processor-marketer to farmer is the play that is being diagrammed for the public. We know better. It is in that middle sector of producer-marketers that power really will be concentrated, and both supply and demand will flow from there.

In the new agriculture, what is produced, how much of it, what quality, and what price will be the decisions of these integrated corporations. In a word, oligopoly.

Breimyer and Barr warned, "If farming were to become so highly concentrated that production and marketing of some farm commodities would be confined to a few firms, any economies of size in production would be partly or wholly denied to the consumer." And that's exactly where Butz and Reagan, have us headed.

In a highly concentrated, noncompetitive food system, prices go up, not down. In 1972 the Federal Trade Commission found that consumers were being overcharged more than two billion dollars a year for food because of monopolies within just thirteen food lines. Things promise to get much worse.

Even less likely is the prospect that any savings from a concentrated food system will be passed along to the consumer. Del Monte, for example, in mid-1973 abandoned its asparagus operations in California, Oregon, and Washington and moved them to Mexico, where cheap labor has reduced the corporation's production costs by 45.1 percent. But of course the price of Del Monte asparagus has not gone down. The savings have stayed with the corporation.

The saddest comment to make on these "changes" is that the direction in which they will lead us is already familiar. Butzculture will mean more of the same—hard tomatoes such as the "Red Rock" variety Rutgers University developed, rubbery chickens, hamburgers without meat, and meat filled with additives, pesticides in our food and in our land. Given a laboratory and an adequate advertising budget, the food manufacturers are willing to reshape the American diet. Given a concentrated food industry, they will be in a position to do just that.

Agribusiness has no room for smallness. Nowhere has that been made clearer than in the USDA's effort to alter the official definition of a farm. Under proposals made to the census bureau in 1972 and 1973, families grossing less than five thousand dollars a year in sales of agricultural products would not be considered farmers—even if their sole occupation and sole source of income was farming. The five thousand dollar cutoff level would rule out 56 percent of the farms in the country. This is likely to mean that these families will be overlooked in future farm programs as well. It is an effort by USDA to put the small producer out of sight and out of mind. The collection of demographic data is a kind of Washington magic—there is no farm income problem if there are no statistics to show it.

Having failed millions of small-scale operations during the past thirty years, USDA now washes its hands of them. Out of the department come recommendations that at least half the farmers in the country must abandon agriculture. Where shall they go?

While USDA ignores the needs of small farmers, it gives rapt attention to the wishes of corporate agribusiness. From genetic research programs to Russian wheat deals, USDA is present to promote agricultural bigness, integration, and industrialization. There is a reason for that:

USDA officials are themselves products of agribusiness. Prior to becoming secretary of agriculture, Earl Butz, for example, served as a paid board member of Ralston Purina, Stokely Van Camp, and International Minerals and Chemical Corporation. Prior to becoming secretary of agriculture, Richard Lyng, for example, was a member of the National Livestock and Meat Board.

The greatest tragedy of agribusiness is that it squanders one of our rare opportunities to make progress a friend of human beings. We've grown so accustomed to thinking of any small enterprises as debris in the path of efficiency that we've assumed that sentiment is the small farmer's only defense. In fact, the family farmers can feed us better than the corporations can, unless agribusiness and its friends in the government drive them off the land.

 LUTHER TWEETEN

Has the Family Farm
Been Treated Unjustly?

Introduction

This paper reviews the impact of federal policies on distribution of income and farm structure including size and number of farms. A later section relates the findings to concepts of justice defined in this paper. Before turning to these issues, it is useful to define terms such as "family farm" and "justice."

Some Definitions

I define a family farm as a crop-livestock-producing unit on which the operator and family provide more than half of the labor, management, and equity capital, and on which a significant share of family income is from farming. This definition is consistent with the one used by other authors in this volume. But for my study, availability of data frequently dictates defining the family farm as a unit with $40 thousand to $250 thousand in gross receipts from farming, which differs from the others by including farms in the $200 thousand to $250 thousand range. Thus I will be including a few, slightly larger units in my discussion of family farm policy and justice.

Justice has been defined in many ways. *Commutative justice* is defined as rewarding each person by the value of his or her contribution to society (Brewster 1961). *Distributive justice* has been defined as assuring each member of society at least minimal basic living standards. It also has been defined as giving to each person equal access to the means of developing creative or earnings potential and of establishing the rules for the economic system (Brewster 1961). Combining commutative and dis-

---

LUTHER TWEETEN is Regents Professor of Agricultural Economics at Oklahoma State University.

tributive justice, some define *economic justice* as providing each person with the value of his or her contribution to society but seeing that that contribution is shared to ensure a minimum level of living for those who lack resources. John Rawls in *A Theory of Justice* (1971) defines a just society as the society individuals would choose if they all had an equally random chance of being placed in any role or position in that society. His principles for justice require (1) equality in the assignment of basic rights and duties and (2) that all positions in society be open to all subject to (1). Rawls's principles would not rule out special benefits to the most productive if such benefits increased output to provide a dividend used to help the less advantaged.

Robert Nozick in *Anarchy, State, and Utopia* (1974) argues that a distribution is just if it has arisen in accordance with justice in acquisition and transfer of wealth. Inequality in income and wealth is acceptable, but if people "steal from others, or defraud them, or enslave them, seizing their product and preventing them from living as they choose . . . the resulting distribution will be unjust."

I prefer concepts of justice more suited to scientific inquiry. *Social justice,* as I use it here, is an allocation of resources, goods, and services providing the most well-being to society. By this definition, justice demands allocations of resources such that marginal utility of resources or income move toward similar magnitude for all members of society. In other words, the incremental unit of income or resources is allocated to uses where it contributes most to satisfactions of the people. This movement does not require arbitrary redistributions but can occur gradually through investment in human resources and through progressive income and estate taxes. Such justice may be achieved in a market economy supplemented by public policy to establish the environment needed for a well-functioning market. (See also Tweeten 1979, chap. 16; 1986.)

Public policy can correct for market imperfections by bringing private incentives (prices) in line with social incentives and by ensuring workable competition among firms and input suppliers. Another role for public policy is to provide public goods (defined as those which are not rival or exclusionary). Public goods will not be provided by the private sector alone because private firms often do not see sufficient benefits in providing them. Such goods must be provided by the public sector or by the private sector guided by public intervention.

In addition, the public sector must reallocate resources, goods, and services from those who have low marginal utility of income and resources to those who have high marginal utility of income and resources — the latter are those who have little income and wealth. Operationally, gauging such allocations requires estimates of marginal productivity and marginal utility of income and wealth, a task that I contend social scientists can perform with a usable degree of objectivity and reliability for public policy. (See Tweeten 1986, references.) Such allocation does not imply equal income for everyone because appropriate allocation of resources must balance utility trade-offs between equity and efficiency.

Three propositions flow from this concept of social justice: (1) Transfer payments from persons of lower income/wealth to persons of higher income/wealth are unjustified. If persons with wealth have temporarily low income, they can draw on wealth or income insurance. (2) The appropriate target of transfer payments is not sectors but individuals or families. Sectors are not uniformly destitute. Transfer payments in cash or kind can assist the poor in whatever sector they are found. (3) Justice for family farms cannot be viewed in isolation from justice for society. For example, a public policy of transfer payments and supply controls to provide justice to family farms may substantially raise taxes and food costs. The injustice to taxpayers and consumers may outweigh the justice to the family farm. This is especially true if transfers are from taxpayers of lower wealth to family farmers with higher wealth.

Federal farm commodity legislation and other public policies have had the stated goal of preserving the family farm. We can debate whether social justice is served by a public policy to preserve the family farm; and we can appraise the performance of past public policies in achieving that goal, whatever its merits. Much of this paper examines this second issue. My conclusion is that whatever the justice of preserving or not preserving family farms, public policy taken as a whole has not diminished the family farm.

Public Programs and Policies

This section looks at the impact on farm structure of five broad public programs: federal commodity programs, monetary-fiscal policy to regulate aggregate demand, credit, taxes, and agricultural research and extension.

FEDERAL COMMODITY PROGRAMS. Federal commodity programs may influence farm size, type, and numbers in several often conflicting ways.

1. Payments and price supports help keep marginal farmers financially afloat or attract them to enter farming. Although large farms received more payments per farm in 1984 than did small farms, the contribution to economies of size and farm structure is better measured by direct payments in relation to receipts and net farm income (Table

18.1). Direct payments averaged 5.1 percent of gross farm income and 31.6 percent of net farm income in 1984. Payments as a proportion of gross farm income ranged from only 2 percent on the largest farms and smallest farms to 7.6 percent on farms with gross income of $100,000 to $249,999 in 1984. Payments as a proportion of net farm income ranged from 7.9 percent on the largest farms to a very large percent (net farm income was negative, hence the ratio of payments to net income was infinite) on the smallest farms shown in Table 18.1. Because direct income benefits of programs are especially large relative to farm receipts (not shown in Table 18.1) and net farm income for small- and medium-sized farms, this effect makes for a greater total number of family and smaller farms on the average.

2. Selected features such as payment limitations, special grazing privileges, and minimum allotments have tended to favor small and medium farms, thereby contributing to more such farms.

3. Income stabilization by programs detracts from investment through the so-called permanent income effect. With a given average income over time, more is invested out of a variable income than a stable income. Commodity programs reduce variability of income. Other things being equal including mean farm income, this permanent income effect makes for more farms because less capital substitutes for labor.

4. Idling of cropland leaves production units with excess machinery and labor that can be budgeted to expand acreage and realize economies of size. The result is fewer and larger farms, other things being equal.

5. To the extent that commodity programs add to the level and stability of farm income over time, they reduce the risk of insolvency as a given equity is leveraged further. This encourages farm expansion and makes for fewer, larger farms.

6. Programs create barriers to entry in the form of land values inflated by capitalization of expected program benefits. The impact makes for fewer, larger farms.

7. Allotments for commodities, most notably tobacco, fragment operations and inhibit opportunities to consolidate production on larger units with labor- and cost-saving technology characterized by economies of size. Despite the lower prices and additional output especially to export markets that would attend the absence of allotment restrictions, the net impact of the restrictions is to make for more and smaller farms.

8. To the extent that commodity programs have raised net farm income and that farmers expand operations to achieve a given net farm income, commodity programs contribute to having more and smaller farms.

9. Commodity program benefits in the aggregate for all farm out-

Table 18.1. Distribution of government program benefits among farms classified by sales, United States, 1984

Item	Total	Farms with sales						
		$500,000 or more	$499,000–250,000	$249,999–100,000	$99,999–40,000	$39,999–20,000	$19,999–10,000	Less than $10,000
Share of all farm				*Percentage*				
Operators	100.0	1.3	3.3	9.9	15.2	10.6	11.6	48.1
Direct government payments	100.0	12.4	18.7	35.4	22.2	6.4	2.8	2.1
Gross farm income	100.0	29.1	16.8	23.5	16.1	5.3	3.3	5.9
Net farm income	100.0	49.5	23.6	27.4	8.0	0.4	−1.5	−7.4
				Dollars per farm				
Net income from all sources	28,659	437,501	93,346	42,568	15,791	21,473	16,253	18,575
Net worth	319,287	2,392,052	1,257,829	735,473	426,735	272,527	190,997	119,053
Direct government payments	3,621	33,418	20,560	13,004	5,309	2,200	865	150
Payments as percent of				*Percentage*				
Gross farm income	5.1	2.2	5.7	7.6	7.0	6.2	4.3	1.7
Net farm income	31.6	7.9	25.1	40.8	87.4	559.8	Large	Large

Source: U.S. Department of Agriculture (1986).

put are greatest per unit of output on medium-sized and small-sized farms as noted in (1) above, but commodity program benefits in the form of payments and additions to receipts are larger per producer and per dollar of output *of supported commodities* on larger farms than on smaller farms for each major commodity program (Lin et al. 1981, 22). The larger payment per dollar arises not because large farms are offered more per dollar of output or per unit of diversion than smaller farms, but because participation rates are higher on larger farms (although not on superfarms because of payment limitations). Larger farms participate more (and hence receive more payments per dollar of output of supported commodities), primarily because they reduce costs per unit more than small farms by participating in programs, a modest benefit that is exaggerated by looking at payments per unit of output alone among farm sizes. Nonetheless, the impact is to contribute marginally to economies of size and to incentives for larger, fewer farms.

10. Program coverage is less for commodities produced on very large farms than for commodities produced on small farms. Data by economic sales classes of farms in Table 18.2 provide insight into commodity program distribution among farms. The enterprises supported by commodity programs (wheat, other grains, soybeans, cotton, tobacco, dairy products) shown in Table 18.2 accounted for 44.5 percent of farm output in 1982 but the figure would be a few percentage points higher if peanuts, wool, sugar, and other minor commodities also supported but not shown in the original data source were included.

Only 21.2 percent of receipts on the largest farms are from covered commodities, in part because payment limitations discourage participation in programs and in part because economies of size favor large-scale operations for production of fruits, vegetables, and nursery products and for cattle, calves, and poultry. Although not covered by commodity programs, some of these commodities receive market protection through marketing orders or other negotiated or administered pricing arrangements.

Because of less program coverage of commodities produced on farms with sales of five hundred thousand dollars or more, such farms with 30 percent of all farm receipts received only 7 percent of government payments in 1982 (U.S. Department of Agriculture 1984). Small farms with sales under ten thousand dollars received 7.2 percent of payments and 3.2 percent of all cash receipts in 1982. Inclusion of additions to farm receipts from program-supported prices would not reverse this overall relationship, because such benefits are proportional to receipts.

The smallest farms also rely less on enterprises covered by commod-

Table 18.2. Shares of products sold by economic class of farms, United States, 1982

		Economic class of farms by agricultural products sold						
Commodity	Total	$500,000 or more	$499,999– 250,000	$249,999– 100,000	$99,999– 40,000	$39,999– 20,000	$19,999– 10,000	Less than $10,000
		Percent						
Supported by programs								
Wheat	5.9	2.4	6.1	7.3	9.0	9.7	8.0	4.5
Other grains	13.8	5.6	16.9	19.2	19.0	17.7	14.3	8.7
Soybeans	7.9	2.2	8.7	10.8	12.0	12.4	11.2	7.5
Cotton	2.4	3.5	3.1	1.9	1.6	1.3	1.0	0.5
Tobacco	2.1	0.3	1.4	2.0	3.1	5.6	8.1	8.9
Dairy	12.4	7.2	12.1	17.9	18.6	9.5	3.7	0.8
Subtotal	44.5	21.2	48.3	59.1	63.3	56.2	46.3	30.9
Not supported by programs								
Fruits, vegetables, and nursery	10.5	19.4	8.7	5.5	5.1	6.2	6.5	6.0
Cattle and calves	24.0	36.3	16.5	14.5	16.3	22.8	31.5	45.2
Poultry and products	7.4	11.8	10.6	6.2	2.2	1.0	0.6	0.6
Hogs and pigs	7.5	4.0	9.9	10.1	8.5	7.5	6.7	5.4
Other	6.1	7.3	6.0	4.6	4.6	6.3	8.4	11.9
Total	100.0	100.0	100.0	100.0	100.0	100.0	100.0	100.0
Total ($1,000)	131,900,223	42,764,189	19,851,024	32,930,351	21,641,795	7,142,112	3,694,306	3,565,838
Percent	100.0a	32.4	15.1	25.0	16.4	5.4	2.8	2.7

Source: U.S. Bureau of the Census (1984, 102, 103).
aTotal across columns adds to only 99.8 percent because abnormal farms excluded.

ity programs than do other farms. It is notable that small farms do not emphasize labor-intensive enterprises such as fruits, vegetables, and nursery products but instead emphasize relatively labor-extensive enterprises such as cattle and calves—perhaps because many are part-time operators with limited time for farming.

Midsize farms rely most heavily on enterprises covered by commodity programs and initially would be relatively most disadvantaged by termination of such programs. The implication is that termination of commodity programs would speed the trend toward a bimodal agriculture comprised of large farms accounting for most output and many small part-time farms accounting for most farm numbers. Even with commodity programs, larger farms produce at less economic cost per dollar of total receipts (including government payments) than medium-sized and small-sized farms; termination of programs could accentuate that pattern (Tweeten 1984b, 106). This conclusion of economies of large size for agriculture in total is backed by studies for specific types of farms. (See Smith et al. 1984; Office of Technology Assessment 1986.)

Emphasis on commodity programs for crops produced on smaller-scale farms contributed to having more and smaller farms. Smaller dairy and tobacco farms were retained; without commodity programs, many of these farms would have become cow-calf producing units with large acreages per farm. Furthermore, beef and hog feeding operations were restrained by the higher feed prices resulting from price supports benefiting smaller-scale grain producers. In short, although commodity programs contributed modestly to having larger and fewer farms within each covered commodity as noted in (9) above, commodity programs contributed to having more and smaller farms by favoring retention of the smaller-scale farms specializing in enterprises supported by government commodity programs.

What is the net impact of these conflicting influences of commodity programs on farm size and numbers? The most comprehensive study to date concluded that, "on net, the mass of data, evidence, and professional judgments provides little basis for any conclusion other than that government price and income payment policy has generally been neutral in its effect on farms of varying sizes producing program commodities" (Spitze et al. 1980, 67). Farm commodity programs have been neutral in their overall impact on farm size and numbers for the farming industry as a whole but not necessarily for farms producing commodities covered by any one program. Of course, programs could be redesigned to favor family and smaller farms.

MONETARY-FISCAL POLICY TO REGULATE AGGREGATE DEMAND. Favorable monetary-fiscal policy is defined as one promoting steady economic progress without sizable inflation. Unfavorable monetary-fiscal policy reduces aggregate demand and creates high real interest and exchange rates, inflation, and instability. Unfavorable monetary-fiscal policy tilts comparative advantage to established family farms; renters, including part-owners; corporate industrial-conglomerate farms; and part-time small farms (Tweeten 1983c, 67). The gains in these categories come partly at the expense of entry-level, full-time, family farms; this eventually means fewer full-time family farms. Unfavorable monetary-fiscal policy may also speed separation of land operation from ownership, with ownership tilting toward nonfarm absentee landlords and corporate stockholders.

On the other hand, by slowing growth in real national income and technology, unfavorable monetary-fiscal policy may slow the general trend to larger and fewer farms. I conclude that there may be fewer total farms on the average with favorable monetary-fiscal policy, but a higher proportion of those farms will be closer to the family farm ideal (Tweeten 1983c, 67).

CREDIT. The expansion of farm size probably would have been slower if credit had not been so abundant. Commercial credit for expansion was, in early 1986, most available to larger, established farms with high income and equity.

The Farmers Home Administration (FmHA) is the major federal lending agency (excluding the Commodity Credit Corporation) serving farmers. It was established to provide lending on concessional terms to limited-resource farmers.

At issue is who FmHA serves: large or small farms? And does its lending result in fewer, larger farms? Farms with gross sales under $40 thousand in 1979 accounted for 60 percent of all farms but for only 22 percent of FmHA lending and for 12 percent of farm receipts (U.S. Bureau of the Census 1982). At the other extreme, farms with sales of over $200 thousand accounted for 7 percent of all farms in 1979 but for 23 percent of all FmHA lending and for 50 percent of farm receipts. In 1979 midsize farms accounted for 33 percent of all farms, for 38 percent of farm receipts, and for 55 percent of FmHA lending. FmHA appears to have deviated sharply from its original mandate of serving limited-resource farmers.

The above data are aggregates of all farms and are not specific to farms with FmHA loans. It is useful to relate FmHA lending to net worth and farm income on farms with FmHA loans. Approximately 90 percent of FmHA lending in 1979 was to farm borrowers with net worth of less than $120 thousand under the 'operating and ownership loan programs, but only half of economic emergency loans went to that category of farmers in 1979 (Hughes et al. 1982, 49). Twenty-one percent of economic emergency lending went to farmers with a net worth of over $300 thousand in 1979. About three-fourths of farm operating and ownership program lending, but only a little more than half of economic emergency lending went to farmers with net farm income of under $22 thousand. (Farm income is for 1978; borrowers probably had less income in 1979 to which the loan data apply. On the other hand, failure to include off-farm income seriously underestimates total buying power of farmers in all categories.)

Less needy borrowers, defined as those with more than $22 thousand income and $120 thousand net worth, received 4 percent of FmHA operating loans and 6.5 percent of farm ownership loans but 27.6 percent of economic emergency loans in 1979. Economic emergency loans represent the sharpest departure from focusing FmHA lending on needy farmers. The economic emergency loan program, which accounted for 18 percent of FmHA farm lending in 1979, was established to assist farmers anywhere in the nation experiencing economic difficulties. Lending distribution data are unavailable for a related program (economic disaster loans), which accounted for 42 percent of FmHA farm lending in 1979 and was available only to farms in counties designated as experiencing an economic disaster. The distribution of lending may be similar under the economic disaster and economic emergency loan programs. The two programs, together constituting 60 percent of FmHA farm lending in 1979, account for much of the departure of FmHA from lending to borrowers with low creditworthiness as measured by income and wealth.

Although FmHA clearly does not emphasize lending to smaller farmers, at issue is whether it contributes to economies of farm size and a more concentrated agriculture by greater lending per dollar of output on large than small farms. Shares of lending are greater than shares of sales on all but the largest farms. Farms with sales of less than $40 thousand per year hold four times more volume of FmHA loans per dollar of sales than do farms with sales of over $200 thousand per year. If embodied subsidies are somewhat proportional to lending, it is not possible to conclude that FmHA abets the trend toward fewer, larger farms.

Concessional FmHA lending for water and other community services as well as housing probably encourages sprawl into the countryside as nonfarm people establish rural residences with just enough "farm" attributes to qualify for FmHA concessional loans. The impact of this policy is to create more small farms.

TAXES. Emphasis in this subsection is on federal income and estate taxes. Implications for farm structure of local and state taxes and services are treated only peripherally.

Federal Income Taxes. Several tax provisions applying to income from agriculture enhance the attractiveness of investment in farming (Tweeten 1984a, 36). By reducing the effective capital-labor price ratio below that of a more perfect market, the deductions and credits for capital and taxes on labor distort incentives and cause resource misallocation. [A perfect market is one in which resources are allocated to uses offering highest returns in a situation where private costs (or returns) to individuals and firms do not differ from social costs (or returns) to society.]

The two most prominent distortions are investment tax credit and rapid depreciation allowances. Tax rules allow larger tax write-offs for depreciation than the decline in the market value of the capital asset, the decline in services forthcoming from the asset, or the cost of maintaining full services (including replacement) of the asset. Tax credits and deductions benefit only those with income and are closely correlated with personal savings that are proportionately greater for high-income persons than for low-income persons. It follows that the incidence of benefits from tax credits and deductions is regressive: the value of these features increases more than proportional to income. Some have suggested lowering the real cost of capital, encouraging substitution of capital for labor, and encouraging investment and growth in farm size and aggregate farm output. But these are objectives of dubious worth in an economy troubled by lack of farming opportunities, demise of the family farm, excess capacity to produce, and underemployment.

The above federal income tax features were designed to encourage investment and firm growth. But an element discouraging growth is progressive income tax rates that reduce advantages of large farms, other things being equal. At issue is the net impact of federal tax features. The impact of taxes can be divided into (1) income or cost effects and (2) substitution effects.

Davenport et al. (1982, 44, 45) summarized results of several studies examining economies of size with and without various federal income tax features. Results were ambiguous regarding the contribution of federal taxes to economies of size and farm firm growth.

Based on economic outcomes for a typical commercial wheat-beef cow farm in southwest Oklahoma simulated over a thirty-year operating horizon under various initial tenure-equity positions, *effective* tax rates were nearly constant over all income levels, under the most realistic of scenarios (Tweeten et al. 1984). Thus, federal income taxes had a relatively neutral *income* effect on growth.

Batte and Sonka (1985) recently reported economies of size before and after federal tax provisions on cash grain farms in Illinois. Results indicated significant economies of farm size and no diseconomies of large size within the range of sizes considered. Federal taxes on the whole were progressive and slightly reduced economies of size. Removal of the investment tax credit diminished economies of size, and the credit was more valuable to large farmers than to small farmers. The investment tax credit did not offset the overall progressiveness of federal taxes, however.

Income taxes have greater absolute impact on savings and investment of the high-income-equity farmer than on the low-income-equity farmer. But the growth rate of net worth as measured by savings as a percentage of net worth is slowed more by taxes for the low-income than for the high-income farmer.

Data from tax records analyzed by Sisson (1982) indicated that taxes for farm families were regressive at low-income levels, were proportional over a wide middle-income range, and were progressive at higher incomes. State and local taxes were more regressive at lower-income levels and less progressive at higher-income levels than were federal taxes. Tax rates for farm families were found to average well below those of nonfarm families for income brackets in excess of eight thousand dollars per year. Overall local, state, and federal taxes not only were lower for farmers than for others, but also were less progressive (U.S. Department of Agriculture 1979, 160).

The following conclusions may be drawn concerning federal income taxes:

1. Federal income taxes simulated by laboratory models of representative farms are mildly progressive over most income ranges when consumption is treated as proportional to income.

2. The basic mild progressiveness of federal income tax rates and other features is offset by behavioral patterns of farmers, especially by lower incentives for and incidence of exploiting tax avoidance features by low-income/wealth farmers.

3. The best available evidence and estimates of taxes actually paid suggest that federal income taxes tend to be nearly proportional to income over a considerable range of income.

4. For farmers aggressively using tax avoidance features, effective

federal income tax rates are very low for farms of various income, wealth, size, and tenure arrangements. Federal income tax provisions have, relatively, most favored part-time small farmers and "syndicates" financing, for example, large cattle-feeding operations because farm losses can shelter off-farm income and provide large savings per dollar of farm output. Many of these farms may not be considered family farms because they derive the vast majority of their income from off-farm sources or, in the case of very large farms, hire most of their labor.

5. Although federal income taxes do not seem to influence farm structure markedly by taking proportionately more income or output from one versus another size of farm or influence economies of size per se, tax provisions do influence structure by encouraging substitution of capital for labor. Federal taxes lower the real cost of capital relative to labor. As more capital replaces labor than would occur in a well-functioning market economy, aggregate farm numbers decrease and size increases over time.

Federal Estate Taxes. Changes in federal estate tax law in 1976 and 1981 were designed to assist transfer of farms among generations by not forcing the breakup of family farms to pay estate taxes. Analysis (Tweeten 1984a, 40) suggests that estate taxes have little impact on farm size and numbers but have some small tendency to restrain growth of large farms and to increase family farm numbers relative to large farms.

Most transfers of farmland are for consolidation with an existing farm, and the average unit of land purchased is of smaller size than the unit formed by the buyer. Even a considerable increase in federal estate tax would affect only large farms. It seems likely that such large farms would be broken into parts forming farming units smaller than the original estate farm. Hence an increase in estate taxes might reduce average farm size and increase the number of family farms. Too little is known about income, wealth, and land holdings of buyers and sellers (including estates) of farmland to resolve this issue.

Property Taxes. Property taxes as a percentage of land prices declined in the 1970s but increased as land prices fell in the 1980s. According to Stam and Sibold (1977), property taxes in the farm sector averaged only .6 to 1 percent of market value compared to 1 to 1.5 percent in the nonfarm sector. The relatively lower farmland tax rate raises land values. This creates barriers to entry, shifting a flow expense (tax) to a stock expense (land price). Long-term net income may not be influenced by relatively lower tax rates on farmland, but higher land values are a barrier to entry. Property taxes probably have had little impact on farm size and numbers.

Concluding Comments. Previous studies emphasizing only tax credits and deductions and largely ignoring the growth retarding impact of progressive tax rates overestimated tendencies for federal income taxes to speed farm consolidation. Federal income taxes are low and nearly proportional to income over a considerable range. Income features have benefited part-time farmers relatively most per dollar of farm output, and many part-time farmers are on family-sized farms. Although direct income effects of income taxes have been somewhat neutral among farms by size, the taxes have encouraged substitution of capital for labor. It seems likely that progressive tax rates slowing growth of large farms are more than offset by features encouraging substitution of capital for labor. The net effect of federal income taxes is fewer family farms.

AGRICULTURAL RESEARCH AND EXTENSION. Research and extension education played a major role in the technological revolution in agriculture. The precise role of public research and extension in structural change is unknown, but some clues are offered below.

Technology may be classified as biological (e.g., improved crop varieties), chemical (e.g., improved fertilizers and pesticides), or mechanical (e.g., improved tractors). Technology also may be classified as output increasing or labor saving. In general, labor-saving technologies tend to be scale biased, increasing the scale of farm providing the lowest cost of production per unit. Output-increasing technologies tend to be somewhat scale neutral.

Most publicly supported agricultural production research is scale neutral and emphasizes development of biological technologies. In 1979 an estimated 85 percent of publicly supported agricultural production research was scale neutral (Carter et al. 1981, 7). Of the 15 percent of research that was scale biased, seven percentage points were directed to benefit farms under twenty thousand dollars in sales, five percentage points to benefit midsize farms, and three percentage points to benefit farms with sales of over one hundred thousand dollars. Mechanization research and engineering research accounted for only 1.5 percent and 5 percent respectively of public agricultural research in 1976.

Private research, which accounts for approximately half the research performed in the United States on agricultural technology, has traditionally emphasized mechanical and chemical technologies. Mechanical technologies especially are scale biased toward large farms. The major single source of structural change in U.S. agriculture has been the tractor and its complements — a mechanical technology developed largely by private firms.

Publicly supported agricultural research and extension were major forces underlying the 2 percent average annual increase in output per unit of aggregate farm production inputs from 1950 to 1982. Meanwhile, demand increased nearly as fast, hence aggregate production input was the same in 1982 as in 1950. If output-increasing technology (as measured by total factor productivity) and demand had increased as they did, and if other things had been equal, then farm size, numbers, and aggregate input would have been nearly the same in 1982 as in 1950. However, increasing output reduced product prices, bringing large gains in social justice for low-income consumers.

Output per unit of labor increased over 6 percent annually from 1950 to 1982. The rapid gains in labor productivity stemming mainly from mechanical technologies caused farm size to increase and farm numbers to fall. Attribution of major structural change to technology generated by private firms does not rule out some contribution of public institutions to economies of size. For example, improvements in disease control techniques by public research helped make large-scale poultry and swine production facilities feasible.

Public research will have the greatest payoff to society if directed to reducing food costs and raising reliability of food supplies. It follows that such research will benefit consumers most by working through farmers who account for most of the food and fiber output. As such, social justice and efficiency are served by continuing to emphasize scale neutral research in public institutions, and by making results available to *all* producers (Tweeten 1983a).

If family farms are to compete with large farms and agribusiness firms in the market, able management is essential. Other things being equal, the party possessing the greatest amount of information also possesses the bargaining advantage. Because large farms have sufficient resources and payoff to obtain information from private sources, the family farm needs help with crop and livestock information and market analysis to be competitive. Public extension, management education, and outlook programs help maintain family-sized farms by supplying them with information and expertise that large farms can afford to purchase from private consulting firms.

In short, public research and extension as a whole have helped family farms relative to large farms. In the absence of such research and extension, large farms would account for a greater proportion of all farm numbers and farm output.

Justice in Agricultural Policy

Social justice is served if public programs promote economic efficiency and income transfer from high- to low-income persons in a manner that raises the well-being of society.

DISTRIBUTION AND EQUITY. My definition of social justice ordinarily is not consistent with redistribution of income from lower-income/wealth taxpayers and consumers to higher-income/wealth family farmers. Data reported by Merrill Oster indicate the net worth of the average taxpayer is $136 thousand, a figure to compare with farm net worth shown in Table 18.1. Net worth of commercial farm families exceeded that of nonfarm families in 1984. On the average over the last three decades, commercial family farmers have had at least as high income and wealth per family as the average nonfarm family. Justice is served if public policy plays an active role in helping establish public or private income insurance and other instruments to reduce the annual and cyclical instability endemic to agriculture. Justice is also served by macroeconomic policies pursued to avoid swings in real interest and exchange rates that have devastated many farmers in the 1980s. Justice is not served by large full-employment deficits that provide greater consumption for the current generation while imposing the burden of debt interest payments on future generations.

On the whole, social justice, as I define it, is not served by sector policies. Public policy best provides for basic needs not because people are family farmers or electricians but because they are needy as measured by low income and wealth.

EFFICIENCY. The major benefit of commodity programs, I believe, is to reduce economic instability in the farming economy, in food prices, and in supplies. (See Tweeten 1979.) A chief source of that stabilizing influence is commodity stock reserves and production capacity in excess of what the private market would hold. That reserve has been useful to society at many times in the past including World War II, the Korean conflict, and food crises of 1967–68 and 1973–74. Reserve production capacity must be held on the farms that account for most output. To concentrate the reserve on midsize family farms would essentially remove them from production—hardly a fitting way to preserve them.

Holding reserve production capacity of benefit to society is like purchasing a commodity that the public values and pays for at a price. It is not an income transfer or welfare issue. It can be purchased on the basis of cost-effectiveness: getting the most diversion per public dollar spent whether on large farms or on family farms. A payment to large farms to idle acres essentially is a business transaction involving a quid pro quo. The public offers a demand price and farmers accept or reject the offer based on their appraisal of the price (cost) of supplying the desired commodity—in this case reserve production capacity. The transaction is no more just or unjust than supply-demand deliberations in wheat, beef, or spinach.

A public policy to preserve family farms and to disadvantage large farms is unjust if it reduces the well-being of society. My research and review of other research (Tweeten 1984a) does not indicate that quality of farm life, production efficiency, protection of the environment (e.g., soil conservation), or assurance of plentiful food supplies of high quality and quantity is fostered more by a system of family farms than by a system of large farms. To be sure, most economies of size are realized by family-sized farms. Family farm systems lean toward inefficiency, however, due to greater costs of federal commodity programs, slightly higher food costs, and less exports. This inefficiency would appear to offset the real benefits to rural communities of having a system of family farms versus a system of large farms.

In my judgment, the strongest justification for special programs to preserve family farms is that they are a part of our national heritage. That intangible or incommensurable benefit is little related to justice and is difficult to measure. How much is society willing to pay in federal transfers and higher food costs to preserve the family farm? The Liberty Bell is preserved despite the crack in it; the family farm may also be deemed well worth preserving despite its flaws. If the only purpose of commodity programs is to preserve the family farm, a major restructuring of such programs is in order. Efficient targeting of benefits to such farms would require termination of loan, storage, and supply control features. Direct payments would be emphasized, and those payments would be directed solely to family farms.

Summary and Conclusions

REVIEW OF SPECIFIC POLICIES. The five broad public programs examined here have unevenly served social justice as I have defined it. Public research, education, and extension programs have provided a massive contribution to social justice by reducing costs of food. Relatively greatest benefits accrued to low-income consumers at home and abroad, because they spend a considerable proportion of their income for food. Such programs have also mightily increased national income through efficiency gains, thereby providing a growth dividend that financed a wide array of social welfare programs to provide distributive justice. Public research, education, and extension may have somewhat decreased the overall number of farms but probably increased the proportion of family farms because the focus of public management and marketing efforts has been on family farms.

Unsound monetary-fiscal policy, which unfortunately has been all too common, has perhaps slowed economic and technological progress that would have speeded trends toward larger and fewer farms. But such policies resulting in inflation in the 1970s and farm financial stress in the 1980s probably reduced the number of family farms. The net effect of these opposing influences on farm structure is difficult to judge, but probably is not large.

I conclude that farm credit policies have increased the number and proportion of family farms where a considerable share of their benefits have been concentrated. Such programs have not necessarily served social justice, however, because they have encouraged overproduction, brought lower farm commodity prices but higher land prices, and have encouraged some operators to remain in farming when they could have contributed more to their own and the nation's well-being in other occupations.

Federal tax policies have slowed growth of large farms through progressive tax rates that create diseconomies of scale. They also have encouraged part-time farming, and many part-time farmers are family operations. Estate taxes have modestly discouraged growth of large farms. Tax credits and deductions on capital coupled with taxes on farm labor have encouraged substitution of capital for labor. This not only has encouraged uneconomic substitution of capital for labor and decrease in number and proportion of family farms, but also has caused uneconomic production and low farm returns inconsistent with my concept of social justice.

Public services such as electricity, telephone, water, education, and school bus transportation tend to be subsidized to farm residents, and benefits are largest per unit of output for small- and family-sized farms. These benefits are sometimes offset by regressive state, local, and (perhaps) federal taxes for low-income farmers but not for higher-income family farms.

Commodity programs have not been responsible for social injustice as defined here either by reducing the number of all farms that are family farms or by creating substantial economic inefficiency. But bigger total benefits to large farms than to family farms and transfer of income from lower-income/wealth taxpayers to higher-income/wealth farmers may be viewed by some as inherently unjust.

IMPLICATIONS FOR JUSTICE. We are now in a position to evaluate better the proposition that federal programs have unjustly favored large farms over small farms. Having larger federal program benefits per large farm than per family farm is not unjust if the purpose of commodity programs is to hold reserve capacity and if the purpose of public research and extension is to increase farm production efficiency to benefit low-income consumers by working through farms that account for most farm output.

Benefits of federal programs are lower per dollar of receipts or income on large farms than on family farms. Federal programs on the whole have not contributed to comparative advantage of large farms or to other pressures for fewer and larger farms. Federal policy has not *unjustly* favored large farms relative to family farms, because such programs have not favored large farms more than family farms in terms of greater benefit per unit of output or per dollar of farm income. The relatively greater federal benefits per unit of output and income on family farms than on large farms is consistent with the social objective of preserving family farms. However, even greater focus of benefits on family farms would be appropriate if the only objective of policy is to preserve family farms.

Trade-offs and multiple objectives are inevitable in public policy. The distribution of benefits per farm reflects efficiency considerations (getting most productivity out of research and extension), family farm preservation considerations, as well as pure political muscle. The distribution of federal benefits among farms also reflects the trade-off between research to benefit low-income consumers (through greater productivity on commercial farms) and targeted assistance to preserve family farms. Shifting public investment from agricultural research to income transfers for family farms could be unjust because family farmers would gain less than consumers would lose. Low-income consumers would experience the greatest relative sacrifice.

The contention is without substance that large subsidies (government payments or price supports) to family farms are necessary to avoid concentration of market power on large farms, which would use that power to unjustly raise food prices. Greater broiler industry concentration has brought production efficiencies and lower poultry prices. In the absence of generous parents passing equity to sons and daughters who farm, the family farm could disappear in a generation. But transfers among generations will continue. The family farm is highly resilient and will be around for generations to come although in slowly diminishing numbers.

Although a summary judgment is difficult indeed, I conclude that, on the whole, public programs have not favored large farms over family farms or had a major negative impact on social justice. A major reason for this conclusion is that social benefits of public research, education, and extension overshadow any negative impact of other policies. This is not to say that public programs cannot be improved to serve social justice. A start would be (1) to phase out the investment tax credit and rapid depreciation allowance and (2) to target public program transfers more heavily on farm families with low incomes. Finally, one should note that the major determinants of farm size and numbers have been technology, national economic growth, and off-farm income. These forces overshadow public programs in shaping farm structure (Tweeten 1984a).

References

Batte, Marvin, and Steven Sonka. "Before- and After-Tax Size Economies: An Example for Cash Grain Production in Illinois." *American Journal of Agricultural Economics* 67 (August 1985): 600–608.

Brewster, John. "Society Values and Goals in Respect to Agriculture." In *Goals and Values in Agricultural Policy,* by Center for Agricultural and Rural Development. Ames: Iowa State University Press, 1961.

Carter, Harold, Willard Cochrane, Lee Day, Ron Powers, and Luther Tweeten. "Research and the Family Farm." Prepared for Agricultural Experiment Station Committee on Organization and Policy. Ithaca, N.Y.: Cornell University, 1981.

Davenport, Charles, Michael Boehlje, and David Martin. *The Effects of Tax Policy on American Agriculture.* Agricultural Economics Report No. 480. Washington, D.C.: ERS, USDA, February 1982.

Hughes, Dean W., Stephen Gabriel, Ronald Meekhof, Michael Boehlje, and George Amols. *Financing the Farm Sector in the 1980s.* ERS Staff Report No. AGES820128. Washington, D.C.: National Economics Division, ERS, USDA, February 1982.

Lesher, William. *Farm Policy Update and Perspective.* Presented to American Farm Bureau Federation meeting in Orlando, Florida, 9 January 1984. Washington, D.C.: ERS, USDA, 1984.

Lin, William, James Johnson, and Linda Calvin. *Farm Commodity Programs: Who Participates and Who Gets the Benefits.* Agricultural Economics Report No. 474. Washington, D.C.: ESCS, USDA, September 1981.

Nozick, Robert. *Anarchy, State, and Utopia.* N.Y.: Basic Books, 1974.

Office of Technology Assessment. "Technology, Public Policy, and the Changing Structure of Agriculture." Report to U.S. Congress. Washington, D.C.: Government Printing Office, February 1986.

Rawls, John. *A Theory of Justice.* Cambridge, Mass.: Belknap Press of Harvard University Press, 1971.

Sisson, Charles A. *Tax Burdens in American Agriculture.* Ames: Iowa State University Press, 1982.

Smith, E. D., J. W. Richardson, and R. D. Knutson. *Cost and Pecuniary Econo-*

mies in Cotton Production and Marketing. D-1475. College Station: Texas Agricultural Experiment Station, Texas A & M University, August 1984.

Spitze, Robert, Daryll Ray, Allen Walter, and Jerry West. "Public Agricultural Food Policies and Small Farms." Paper 1 of NRC Small Farms Project. Washington, D.C.: National Rural Center, 1980.

Stam, Jerome, and Ann Sibold. *Agriculture and the Property Tax.* Agricultural Economics Report No. 392. Washington, D.C.: ERS, USDA, 1977.

Tweeten, Luther. *Causes and Consequences of Structural Change in the Farming Industry.* NPA Report No. 207. Washington, D.C.: National Planning Association, 1984a.

_____. "Diagnosing and Treating Farm Problems." In *Trade Policy Perspectives: Setting the Stage for 1985 Agricultural Legislation,* 95–123. Committee on Agriculture, Nutrition, and Forestry, U.S. Senate, 98th Cong., 2d sess. Washington, D.C.: Government Printing Office, December 1984b.

_____. "Domestic Food and Agricultural Policy Research Priorities." In *Interdependencies of Agriculture and Rural Communities in the 21st Century: The North Central Region,* edited by William D. Heffernan and Rex R. Campbell. Ames: Iowa State University Press, 1986.

_____. "The Economics of Small Farms." *Science* 219 (March 1983a): 1037–41.

_____. "Farm Financial Stress, Structure of Agriculture, and Public Policy." In *U.S. Agricultural Policy, the 1985 Farm Legislation,* edited by Bruce Gardner, 83–112. Washington, D.C.: American Enterprise Institute for Public Policy, 1985.

_____. "Food for People and Profit: Ethics and Capitalism." *FS5* in *The Farm and Food System in Transition, Emerging Policy Issues.* Resource paper sponsored by the Extension Committee on Policy, USDA-Extension and the MSU Cooperative Extension Service. East Lansing: Cooperative Extension Service, Michigan State University, 1983b.

_____. *Foundations of Farm Policy.* Lincoln: University of Nebraska Press, 1979.

_____. "Impacts of Federal Fiscal-Monetary Policy on Farm Structure." *Southern Journal of Agricultural Economics* 15 (July 1983c): 61–68.

_____, Tom Barclay, David Pyles, and Stanley Ralstin. "Simulated Farm Firm Growth and Survivability." Research Report P-848. Stillwater, Oklahoma Agricultural Experiment Station, 1984.

U.S. Bureau of the Census. *1982 Census of Agriculture,* Vol. 1, Pt. 51. AC82-A-51. Washington, D.C.: Government Printing Office, October 1984.

_____. "1979 Farm Finance Survey." *1978 Census of Agriculture,* Vol. 5, Pt. 6. AC78-SR-6. Washington, D.C.: Government Printing Office, July 1982.

U.S. Department of Agriculture. *Economic Indicators of the Farm Sector: Income and Balance Sheet Statistics, 1983.* ECIFS 3-3. Washington, D.C.: ERS, USDA, September 1984.

_____. *Economic Indicators of the Farm Sector: National Financial Summary, 1984.* ECIFS 4-3. Washington, D.C.: ERS, USDA, January 1986.

_____. *Structural Issues of American Agriculture.* Washington, D.C.: ESCS, USDA, November 1979.

RUSSELL C. PARKER
AND JOHN M. CONNOR

Consumer Loss Due to Monopoly in Food Manufacturing

THE ESSENCE OF MARKET POWER is the ability of a firm to raise the price of its product above the level that would obtain if its market were more competitive. Of the numerous, still tenable theories of oligopoly developed by economists over the last 140 years, all predict higher long-run average prices where seller concentration is high (Weiss 1974), product differentiation is effective (Comanor and Wilson 1974), and/or barriers forestall the entry of new businesses into the market (Bain 1956). The major empirical tests of these hypotheses have generally established statistical relationships between the aforementioned market imperfections and higher profits, an indirect effect of raising prices.

Most major industrial groups are composed of industries that produce mixtures of high- and low-priced products of varying durability and that employ diverse channels of distribution. Food and kindred products manufacturing, on the other hand, is characterized by low-unit-priced, high-turnover consumer products sold mainly through grocery stores. To the extent that some of the food industries sell to other industries such as flour and sugar, customers are mainly other food industries. These characteristics, plus the fact that imports and exports are slight, mean that the food and kindred products industries as a group are largely self-contained and do not present the empirical researcher with the need for complex model specification.

A review of the elements of market structure and performance of the food-manufacturing industries provides ample support for the likelihood of substantial departures from pure competition (Connor 1979; Parker 1976). Not only are market shares, sales concentration, advertis-

RUSSELL C. PARKER was a member of the Food System Research Group at the University of Wisconsin when this essay was written. JOHN M. CONNOR is assistant head for research in the Department of Agricultural Economics at Purdue University. This essay is taken from "Estimates of Consumer Loss Due to Monopoly in the U.S. Food-Manufacturing Industries," *American Journal of Agricultural Economics* 61 (Nov. 1979): 626–39, and is reprinted with permission of the authors.

ing, other product differentiation expenditures, and profits high, but also they have been increasing steadily over the last three decades. The processing and manufacturing of food products occurs in some forty-seven four-digit industries whose competitive structures are quite varied; but in the main, they are consumer products industries dominated by companies with high market shares. Increasingly, these leading firms have become consolidated, through merger, into the hands of a few, very large, conglomerate enterprises. Just fifty firms accounted for 64 percent of food manufacturers' assets in 1978, up from 41 percent in 1950. Concentration of profits, sales promotion activities, and the holding of leading positions by these fifty firms is substantially higher, ranging upward to 90 percent.

The weighted average four-firm concentration ratio for all national-market food-processing industries in 1972 was 43 percent, which is slightly higher than the average of national market industries in the remainder of manufacturing. Adding local market industries (fluid milk, ice cream, animal feeds, bread, and soft drink bottling) and substituting average local for national concentration, the shipments-weighted-average food industry concentration ratio was 52 percent, indicating a significant degree of oligopoly and potential for competitive problems. Scherer (1970), in his review of the effects of concentration on market performance, concluded that "when the leading four firms control 40 percent or more of the total market, it is fair to assume that oligopoly is beginning to rear its head" (p. 60).

Promotional activities have accounted for a significant increase of the costs of food manufacturers since World War II. Some methods of sales promotion such as coupons and free samples provide consumers with price discounts while creating consumer loyalty. Other promotional devices such as cooperative advertising allowances, special in-store displays, shelf-stocking services, delivery, and information reduce the costs of retailers or wholesalers. If passed on, these devices would also represent price reductions to consumers. But the bulk of promotional activities currently used for food products (market surveys and test marketing, package design and elaborate packaging materials, point-of-sale advertising, direct mailings, and media advertising) are largely self-canceling and simply add to distribution costs. IRS data indicate that in 1975 total U.S. advertising expenditures (which include some nonadvertising promotional outlays) by food manufacturers were $4.1 billion. (This figure includes nonfood advertising by the companies but excludes a substantial amount of food advertising by soap and drug companies, other manufacturers, conglomerates, food wholesalers and retailers, and others.)

Media advertising (TV, radio, magazines, newspapers, and outdoor) of food typically accounts for more than 60 percent of total food advertising expenditures (Connor 1979). Media advertising of processed foods has risen by an average of 10 percent per year from 1950 to 1975. Because advertising expenses have increased at a much faster rate than sales, the advertising-to-sales ratio of food processors has increased also, from only 1.1 percent in 1947 to about 2.5 percent in 1975, with most of the increase accounted for by television advertising.

Persistently high profit rates are another index of monopoly pricing. The profits after taxes, as a percentage of stockholder equity of food manufacturers, have experienced a quarter-century increase of more than 50 percent. They also have risen relative to profit rates in the rest of manufacturing, itself hardly a model of competition.

The central purpose of this paper is to compute and compare annual "consumer loss" estimates for the U.S. food-manufacturing industries using three different and independent methodological approaches. (We use "monopoly" to encompass not only the one-firm case but oligopoly as well.) These procedures also make use of widely different data bases, two of which have never previously been employed for computing monopoly losses. The estimates display a significant convergence, twelve to fourteen billion dollars for 1975. These large dollar amounts suggest a high payoff for increased public policy attention to competitive problems in the food-manufacturing industries. . . .

• • •

Summary and Conclusions

A summary of the results of the three statistical estimates (as described in our full 1979 study) is shown in Table 19.1. We believe these estimates of consumer loss due to monopoly are conservative. However, because of the data sets used and because of the estimating procedures employed, a considerable degree of error is likely for each of the estimates. Scherer, for the whole economy, thought that the maximum error would fall within the range of 50 percent less to 100 percent greater than his estimate. The present authors have used statistically fitted structure-performance relationships and data specific to food manufacturing to estimate Scherer categories comprising about 60 percent of the total consumer loss estimate. The estimated values of these components are believed to be more reliable than Scherer's original values and should tend to reduce the error range of the overall loss estimate. The authors have no method for estimating the likely error range of their two additional methods. They feel, however, that a 25 percent error on the two estimates is the most that would reasonably be expected. The extent of convergence of all three essentially independent estimates gives strength to the conclusion that consumer loss due to monopoly in the U.S. food-manufacturing industries in 1975 was at least $10 billion but possibly as high as $15 billion.

To put these estimates in perspective, note that Scherer ventured that the monopoly loss to consumers in the whole economy was 9 percent of GNP; with U.S. GNP running at $1,500 billion in 1975 (Council of Economic Advisors 1977), this implies an overcharge of some $135 billion (or $120 billion if he meant to include only the private sector of the economy). For food manufacturing, our estimate is equal to one-fourth of the total GNP (value added) attributed to the sector. However, put another way, our monopoly loss estimate for processed foods represents 1.1 percent of U.S. personal disposable income and about 5.7 percent of average U.S household food expenditures in 1975.

There are significant implications of our monopoly loss estimates for public policy. The annual loss to consumers in food manufacturing alone is 250 times the combined antitrust budgets of both U.S. antitrust agencies, and is several thousand times that part of federal antitrust expenditures.

Besides demonstrating that food processing ought to have a high priority for the antitrust agencies, the findings of our price-cost margin and national brand-private label price models suggested that advertising represents an important problem area for consumer products. Furthermore, the problem is most serious when TV is the primary medium and when the advertisers are large firms. This suggests that consideration be given to limiting advertising in industries where it is already intense and to formulating stricter policies that would discourage product extension mergers where differentiated consumer products are involved. This latter suggestion is further supported by evidence that food-manufacturing mergers, on average, exhibit a subsequent doubling of advertising outlays (NCFM 1966), often accompanied by a shift toward greater use of

Table 19.1. Summary of monopoly loss estimates in U.S. food manufacturing, 1975

Type of consumer loss	Loss components (adjusted)	Price-cost margin	Private label differential
		($ millions)	
Monopoly profit	3,613	} 12,933	11,877
X-inefficiency	8,480		
Allocative	430	823	559
Total	12,523	13,756	12,436

TV. An antitrust policy or a new law that reduces food company mergers, especially takeovers by leading grocery product firms and other conglomerates, would be expected to moderate the market power of sellers by its effect on both concentration and advertising.

However, we recognize that neither advertising restrictions nor merger prohibitions may erode this existing market power at sufficient speed to achieve workable competition in all food-manufacturing industries. Under these circumstances, therefore, more direct restructuring may be necessary. Such restructuring could take the form of divestiture of portions of the physical assets of leading firms, compulsory licensing of major trademarks, or other affirmative programs to encourage the entry of firms into the affected markets.

References

Bain, Joe E. *Barriers to New Competition.* Cambridge, Mass.: Harvard University Press, 1956.

Comanor, William S., and Thomas Wilson. *Advertising and Market Power.* Cambridge, Mass.: Harvard University Press, 1974.

Connor, John M. *Competition and the Role of the Largest Firms in the U.S. Food and Tobacco Industries.* Washington, D.C.: ESCS, USDA, 1979.

Council of Economic Advisors. *Economic Report to the President 1976.* Washington, D.C., 1977.

National Commission on Food Marketing (NCFM). *The Structure of Food Manufacturing.* Tech. Stud. No. 8. Washington, D.C., 1966.

Parker, Russell C. *The Status of Competition in the Food Manufacturing and Food Retailing Industries.* Work. Pap. No. 6. NC-117, 1976.

Scherer, F. M. *Industrial Market Structure and Economic Performance.* Chicago: Rand-McNally & Co., 1970.

Weiss, Leonard. "The Concentration Profit Relationship and Antitrust." In *Industrial Concentration: The New Learning,* edited by Harvey J. Goldschmid, H. Michael Mann, and J. Fred Weston. Boston: Little, Brown & Co., 1974.

20

CATHERINE LERZA
AND MICHAEL JACOBSON

Food for People, Not for Profit

FORMER SECRETARY OF AGRICULTURE Earl Butz was not often credited with making particularly sage remarks. In the summer of 1974, however, he made a statement that rang with truth: "Food is power," he said. But we ask: who *has* that power?

Corporations have it. In the United States and other industrialized nations, agribusiness firms have adopted an economic philosophy that economist Herman Daly calls "growthmania." According to Daly, growthmania assumes "there is no such thing as enough." This quest for perpetual economic growth has created, among other items, gasoline-guzzling automobiles, no-deposit, no-return beverage containers, and artificially flavored and colored, sugar-coated breakfast cereals. It has also sent American corporations abroad in search of new and bigger markets for their goods. Japan and Western Europe, the most lucrative markets, have been carefully courted and cultivated, but underdeveloped nations have not been overlooked as potential markets. Coca-Cola, Kellogg, and other brand names familiar to Americans are seen in even the remotest regions of the world. These corporations literally cash in on the desire of poor people around the world to emulate the "modern" lifestyle of affluent nations.

Food corporations have the power Butz described. Firms have significantly altered traditional infant feeding practices in Third World countries where some mothers were persuaded that canned formula was a better infant food than their own breast milk. Although mother's milk is the least expensive, most healthful way to feed most infants, Western-owned food corporations encouraged mothers to buy formula, even go-

CATHERINE LERZA is former editor of *Environmental Action*. MICHAEL JACOBSON is codirector of the Center for Science in the Public Interest. This essay is taken from the introduction to *Food for People, Not for Profit*, eds. Catherine Lerza and Michael Jacobson (New York: Ballantine, 1975), reprinted with permission from Center for Science in the Public Interest, copyright 1975.

ing so far as to dress their sales representatives as nurses. Poorly educated mothers frequently overdiluted formula with water—water that was often contaminated. Despite the economic hardships and public health risks involved in formula feeding, food corporations continue their global sales pitches.

Yet another symbol of "modernity" is the soft drink, now found in the farthest corners of even the poorest countries. Soda pop invariably displaces milk and other more nutritious beverages and foods from indigenous diets. Like canned formula, soda is expensive, and usually produced by foreign-owned corporations. Thus, even the profits from sales accrue to foreign rather than local manufacturers and are lost to the domestic economy.

Food aid notwithstanding, the international food gap can be closed only if hungry nations increase their domestic food production and strive for self-sufficiency. China has shown that this can be done, even though it is the most populous nation on earth. France and Bulgaria have also substantially increased food production. The experiences of these three nations, as well as those of countries that have failed to improve production, demonstrate that meaningful advances in production are probably impossible without land reform, cooperative efforts, and other significant political changes. Traditionally, however, the interests of multinational corporations have been served by the status quo rather than by efforts toward political change. Global corporations are glad to help Third World countries increase agricultural production by selling them seeds, tractors, fertilizer, and agricultural chemicals—the trappings of large-scale, mechanized agribusiness—but they continue to present serious obstacles to the political and social changes necessary before technical expertise can be used to increase food production in a way that will benefit native populations.

The economic well-being of large corporations and landowners is not necessarily tied to the economic well-being of the native population. Copying American methods, agricultural production is being mechanized and has become highly dependent on the costly fossil fuels needed to power tractors and produce pesticides, herbicides, and fertilizers. These large agricultural concerns usually do not grow crops that are consumed by the native population. Instead, they produce "cash crops" such as bananas, coffee, tea, spices, sugar, and cocoa, all of which are sold to wealthier nations. Of course, the profits of cash crop production do not "trickle down" to the poor, but instead remain with the wealthy or find their way into the pockets of foreign-owned corporations.

The rise in mechanized, capital-intensive agriculture and the increasing influence of global corporations often results in peasant landowners or tenant farmers being forced off the land. If they have been mechanized out of an agriculture-related job, they often wind up in cities. In some cases filthy urban conditions breed disease; and lack of social services breeds continued illiteracy—and little hope of ever escaping.

American foreign policy has also been shaped by the influence of multinational corporations. The United States has "intervened" militarily in foreign countries on an average of at least once every two years since World War II, and a number of these interventions—as in the CIA-instigated overthrow of a democratically elected government in Guatemala in 1954 or the abortive attempt to oust Castro's regime from Cuba in 1961—were against governments pledged to undertake extensive land reform at the expense of American corporations (that of the United Fruit Company in the case of Guatemala). U.S. involvement in the ouster of Salvadore Allende in Chile in 1973 is an example of armed opposition to a government committed to land reform.

If the power of global food corporations is great abroad, it is enormous here in the United States. Before World War II, our food was produced by independent, family farm operations and eventually marketed in small mom-and-pop-style grocery stores. Today, food corporations have displaced small-scale entrepreneurs all along the production/marketing chain.

Although big is not necessarily bad, numerous instances of corporate abuse of power have raised the suspicion that American consumers are being shortchanged by these industry giants. To decide when the power of a corporation or corporations is too great within an industry, economists use a simple guideline: if four (or fewer) corporations control 50 percent or more of a given market, they are termed an oligopoly—a shared monopoly. In such situations, common in the food industry, cooperative "competition" results in price-fixing, and prevents newcomers, who may offer real competition, from entering the market. In 1974, then Attorney General William Saxbe announced that the justice department would look into at least eight different food industries to uncover possible violations of the law. In the same year, the justice department indicted a sugar refiner for price-fixing in the Midwest and West, while the Federal Trade Commission (FTC) issued a formal complaint against ITT-Continental Baking for attempting to monopolize the wholesale baking industry. In a 1972 antitrust complaint, the FTC charged Kellogg, General Mills, General Foods, and Quaker Oats with monopolizing the breakfast cereal industry.

Corporate concentration costs consumers money. (See Parker and Connor's "Consumer Loss Due to Monopoly in Food Manufacturing,"

Chap. 19.) The FTC estimated in 1972 that overcharges resulting from monopoly in seventeen food lines cost consumers $2.6 billion out of $65 billion worth of sales. Another earlier FTC study found that, during the 1950s, Seattle residents had been bilked out of $30 million as a result of price-fixing by local bakers. It is difficult to estimate precisely the total cost to consumers of industry gouging, but it is probably at least $100 per family each year. It is ironic that, whenever the federal government prepares an action to eliminate such practices, industry executives speak sanctimoniously about the free enterprise system and the spirit of competition.

Farmers also suffer economically from the widespread, concentrated power of food monopolies and oligopolies. When one processor or wholesaler dominates a regional or product market, farmers have little leverage in determining the sale price of their livestock or crops. While consumers are forced to pay higher prices for food, farmers are being squeezed out of their livelihood.

Those family farmers who remain on the land are not immune to the economic power of agribusiness corporations. With increasing regularity, once-independent farmers now work under contract to agribusiness corporations. The big food producers can dictate terms to farmers who must toe the corporate line in order to find a market for their crops. The family farmer, still operating under the myth of "free enterprise," cannot expect to compete in the marketplace.

A prime example of the way in which agribusiness corporations have relegated once-independent farmers to the role of tenant is the corporate takeover of chicken production. Chicken processors such as Holly Farms and Safeway have removed competition from the chicken industry, while maintaining what is, for them, the most profitable element: cheap human labor in the form of farmers who do the growing for them.

Agribusiness enjoys other advantages not available to family farmers. Interlocking corporate directorates mean that a food company executive may also sit on the board of a bank, a feed company, or an agrichemical company. Such cozy arrangements mean that these companies will be more cooperative than competitive. These "interlocks" extend beyond the corporate level. In 1973, Robert Long, former agricultural loan officer of the mammoth Bank of America, a major agricultural landholder in California, became an assistant secretary of agriculture. Soon after arriving at USDA, Long predicted that agriculture would progress toward fewer and fewer units, and those remaining would find it easier to work together for greater profits. If such a blatant instance of incest occurs within the upper echelons of the USDA, it is clear that the interests of the family farmer play no part in governmental and economic decision making.

An agribusiness corporation can also afford to undercut independent farmers by operating just one of its divisions at a loss over an extended period of time. These losses are covered by revenues generated in other parts of the corporation. Such undercutting either forces small farmers completely out of the picture or into contractual peonage. Thus, the myth perpetuated by agribusiness that its operations are more "efficient" than those of the small farmers, is a function of economic power rather than actual fact. A 1967 USDA report, *The Economies of Scale in Farming,* concludes that small, one- or two-person farms are just as, and often more, efficient than massive corporate farms.

Agribusiness is the recipient of much taxpayer-funded agricultural research via the land grant college system. Jim Hightower's important study, *Hard Tomatoes, Hard Times,* documents the way in which the land grant colleges, along with companion organizations like the agricultural experiment stations and state agricultural research services, have served corporate interests rather than the needs of independent farmers and consumers. (See Chap. 12.) According to Hightower, in 1969 the land grant system invested only 289 of 6,000 scientific person-years into what the USDA terms "people-oriented" research. For instance, much land grant research has gone into developing highly mechanized farming methods, as well as crops (e.g., the "Red Rock" tomato), which can be picked by machine instead of by hand. "The primary beneficiaries of land grant research are agribusiness corporations. These interests envision rural America as a factory that will produce food, fiber, and profits on a corporate assembly line," Hightower wrote.

There is no way to compute the social costs of the loss of agriculture as a way of life for a significant sector of the American population. Likewise, there is no easy way to translate into economic terms the environmental degradation resulting from agribusiness farming methods. Natural systems and their inherent capacity to recycle nutrients from one system for use in another are often disrupted by agricultural technologies. For instance, when cattle are allowed to graze on pastureland, their waste becomes manure that replenishes the soil and insures the growth of future forage. When the same cattle are crowded into a feedlot and fed high-protein cereals, the accumulation of their excreta creates disposal and pollution problems. The concentration of cow manure results in the contamination of both groundwater supplies and streams and rivers. In fact, feedlots produce about 750 million tons of waste every year, and one 10,000-animal lot produces an amount of sewage equivalent to a city of 164,000 people.

Modern farming methods rely heavily on agrichemicals: pesticides, herbicides, fertilizers. Although these substances have helped to improve yield-per-acre in the short run, their long-term effects on ecosystems, soil quality, and human health have not yet been determined. Extensive use of chemical fertilizers, especially nitrogen fertilizers, results in reduced amounts of organic material in the soil, and thus, lower soil quality. U.S. fertilizer use in 1970 averaged 150 pounds per acre, while in some regions of the country, Texas's Rio Grande Valley, for example, application reached 800 pounds per acre. Chemical fertilizers were virtually unused before 1940.

Use of insecticides and herbicides locks farmers into an ever-mounting spiral of application. As target pests develop resistance to these chemical poisons, farmers must use larger and larger amounts of the products. These increased doses often mean that other, nontarget pests will be damaged and the human beings who apply agrichemicals or who work in recently sprayed fields may suffer illness or death. (The Department of Health, Education, and Welfare estimated in 1969 that eight hundred farm workers died and some eighty thousand were injured due to pesticide poisoning yearly.) Many of these chemicals have a long biological life and, once they enter an ecosystem, will remain there for long periods of time.

Finally, we must ask *why* agribusiness corporations have the power they do. The primary responsibility of private corporations is to earn money for their stockholder-owners. Their responsibility to the public, be it eliminating air or water pollution or contributing to worthy charitable causes, is seen as secondary. But government, the institution created by society to serve its best interests, is supposed to place those interests first. It should also insure that corporations respect the public interest, while they satisfy their own special interests.

Unfortunately, the lesson of the past has been that our government is not protecting the public interest, and is, in fact, promoting private interests at the public's expense. Watergate, Vietnam, special favors to ITT, illegal campaign contributions, dairy subsidies, and domestic spying offer proof that personal gain, to both government officials and corporations, has been a major force behind much government policy making.

A symptom of the "you-scratch-my-back-and-I'll-scratch-yours" attitude that exists between corporations and the federal government is the extent to which former corporate officers serve as officials in agencies that regulate the activities of the industry of which they were once a part, and to which they may eventually return. This cross-pollination process works both ways, as many federal officials leave their posts for jobs in the industries they oversee. The top positions of the Food and Drug Administration (FDA), the agency charged with regulating food safety and labeling, have been rife with such conflicts for years. A 1969 congressional study showed that thirty-seven of forty-nine of the top FDA officials who had recently left the agency went on to take jobs with food or drug companies. In 1974, Virgil Wodicka, director of the Bureau of Foods, quit to become a private consultant to industry; prior to his five-year sojourn with FDA, Wodicka worked for Ralston Purina; Libby, McNeill and Libby; and Hunt-Wesson (where he was a vice-president). The nutrition director of the FDA also left his post in 1974 and joined the Hershey Corporation. In prior years, the same pattern was repeated. The FDA's general counsel previously represented ITT-Continental Baking, dairy interests, and the chewing gum industry. The man he replaced as general counsel, William Goodrich, left the FDA to become the president of the Institute of Shortening and Edible Oils, an industry lobby. The man Goodrich replaced at the institute was a former assistant commissioner of the FDA.

The situation at USDA has always mirrored that of FDA. The secretary of agriculture from 1971 to 1976, Earl Butz, was formerly a director of several agribusiness corporations including Ralston Purina and Stokely Van Camp. Several of the top officials in USDA jumped ship in the early seventies for the lusher environs of Continental Grain, Miles Laboratories, and other corporations. In 1985 Ronald Reagan appointed Richard Lyng as secretary of agriculture. Lyng was formerly a member of the National Livestock and Meat Board.

Long employment in industry can be expected to etch deeply in anyone's mind a special respect for the problems and needs of private corporations. Knowing the employment history of many federal agency officials, it is not unrealistic to expect that on the day an official receives, for instance, information about the health problems inherent in eating excessive amounts of refined sugar, the officials may be negotiating with sugar refiners or industrial users about a future position within those industries or perhaps even accepting such a position. The chance that responsible regulatory action will ensue from that agency diminishes considerably. It is also not unrealistic to believe that such an individual has less than a total commitment to defending the public interest.

This is not to say that people formerly employed by industry should be barred from government employment. Rather, their expertise should be called upon, as should that of people who have been involved in public interest work.

Even more significant than the questionable allegiance of government decision makers to the public interest is the basic political and economic power wielded by food corporations. In the early seventies total sales of the food and beverage (including alcoholic beverage) industries amounted to approximately $154 billion a year. This obviously gives industry representatives political leverage that can be used on countless members of Congress, in the White House, and in federal administrative agencies.

If elected and appointed government officials have not been speaking the unvarnished truth or have not been fighting for the consumer's interests, one can always turn to professors at the nation's great universities for unbiased analyses of complex problems — or so one might think. Sad to say, corporate power reaches deep into universities, influencing curricula, the contents of textbooks, and the views of professors. Food and chemical corporations have a clear interest in what nutrition, chemistry, food science, and agriculture students are taught. Companies hire professors as consultants, pay them retainers so they may ask them for routine advice, offer their students summer and permanent jobs, provide grants to support research (these are especially important at times when government and foundation money is scarce), and offer every kindness at scientific meetings and conventions. Students at the Harvard School of Public Health are reminded of General Foods' benevolence every time they pass a plaque, mounted in the main stairwell, that thanks that company for helping pay for the research laboratories.

While taking research money or a retainer from industry may not cause professors to falsify experimental results or prostitute their beliefs, it does seem to prevent academics from criticizing "benevolent" companies or finding fault with the corporate philanthropist's products. It may be that the neutralization of potential criticisms is the public's greatest loss.

With losses of this sort, the people lose power, and food becomes a tool in the hands of the wealthy few. The result? Food is grown for profits, not for people.

LUTHER TWEETEN

Food for People and Profit

CRITICS HAVE CHARACTERIZED the food system as being made up of soulless corporations that reap obscene profits while providing unfair returns to family farmers and exploiting workers. Describing our food supply system, Lerza and Jacobson state in Chapter 20 of this volume that food corporations "literally cash in on the desire(s) of poor people," adding that "while consumers are forced to pay higher prices for food, farmers are being squeezed out of their livelihood." These critics of the food system say food should be for people and not for profit. (See also Robbins 1974.)

These charges of immorality must be answered on grounds of moral philosophy and not just of economics. That is the purpose of this paper. Few critics are saying the *people* supplying food are immoral. Rather, the food *system* is on trial. That system includes input supply, farm and marketing firms, and industries, along with the public sector that serves the system. A food system out of step with the moral philosophy of the nation will not survive. But is it out of step?

In the following pages, I first briefly review a norm of moral philosophy used to judge performance of the food system. The remainder of the report examines two systems, the market sector and the public sector, for their ethical promise and performance in serving people.

Moral Philosophy

To judge morality, a norm is essential. I contend that the appropriate norm to use in judging the food system is utilitarianism, perhaps the

LUTHER TWEETEN is Regents Professor of Agricultural Economics at Oklahoma State University. This essay, with the subtitle "Ethics and Capitalism," was originally published as one of the resource papers in a series called *The Farm and Food in Transition* sponsored by the Extension Committee on Policy, USDA-Extension, Michigan State University Cooperative Extension Service, and other universities and organizations.

most widely shared ethical system in America. Utilitarianism holds that judgments of right or wrong ought to be based on what is variously termed well-being, satisfaction, quality of life, welfare, avoidance of pain, pursuit of happiness, or the greatest good for the greatest number of people.

As a norm of food system performance, utilitarianism raises several questions.

1. Whose utility should be increased? Utility applies to "society," but who or what is society? The answer is critical because utility can be increased for one individual, interest group, or nation at the expense of others. Does "society" include foreigners and illegal aliens, unborn generations, and animals? The last is not a frivolous addition as evidenced by the emerging animal rights movement. The issue of whom to include is not resolved here, but I proceed on the assumption that society includes all Americans. Later I will turn to the issue of balancing interests of individuals with those of society.

2. Can utility be measured? If utilitarianism is to be used to judge whether actions are right or wrong, then satisfactions must be measurable. Measurement need not be precisely quantitative, however. In general, individuals are presumed to be rational in determining and acting upon what gives them pleasure or pain. A more troublesome issue is whether one individual or group can measure the utility of another individual or group. Elected officials repeatedly must make such judgments. Implicitly, so does the market price system. Psychologists and sociologists have devised sophisticated attitudinal scales to measure well-being, but the predictive power of such scales is limited. Although useful comparisons can be made among groups of individuals, the conclusion is that utility can be measured only imperfectly.

3. Is utilitarianism inconsistent with our humanist and Judeo-Christian heritage and national civil codes? Religious and civil laws providing moral imperatives to judge right or wrong would seem to conflict with the utilitarian philosophy that acts are not intrinsically moral or immoral but must be judged according to their consequences for the well-being of people.

Our nation's laws, formulated under the presumption that they enhanced well-being (avoided dissatisfaction), have often been changed when they became inconsistent with the perceived welfare of society. Examples are changes in laws to end slavery and discrimination against females and minorities.

Some regard the laws from Mount Sinai against lying, stealing, and killing as absolute moral imperatives, but most Americans sanction just wars, white lies, or stealing a loaf of bread to save the life of a starving child—presumably because such actions promote a greater good or prevent a greater evil. Jesus stressed that the Old Testament law of the Ten Commandments could be fulfilled only through obedience to the higher law, "Love thy neighbor." Because concern for one's "neighbor" extended to more than just the person next door (as taught in the parable of the Good Samaritan), it seems clear that Christian moral law is not inconsistent with utilitarianism.

There is almost universal agreement that above all else the food system must serve the needs of people. The major participants in the American food system, producers and consumers alike, probably accept in principle this utilitarian ethical philosophy.

The issue is not, as Lerza and Jacobson have it, food for people versus profit but public sector versus private sector. I shall contend that the private sector is generally more efficient than the public sector in converting resources into output. Government takeover of production and marketing activities now performed by the food-for-profit sector would not serve the public interest. I also contend that laissez-faire is inappropriate: an activist government is essential to preserve competition, provide essentials for those with inadequate resources, improve technology through basic science, and perform other functions the market cannot do well.

The controversy in ethical systems is not so much over this dominant philosophy as over the best means of serving utilitarianism. Some system must determine what, when, where, and how to produce and distribute food. Billions of decisions must be made each year. Without a coordinating framework, decision makers acting individually could not promote the general welfare. By pursuing self-interests, individuals would bring greater harm to society than benefit to themselves. Accordingly, society establishes institutions to coordinate food production activity and to balance competing interests for the benefit of society. Two major institutions perform this role, the market system and the public system. The "public system" here broadly refers to all nonmarket allocating institutions including family and government. It also includes individuals or groups acting for others and chosen by tradition, election, or appointment.

Theory and performance help to appraise how well each system serves utilitarian ends. We shall note that in theory either the public or the market system can be consistent with utilitarian ethical philosophy. Another test is how well the assumptions are satisfied for each system to work. The most important test of all, however, is how well each system performs.

The Market System

Adam Smith, who held an academic chair in moral philosophy, early showed that under certain conditions the market price system would maximize utility. Each person "intends only his own gain, but he is in this . . . led by an invisible hand to promote an end which was no part of his intention." I contend that under strict assumptions the allocations of what economists call "perfect competition" maximize utility of society. (See Tweeten 1979.) It is critical to note that this optimal allocation of resources and products is the same for a barter, socialist, capitalist, or any other economic system. Before critically examining assumptions and performance, let us briefly review some outcomes of profit-seeking behavior within this ideal competitive framework.

Profit and the competitive market price system are not ends in themselves but means to serve the needs of people. Under competitive markets, price is a measure of utility value per unit of a resource or product. Because the search for profit causes firms to allocate resources and products to highest value uses, profit-seeking provides allocations consistent with utilitarian moral precepts.

To see why a market serves utilitarian objectives by responding to needs of people, consider the case of hard tomatoes, an issue Jim Hightower has raised in the past and raises again in this volume (Chap. 12). If consumers are dissatisfied with hard tomatoes, they can vote with their dollars for tasty, thin-skinned tomatoes and depend on a profit-seeking capitalist to supply them. If this capitalist's profit rate is "obscene," they can count on other self-serving producers to enter the market with tasty tomatoes. The increased supply of tomatoes will reduce profit rates.

Profit is for people because it provides incentives for firms to supply food that meets people's wants. Profit is also for people because it either is invested to produce more or is distributed to the persons who own the firm. Profit is for people because it compensates the investor for risk and for consumption sacrificed while his or her capital is used to produce output for others.

But to make more profit don't private firms in a competitive market economy exploit their workers, paying them less than they contribute to the firm? Most workers have skills that can be used by more than one firm. A firm initially paying less than what workers contribute to output will lose profits as these workers are bid away to other firms. Similar reasoning applies to race and sex discriminations. If women and blacks are paid less than they contribute to output value, firms increase profit by hiring them. If people value information on safe and nutritious foods, it will pay a profit-making firm to draw customers by advertising food benefits.

Critics contend the market system wastes fossil fuel, spends too much on transportation, and provides too little regional self-sufficiency. But if firms seek profit and if energy is properly priced, firms can be depended upon to conserve energy and provide appropriate regional self-sufficiency.

Those who favor reliance on the market often hold several value judgments. They hold that people are basically self-serving and require an extensive check and balance system if individuals are to act in the public interest. They hold that each individual is owed "commutative justice" (defined as the value of his or her contribution to society). They hold that the best way to achieve a just society is to rely on the market price mechanism with each producer freely allocating resources in search of profit and each consumer freely allocating resources to satisfy his or her needs. They hold to the enterprise creed that each individual (or immediate family) ought to be primarily responsible for his or her economic security and that a prime function of the government is to ensure a healthy climate for exercise of business enterprise and consumer sovereignty. They hold that the system of property rights is fair and equitable. They hold that "the government is best that governs least."

These values notwithstanding, more solid grounds must exist to choose an economic system. Critical choices depend on conditions required for the market price system to work and how in fact it does work. The market price system brings outcomes in the public interest if (1) resources are equitably distributed, (2) costs of goods and services to firms and individuals coincide with costs to society, (3) there are enough buyers and sellers to provide effective competition, and (4) firms act to increase profits. How well are these conditions met?

EVIDENCE OF THE PROFIT MOTIVE. Subjective and objective information suggests that firms do act to increase profits. On the subjective level, firm decision makers say they do. On the objective level, economists have found that firms act as if they seek profits. Indeed, because profit is essential to survive in the market system, it is difficult to believe that firms could behave otherwise. Firms also pursue growth, security, and other goals. But profit must be a key objective if firms are to serve utilitarian ends in a market economy.

EQUITABLE DISTRIBUTION OF RESOURCES. Even if firms act to increase profits, the market price system will not serve society well if resources are inequitably distributed. "Distributive justice" refers to fairness in access to economic opportunity, in sharing of economic outcomes, and in making rules of the economic game.

The issue of optimal distributive justice in a utilitarian framework need not be solely a value judgment. Marginal products and marginal utilities required to calculate appropriate transfers have been measured crudely. Results indicate that maintaining incentives and fostering well-being in society require a somewhat uneven distribution of income rather than equal income for each person.

Economists call the outcome of a well-functioning market a *Pareto optimum* (defined as an allocation of resources and products where one individual cannot be made better off without making someone else worse off). Efficiency is at a maximum with this optimum if we naively assume everyone receives the same satisfaction from another dollar of goods or services. Commutative justice is also served because each worker is paid what he or she contributes to society.

The problem is that this outcome depends on the initial distribution of resources. If one person begins with few human and material resources, that person may starve while others are sated with food under Pareto optimum efficiency. Thus commutative justice is served by the market price system but distributive justice is not. Utility could be increased by transferring income or resources from the sated person to the starving person.

The public intervenes in the market in numerous ways to provide distributive justice. Examples are food stamps, public education, social security, and public assistance. Redistribution of income requires taxing some producers' output, a violation of commutative justice. Distortion of incentives reduces national dollar output. To be consistent with utilitarianism, such reallocation is justified only if the utility gained by transfer recipients outweighs the utility lost by others.

COSTS (BENEFITS) TO FIRMS COINCIDE WITH COSTS (BENEFITS) TO SOCIETY. For the profit motivated market to serve society well, prices must signal private firms to act in the public interest. If incremental costs or benefits to firms differ from those to society, the market price system will not bring desired allocations. Prices and markets do not even exist for many goods and services. For example, individuals and families allocate human emotions without benefit of markets or property rights.

The lack of markets to allocate air and water causes problems.

Individuals and firms paying no charge to dump wastes in the air pollute the atmosphere to the detriment of society. Soil erosion may occur because the farm operator is unaware of the problem, does not care, or is concerned with immediate output only. Losses are then incurred by other farmers "downstream," by future operators of the eroded farm, and by consumers. A related example is the cost to consumers who are unaware of chemical pesticide residues or harmful microorganisms in purchased food. When benefits of safe, nutritious foods are not perceived by consumers, they will pay no premium for such foods. Thus firms will lack incentives to supply them.

Benefits to society exceed benefits to private firms engaged in basic agricultural research. Because an improved wheat variety can easily be propagated by one farmer without the knowledge of the seed developer and sold to other farmers, the firm that develops the new variety is unable to appropriate enough benefits to cover research and development costs. Consequently, if research is left to the private sector it will be underfunded. Risks also dampen incentive for private firms to develop new varieties. A public research facility operating on a large scale can average out these risks. It can continue to produce new technologies, which, on the average, provide a high rate of return, but which a private firm will forego because it cannot chance bankruptcy producing technologies that are mostly losers. Only rarely does a big winner compensate for the many losers in basic research.

Public education and incentive programs improve the efficiency of the private market by aligning private and social costs (benefits). The government has intervened in markets to establish food grades and standards, information systems, health and safety requirements, and labeling regulations. The federal government provides soil conservation cost-sharing, technical assistance, and educational programs to supplement the market. Taxes on tobacco and alcohol consumption help bring costs to individuals who use these products into line with the costs that society bears as a result of their use.

COMPETITION IN THE MARKETPLACE. The price system ideally provides people options to enter the marketplace without arbitrary constraint, trading goods and services only if a transaction benefits both buyer and seller. But all parties do not enter the marketplace with equal bargaining power. A sole buyer may exercise bargaining power to absorb the gains from trade with many unorganized sellers. Or a single seller may exercise bargaining power to absorb the benefits from trade with many unorganized buyers. Concentration of economic power can lead to high prices, reduced output, and benefit to the few at the expense of the public at large. Do such concentrations and associated costs characterize U.S. farm and food industries?

Farm production is increasingly concentrated on fewer farms. In 1981, the 298 thousand farms with sales of one hundred thousand dollars or more accounted for 12 percent of all farms and for two-thirds of all farm output (sales). By the year 2000, only 50 thousand large farms will account for two-thirds of all sales if current trends continue. However, farm production in the foreseeable future is in no danger of being concentrated in so few firms that monopoly pricing and profit will be a problem.

A number of studies indicate that market concentration does not significantly influence profit rates until eight firms control 65–80 percent or four firms account for 50–65 percent of the output of an industry. No major farm commodities are characterized by such degrees of industry concentration. The most concentrated major component of agriculture is broiler production and processing. An estimated 97 percent of all broilers are produced under vertical coordination—90 percent under production contracts and 7 percent directly by integrated broiler processing firms. In 1975, 30 percent of broilers were processed by the eight largest firms and 50 percent by the eleven largest firms. Through production contracts with growers or company-owned growing operations, these firms could effectively control broiler production for their processing facilities. Despite this high degree of concentration or because of it, broiler production efficiency has increased more rapidly than beef and pork production efficiency. Cost savings have been reflected quite fully in wholesale broiler prices. From 1955 to 1970, for example, the wholesale price per pound of broilers fell from forty-two cents to twenty-five cents while the cost of production fell from thirty-six cents to twenty-seven cents per pound.

Concentration of economic activity in the hands of a few dominant firms is more prevalent in the farm input supply and product marketing sectors of agriculture. Turning first to inputs, the large numbers of private and cooperative firms provide workable competition in the feed, fertilizer, and credit industries supplying farmers. The farm machinery industry is highly concentrated in a few private firms, but competition among these firms is intense and innovation is rapid. Industry profit rates have not been high on the average and some firms are on the brink of bankruptcy. If the demand and price situation warrants, foreign manufacturers are positioned to enter the machinery industry and provide effective competition.

As Parker and Connor (1979) suggest, market power is of greatest concern in the food marketing industry, which performs wholesaling, processing, transportation, storage, and retailing functions from the farm gate to the consumer. (See Chap. 19.) With few exceptions, profit rates of marketing firms have been no higher than those of other industries. Because profit rates indicate monopoly power inadequately (equity values of firms with monopoly power tend to be bid up by investors until profit rates are comparable to those of firms in other industries), economists use other approaches to estimate the costs of monopoly.

Parker and Connor estimate that "consumer loss due to monopoly in the U.S. food-manufacturing industries in 1975 was at least $10 billion, but possibly as high as $15 billion." At 6 to 8 percent of the value of shipments, this is the highest relative loss estimated for any major food marketing industry. The Parker-Connor estimates have been disputed — other economists measuring the same phenomenon derived estimates only 3 to 6 percent as great as those from Parker and Connor. In 1982 Gisser found a statistically positive and highly significant association between food manufacturing concentration and productivity, a relationship that economists had found earlier for the entire U.S. industrial sector and had accounted for by economies of large size. Benefits accruing to consumers from increased productivity were omitted by Parker and Connor, but according to Gisser they were sufficient to offset their estimates of welfare losses from monopoly.

Even if the Parker-Connor results are taken as fact, they do not necessarily constitute a case for a major restructuring of the industry into many small private firms or into cooperative or public firms. The inefficiencies of alternatives could exceed those of the current market structure. Nonetheless, the situation needs to be monitored. Antitrust and other laws to maintain competition and stop unfair business practices must be applied when warranted.

The conclusions from these studies, as well as from a large number of other studies of food marketing industries, is that efficiency losses from market power in food industries are of modest proportions.

Overall Food System Performance

The American food system, relying primarily on the market price system but supplemented by public intervention to correct for market inadequacies, has compiled an impressive record. Americans receive a high quantity, quality, and variety of food at lower real cost in terms of income than do consumers in any other country.

From 1920 to 1981, farmers increased output 140 percent with approximately a constant volume of total production inputs. They accomplished this feat on fewer harvested acres in 1981 than in 1920. In the absence of productivity gains, millions of acres of fragile lands subject to soil erosion would have been cropped. Progress in soil conservation also is notable because farmers are rapidly adopting reduced-tillage techniques that protect soil to increase their profits.

In the final analysis, the case for the market system rises or falls on the issues of whether competitive markets (1) reward labor and other factors of production according to their contributions, (2) direct output to markets of highest returns, and (3) move apace toward equal (equilibrium) rates of return on resources. The empirical evidence in support of each of these issues is compelling. My calculations for 1960, 1965, 1970, and 1981 indicate that production resources on adequate-sized farms have earned returns at least comparable to returns on resources in the nonfarm sector. (For cost/returns for 1981 see Tweeten 1986.) Other rates of return are reported in *A Time to Choose* by the U.S. Department of Agriculture (1981, 51).

The impressive performance of agriculture is in no small part the result of a highly successful synergism between the private sector and public agricultural research and extension—one of the few public sector performances that is truly exemplary. But even here the record is flawed by too little public investment. For a more complete treatment see Ruttan (1982). Public sector research and extension constitute only 1 percent of the nation's farm resources, hence private sector decisions dominate economic performance.

The level and distribution of returns among land, labor, and capital have behaved about as predicted by economic theory. As resources and products moved to highest value uses, the structure of the farming economy changed massively, which resulted in utility gains to society, especially to low-income consumers who spend a high proportion of their income for food. How those who left farming improved their circumstances has also been documented (Tweeten 1983a).

After lagging sharply for decades, income per person in farming gradually improved to parity with nonfarmers' income per person by the late 1970s. With net farm income unusually low in 1981 (the second lowest in real terms since the early 1930s), farm income per capita from all sources still averaged nearly 90 percent that of nonfarmers! The incidence of farm poverty has dropped from about 50 percent as recently as 1960 to approximately 14 percent today, about the same incidence as nonfarm poverty. Many of the once-massive family income differences among farm size classes and geographic regions have been eliminated. The major reason is rural industrial development induced by market incentives. This development created jobs, giving farm families — especially those on small farms — off-farm income to supplement their inadequate farm incomes.

The Public Sector System

Americans widely favor at least some regulation, taxes, and subsidies by the government in order to align private incentives with social ones, to avert restraint of trade, and to provide for those who are incapable of providing for themselves. Central to the food for profit or people controversy is the question of whether the public sector should undertake the basic production-marketing activities now performed by the private sector for profit. To examine that issue and appraise overall performance of the public sector, this section draws heavily on international experience.

Public sector food systems range from cooperative farms such as the Israeli kibbutz (where residents make decisions more or less democratically and sell to the private market) to the centrally planned and administered agricultures of communist countries. The key feature of such systems is that principal decisions regarding what, when, how, and where to produce and distribute are guided by individual or collective judgments rather than by the market price system.

Those who advocate public allocation of food system resources and output frequently hold the belief that people are basically good, although they are corrupted by institutions. They hold that the good society must provide a just wage for labor, a just interest rate on financial capital, and a just rent for physical capital. They hold that people have a right to food, a right to a clean environment, and a right to a fair return on investment. They hold that a just society requires production from each according to his or her ability and allocates to each consumer according to his or her needs. They hold that, above all, society owes distributive justice to each individual.

In theory, informed managers committed to utilitarian ethical philosophy and operating a public sector food system could allocate resources and products to maximize utility of society. For public sector allocations to bring outcomes in the public interest, those making allocative decisions must (1) desire to serve the public interest, (2) recognize preferences, and (3) have the administrative and analytical capabilities to serve the public interest.

DO INDIVIDUALS SERVE SELF OR SOCIETY? Great thinkers have long debated whether people are basically selfish or altruistic. The issue is not resolved here, but worldwide experience is compelling: impersonal men and women cannot be depended on to act in the best interest of others. Nonmarket allocation seems to work only in groups with strong affective bonds such as families or small communes. Large groups have succeeded only if they have been bonded and motivated by strong religious or political ideology.

RECOGNIZING PREFERENCES. Public decision makers, however well intentioned, cannot serve society without knowing people's wants. Decision makers rely on various institutional systems to gauge the preferences of those they serve. In a democratic society, the voting process is used to reveal preferences and maintain accountability. The one person-one vote system may seem intuitively fairer than "dollar voting" in the market system. But democratic systems are notoriously clumsy, costly, and subject to manipulation by special interests. Representative systems, as opposed to "town hall" democratic systems, give rise to fragmented jurisdictions—politicians find it difficult to resist programs for supporters paid for by others. Benefits of public programs are frequently concentrated among the few, while costs are widely distributed. In the political arena, millions of indifferent losers are no match for a few determined gainers, even when aggregate utility losses far exceed gains.

Those successful in forming power collectives aggrandize themselves at the expense of the unorganized. A responsive public sector is likely to serve multiple objectives, each supported by narrow constituencies with no recognition of the cost to society of serving self. In contrast, a well-functioning private market serves society by pursuit of the singular profit objective. An effective, efficient check and balance system cannot be built into large democratic systems; no invisible hand guides government decision makers to serve the public interest.

ADMINISTRATIVE AND ANALYTICAL CAPABILITIES. Even if the political process accurately signaled the highest and best uses of resources and products of the food system and if all public decision makers were well intentioned, the administrative decision processes would require so many resources that society's welfare would be sharply reduced. Faced with more decisions than can be made, bureaucratic systems simplify administration. Because each producer and consumer has unique wants and needs, simple rules will not do—for example, that each individual should receive the same food.

Even the most complex econometric models of the U.S. food economy predict economic outcomes very imperfectly. Highly trained and experienced analysts frequently err in projecting implications of proposed public policy. The analyst needs to foresee impacts on the general welfare that result from complex policy options working through millions of producers and consumers, but he or she cannot do so. Given the formidable obstacles to competent planning, it is not surprising that public sector political-administrative processes frequently produce outcomes opposite to those intended.

Public Sector Performance

Centrally planned and administered food systems have performed better in distributing food than in producing it. Food presumably produced for people and not for profit has been lacking in quality, variety, and often in quantity. Consumers pay a high percentage of their income for food and often spend hours waiting in line for it.

As food production systems, the centrally planned economies have failed dramatically with the Soviet Union the most obvious example. Chronic food deficits in socialist economies contrast sharply with chronic food surpluses in the United States. Producers paid according to their needs, rather than according to their abilities or contributions, receive inadequate incentives to produce. The ideology of self-sacrifice for the good of society seems not to endure long after the revolution. Producers contribute less than what they are able to produce, consumers take more than they need. As altruism fades in the absence of adequate checks and balances, the pursuit of self-interest emerges in the form of waste, mismanagement, and corruption. Centrally planned production systems seem to work only by introducing economic incentives (profits) — with notable recent examples of modest success in Hungary and the People's Republic of China. Western countries have similar experiences with socialism — state-owned firms operate efficiently and compete in international markets only if they are allowed the freedom to operate like profit-seeking private firms. Where exact operational management and worker dedication are essential, as in planting and harvesting crops on time or in overseeing sow farrowing at midnight, no substitute has been found for the market system where the owner-operator has a direct stake in the outcome.

Undertaking food production and marketing may not be viable options for the public sector in this country. How well has the public sector performed its more traditional role in market economies, the role of providing distributive justice, stabilization, and correction for market distortions? Here attention is focused on a range of public activities influencing the food system directly and indirectly.

STABILIZATION. Performance of government monetary-fiscal policies to dampen business cycles while promoting growth without inflation has been disappointing. Monetary and fiscal policies have been erratic and sometimes at cross-purposes, adding instability and uncertainty to the economy. Periodic inflation, high real interest rates, and other products of failed monetary-fiscal stabilization policies have a debilitating impact on the farming sector (Tweeten 1983b).

DISTRIBUTIVE JUSTICE. Public efforts to force private firms to pay a "just" wage and to promote organized labor have caused unemployment for disadvantaged workers, reduced output, and interfered with the ability of U.S. firms to compete with international firms (notable recent examples are the steel and auto industries). Regulation forcing low interest rates on savings has reduced savings, investment, and economic progress. Approximately three-fourths of the nation's massive transfer payments go to the nonpoor. [This estimate originates with data published by Robert Lampman extended to include farm commodity programs and updated (Tweeten and Brinkman 1976, 165).] Public assistance payments encourage family disintegration. The working, intact-family poor found disproportionately in rural areas are often not eligible for payments.

CORRECTION OF MARKET DISTORTIONS. Government programs to control soil erosion and other environmental problems have not been well targeted or cost-effective. In the food sector and in sectors influencing the food sector, government has been responsive to special interest groups such as labor and trade associations. The results have been tariff and other international trade barriers, unnecessary licensing requirements for entry into trades and professions, along with featherbedding and other arbitrary rules that restrain productivity gains. Government agencies to regulate economic activity have often become the tools of the regulated. The elimination of many banking and trucking regulations gives the promise of large benefits to society.

Shortcomings of government attempts to regulate are not confined to the United States. Olson (1982) provides an absorbing historical exposition of how failure of government to block labor and trade association activities in restraint of trade caused an institutional sclerosis that stagnated economies the world over. One needs only to observe evidence from centrally planned countries to realize that capitalists have no monopoly on the suborning of public officials.

Although government intervention may actually rarely be needed to correct farm product price distortions, it is, in fact, commonplace. Such intervention frequently is made not to align private incentives with social incentives, but to distort private actions that initially were in line with social incentives. Schultz and others (1978) have documented research that reveals a pattern of governments in developing countries artificially holding down farm prices, thereby causing hardship to producers and reducing output. On the other hand, developed countries frequently hold farm prices artificially high. In the case of the European Community, domestic farm commodity price supports generate excess production, which is exported at considerable cost to European taxpayers and chagrin to American farmers. The distorted price incentives are not an attempt to correct for private market failures but rather the result of successful public lobbying by the urban elite in developing countries and a few powerful producer organizations in developed countries.

Many criticisms of markets, when analyzed in depth, turn out to be criticisms of government actions that encouraged or forced markets to work poorly. Perhaps "leftover" tasks assigned to government are more difficult to perform than are the tasks performed by the market sector. But an increasing body of economic analysis and world experience reveals that government failure to perform what is expected of it is far more pervasive than is market failure.

Summary and Conclusions

The American food system is a mixture of a public sector guided largely by political-administrative processes and a private sector guided largely by the price system and profit motive. The private market dominates, however, and in recent years the entire system has been severely criticized along the lines taken in this volume by Hightower, Lerza and Jacobson, and Parker and Connor; the system produces food for the profit of corporations rather than for the needs of people. Judging whether the food system is for people or for profit requires evaluation of ethical standards as well as the performance of the food system in serving these standards.

The limitations of the market must be recognized. It will not serve the needs of those who enter the market with no resources; it will not serve where social needs have no market or where private price incentives fall short of social incentives. Acting alone it will concentrate economic activity temporarily (business and commodity cycles) and interpersonally (case poverty). If the market is evaluated only in regard to what it can be expected to do, its performance has been exemplary in directing resources and products to highest value uses, and hence, in serving utilitarian ends. It is quite robust in providing utilitarian outcomes even when the number of domestic firms is not large—especially if international trade channels are kept open.

The public sector supplements the market sector to correct incentives and to promote stability and distributive justice. As government efforts have been extended, weaknesses of public allocation have surfaced. Public decision makers and administrators often do not act in the public interest. Worldwide experience gives no evidence that a check and balance or incentive system can be designed to ensure that they do. Analytical systems cannot be devised to enable these decision makers to foresee consequences of complex and far-reaching policies with sufficient clarity for even well-intentioned, wise, responsive public officials to act in the public interest.

Public involvement is essential for environmental protection, poverty alleviation, and national economic stabilization; but in the light of history, one cannot predict that such intervention will be well done. Outstanding examples of effective and efficient public sector performance occur but they are not the pattern. The optimal food system consistent with utilitarian moral philosophy utilizes the market system to the extent possible and confines the public sector to functions the private sector cannot perform. The widespread evidence of government failure suggests caution in extending the public sector into activities now performed by the private sector—the costs of government intervention may be greater than those of the market failure that it was intended to correct.

Impersonal markets alienate many by unceremoniously releasing workers from employment or by pricing consumers out of the market without regard for personal consequences. American farmers have long viewed banks, railroads, and grain exchanges as diabolical and sinister. Alienation and resulting populist and protest movements arise in part because producers do not understand or trust an impersonal market system operating with an invisible hand. Alienation is not confined to market economies, as illustrated by the widespread social unrest in Soviet satellite countries and by the high rates of alcoholism and the heavy-handed measures required to squelch dissent in the Soviet Union.

In the past, the promise of socialistic economic organization was frequently contrasted with the reality of the market price system, with the latter often coming out the loser. The market system looks much better when it is compared with the reality of the performance of centrally planned economies and of public sector performance in the United States. While in theory each system can be consistent with the utilitarian ethic, in reality the market system performs much more efficiently. We should take a lesson from the fact that it is being injected into food systems in centrally planned economies. At the same time, a pure market system does not alone serve utilitarian needs of society. A mixed public-private food system is optimal. The search for proper proportions of the two systems will perennially create conflict, in part because of the inherent tension between distributive and commutative justice.

References

Gisser, Micha. "Welfare Implications of Oligopoly in U.S. Food Manufacturing." *American Journal of Agricultural Economics* 64(1982): 616–24.

Olson, Mancur. *The Rise and Decline of Nations*. New Haven, Conn.: Yale University Press, 1982.

Parker, Russell, and John M. Connor. "Estimates of Consumer Loss Due to Monopoly in the U.S. Food-Manufacturing Industries." *American Journal of Agricultural Economics* 61(1979): 629–39.

Robbins, William. *The American Food Scandal*. New York: William Morrow, 1974.

Ruttan, Vernon. *Bureaucratic Productivity: The Case of Agricultural Research*. St. Paul: Economic Development Center, University of Minnesota, 1982.

Schultz, T. W., et al. *Distortions of Agricultural Incentives*. Bloomington: Indiana University Press, 1978.

Tweeten, Luther. "The Economics of Small Farms." *Science* 219(4 March 1983a): 1037–41.

_____. *Excess Farm Capacity: Permanent or Transitory?* Proceedings, National Agricultural Policy Symposium, 28 March 1983. Columbia: University of Missouri, 1986.

_____. *Foundations of Farm Policy*. Lincoln: University of Nebraska Press, 1979.

_____. "Impact of Federal Fiscal-Monetary Policy on Farm Structure." *Southern Journal of Agricultural Economics* 3(July 1983b): 61–68.

_____, and George Brinkman. *Micropolitan Development*. Ames: Iowa State University Press, 1976.

U.S. Department of Agriculture. *A Time to Choose*. Washington, D.C.: Government Printing Office, 1981.

SUGGESTED READINGS

Beauchamp, Tom. "The Ethical Foundations of Economic Justice." *Review of Social Economics* 40(1982): 291–99.

Buttel, Frederick H., and William L. Flinn. "Sources and Consequences of Agrarian Values in American Society." *Rural Sociology* 40, no. 2(1977): 134–51.

Dillman, Don, and Daryl J. Hobbs, eds. *Rural Society in the U.S.: Issues for the 1980's*. Boulder, Colo.: Westview Press, 1982.

Haynes, Richard, and Ray Lanier. *Agriculture, Change, and Human Values,* Vols. 1 and 2. Gainesville: University of Florida Press, 1982.

Jacob, Jeffrey C., and Merline B. Brinkerhoff. "Alternative Technology and Part-time, Semi-subsistence Agriculture: A Survey from the Back-to-the-Land Movement." *Rural Sociology* 51, no. 1(1986): 43–59.

Rawls, John. *A Theory of Justice.* Cambridge, Mass.: Belknap Press, 1971.

Schertz, Lyle P., et al. *Another Revolution in U.S. Farming.* Agricultural Economics Report No. 441. Washington, D.C.: ESCS, USDA, 1979.

U.S. Congress, Office of Technology Assessment. "Technology, Public Policy, and the Changing Structure of American Agriculture." OTA-F-285. March 1986.

Christian Theology and the Family Farm

VI

CAN RELIGION HELP us decide how much to value the family farm? In Part Five one viewpoint said that agriculture should serve people first, while another argued that it can serve people and profits at the same time. Can Christian theologians help us settle the debate?

In 1981 Michael Novak published "The Parable of Iowa" in which he suggested that Iowa was a model of the free market system. When the political system stays out of the way, allowing the economic sphere to operate on its own, "free enterprise," wrote Novak, "can flourish." This, according to the Catholic theologian, is what happened in "poor, primitive Iowa." Within the span of just a few generations, settlers created wealth "by invention and by discipline, turning the prairie into a 'wealthy, developed nation.'"

Five years after Novak penned these words, Iowa was no longer the wealthy state it once was, and Maurice Dingman, Catholic bishop of Des Moines, was condemning the effects of the free market. Calling Iowa an example of the "El Salvadorization" of the

Midwest, Dingman charged that large agribusiness firms and corporate farms had taken over the farm economy (Shanley 1985, 5). His views were strongly reflected in the third draft of the United States bishops' pastoral letter on the economy, excerpted here.

While Christians of all stripes embrace the family farm, they are not of a single mind in their approach to the present crisis. We find both Catholics and Protestants defending the principles of free market capitalism, and both Catholics and Protestants sounding more socialist views. The essays on religion, ethics, and agriculture in Part Six represent this diversity. Consider the democratic capitalism of Michael Novak, a Catholic, and Merrill Oster, a Protestant. Both are firm in their resolve to get government out of the agricultural sector, allowing farmers to stand on their own feet and compete with others. Then compare the statements of the Catholic bishops and David Ostendorf, a United Church of Christ minister. Both are critical of the free market, calling for extensive government intervention and the formation of a farmers' union.

The various sides in this debate are represented here. Chapter 22 opens with Novak's tribute to the heartland, "The Parable of Iowa," in which the author suggests that "a theology of development" rests ultimately with our response to the divine being, not with our response to our own self-interests. Early Iowans provide a good example of devout people who "invented the possibility of economic development not solely for themselves . . . but in the name of God, whose precepts guided their choices." Novak's second essay looks at corporations in theological terms and is a unique Christian defense of those enterprises. It is significant here since it can be related to agribusiness. To my knowledge a Christian defense of agribusiness corporations is yet to be written. However, since Novak's argument does not exclude farm businesses, one might substitute "agribusiness corporation" where he writes "corporation."

Next is an excerpt from the Catholic bishops' pastoral letter on the U.S. economy (third draft). The bishops express some skepticism about the market and suggest that we consider raising farm prices through mandatory production control programs. They find fault with present agricultural policies that allow big farmers to reap huge government price-support payments while smaller farmers go broke. Their support goes to a progressive land tax on farm acreage in order to discourage farmers from amassing huge land-holdings at the expense of their neighbors.

In contrast to the populist Catholic bishops, Protestant Merrill Oster speaks from a capitalist perspective similar to Novak's. Farming is "risky business" and those who manage the risk well will succeed, those who do not will fail. That is how capitalism operates and that is as it should be; God favors this system. Thus, attempts by clergy "to establish a national land reform or a policy favoring the 'family farm' at the expense of the 'corporate farm' " come in for some of Oster's harshest criticism. These programs "would require nationalized theft of property from one group to give to another. This would be a great affront to God, biblical principle and the rights of free people." Although Oster suggests that national and individual repentance is needed, he sees attempts to redistribute land as misguided. He concludes "there is no biblical mandate favoring farms of any size," and pleads for the right to individual success or failure.

Oster finds no biblical warrant for trying to spread out landownership. He adds that the Old Testament idea of a Jubilee year (in which land returns to its original owners every fifty years) is no longer applicable. In the New Testament, moreover, Jesus shows no concern for land redistribution.

David Ostendorf's essay could hardly present a more contrasting view. Ostendorf, a Protestant like Oster, argues that the biblical Jubilee is indeed applicable if original landowners are viewed not as individuals but as the *people* of God. To him, the New Testament only confirms an Old Testament point. Individuals trying to appropriate land against the will of God should heed the fate accorded to King Ahab or Sapphira; both were struck dead. Ostendorf considers family farms more socially and religiously desirable than large operations. Like Oster, he appeals to the Bible, but not as a justification for the inviolability of individual freedom, rather for the sanctity of distributive justice.

In the last essay in this part, Charles Lutz claims that, for the farm crisis, neither Christian theology nor its ethical tradition offers "any specific solutions in terms of public policy decisions." Christians can act, however, by focusing concern on farmers as human beings who are now experiencing crises. Christians should do this because God asks them "to be in ministry with people who are hurting, displaced, perplexed, anxious, or oppressed *wherever* that is true." Lutz warns that the Church should not align itself with any specific organization, be it the Farm Bureau, the American Agriculture movement, or the NFO. The churches must be political, he says; they must not be partisan.

Reference

Shanley, Mary Kay. "Meet the Bishop." *The Iowan,* Fall 1985, 5.

MICHAEL NOVAK

The Parable of Iowa
and
Toward a Theology
of the Corporation

The Parable of Iowa

Having just completed a manuscript on the spirit of democratic capitalism, I have been preoccupied with the next step, a theology of development. What courses of action should the churches support in order to help end the manifest tyranny and poverty of most of the inhabited world? Simultaneously, life so arranged matters that my family went back to Iowa to a farm inherited from my wife's family for a three-week vacation.

I have been browsing in Iowa's history in between catching my first three trout in Coldwater, hacking away on a golf course built over the ruins of a ghost town that the railroads skipped eleven decades ago, throwing sawdust mulch around some new trees, and other assorted sweat-inducing activities in the heavily scented air of northeastern Iowa. The corn is not so high as it used to get by this date, but this 1981 crop should be the second biggest in history. The hay is lying in rows for more days than it should, for each night sweet rains have swept the land. The sky is as big as ever, far bigger than eastern skies. From the rolling hills one can see beautiful farms and darkly wooded lands for miles around.

This gorgeous part of the world was an undeveloped land hardly more than a century ago. About 1836 Chief Black Hawk ceded the strip of land where our farm is located to the United States. There were never, experts say, more than about ten thousand Indians in all of Iowa at any one time. Most were hunters, although Black Hawk's tribe kept some eight hundred acres planted in maize down around Rock Island. The

MICHAEL NOVAK is resident scholar at the American Enterprise Institute in Washington, D.C. "The Parable of Iowa" originally appeared in *National Review,* 21 August 1981, and is reprinted with permission. "A Theology of the Corporation" originally appeared in *The Corporation: A Theological Inquiry,* edited by Michael Novak and John W. Cooper (Washington, D.C.: American Enterprise Institute for Public Policy Research, 1981) and is reprinted with permission.

Indians' implements were of the Stone Age; they did not use the wheel. Horses had come with the white man.

The Sioux warred with the Fox and the Sac so bitterly that the land where our farm lies, not far from the Upper Iowa and the Turkey rivers, was by tribal consent made into a neutral zone. Fort Atkinson, not far away, was established at tribal invitation to keep the neutral zone free of warfare between the tribes.

When a certain Baptist missionary, of whom my wife is a collateral descendant, reached Iowa in 1842, there were scarcely ten thousand whites in the territory, which stretched out at great distances. For much of the year the Indians were away on long hunting expeditions, but they came back to this place for its hills, woods, streams, bluffs, for its sheer beauty.

The town of Cresco ("I grow") was incorporated by a handful of pioneers in 1866. The Reverend Charles E. Brown built the First Baptist Church here (after having earlier built another at the doomed ghost town a few miles south). His son, Charles, Jr., was the first Howard County man (among scores from the first few families) to volunteer for the Union to help end slavery—his father having preached on the subject in a notable July 4 address in 1844. Charles, Jr., served at the battles of Corinth and Shiloh, among others.

How did poor, primitive Iowa become a "wealthy developed nation" within something like three generations? Early writers in the pioneer generation placed great stress on the character and learning of people. They carried books in their packs—Mill and Locke, for example. But government and law played an extremely important role. The Homestead Act opened up the territory to land- and home-ownership for all. The land grant colleges, from the beginning, founded agriculture upon science, in the belief that even the fields most favored by nature could, by human invention, be made to yield a hundred times greater wealth than nature itself provided. The people set out to create wealth by invention and by discipline.

They needed strong communities and a stable system of law. They needed transport, and they bent ingenuity in every direction to attain it, by water and by land. They built roads, bridges, dams, mills. All this they did communally, either through volunteer help or through government conscription.

Eventually the state experiment stations helped farmers in every community to diagnose and correct the problems of soil management, seed quality, blight control, irrigation, drainage, and the rest. Eventually, rural electrification—a federal program—brought electricity, as telegraph, telephone, and radio brought swift contact with distant worlds.

Was this free enterprise alone? Of course not. The political system took on many tasks that the economic system alone could not accomplish. But the political system also tried to stay out of the way, not to take over the economic system. And the moral-cultural system—rooted in institutions of family, church, schools, the press, and voluntary associations of every sort—watched over the foundations of character required by such a political system in such an economic system.

The reality of Iowa today is a parable for development. Like all parables, it must be applied with appropriate changes for each situation. Yet a theology of development that ignores this parable is bound to miss some wisdom, human and divine. Such peoples as the early Iowans invented the possibility of economic development not solely for themselves—they still feed millions far from their site—but in the name of God, whose precepts guided their choices.

To my dim-witted eyes, this parable of development stresses (1) the character of a people, (2) the economic system, (3) the political system. All three sets of institutions are critical. Their order is also critical. And their mutual checks and balances, complementary strengths and weaknesses, and cooperative functioning are also critical. Two further facts—that agriculture is the basis of development and that individual farmers had incentives to do what they decided to do and did best—also seem important.

I should mention that I was not long ago reading Ronald C. Nairn's *The Wealth of Nations in Crisis,* which deals with many of today's developing nations. It makes sense of what I see and hear in Iowa. It also makes sense of the parable of Iowa, on a worldwide canvas. Last Sunday, I also picked up Arthur Simon's *Bread for the World* in the vestibule rack of Assumption Church. It seems to miss every significant point.

References

Nairn, Ronald C. *The Wealth of Nations in Crisis.* Houston: Bayland, 1979.
Simon, Arthur. *Bread for the World.* New York: Paulist, 1975.

EDITOR'S NOTE: In 1980, Notre Dame's Business School and Theology Department asked Michael Novak to present a lecture on the subject "Can a Christian Work for a Corporation?" Although the original essay did not address specific industries or particular corporations, Mr. Novak has kindly, at the editor's request, added a few new sentences to his original text of 1981. In addition, he calls the attention of readers to the volume of essays he edited in 1981,

The Corporation: A Theological Inquiry. Both that volume and *Toward a Theology of the Corporation* were originally published by The American Enterprise Institute and are obtainable from the University Press of America (4720 Boston Way, Lanham, Md. 20706).

Toward a Theology of the Corporation

My task is to set forth some steps toward a theology of the corporation. We need such a theology so that the ministers who serve men and women of business and labor might be able to preach more illuminating and practical sermons, and so that critics might have at their disposal a theologically sound standard to behavior for corporations.

Among the intellectuals and journalists with whom I have grown up—and in my own earlier writing—"big business" and "big corporations" are terms often voiced in a condemnatory tone. Most of us who have been trained in the humanities have been taught to look down upon capitalism as vulgar and inferior. Our training in the social sciences has also taught us to prefer a more social and scientific ideal (more like scientific socialism, actually) and to find in capitalist economies mainly alienation, exploitation, conspiracies, greed, and self-interest. The embodiment of these evils is, we have often been led to imagine, the corporation.

When I first wrote this essay in 1980, I thought it would have use as an instrument for criticizing the corporation. To do that fairly, though, one needs to see the corporation in the light of *its own* ideals: there is no use criticizing it for not being socialist, or traditionalist, or precapitalist. The business corporation—free, voluntarily formed, not a creature of the state but of independent individuals—is one of the chief social inventions of the political economy composed of political democracy and a capitalist economy. It is one of the distinctive inventions of the liberal society, and flourishes in number and in variety only in the liberal society. (There are illiberal, authoritarian societies that permit some few corporations; but these are in many cases rife with political cronyism.)

Moreover, the corporation, as an institutional form invented in relatively recent times, owes a great deal to Jewish and Christian concepts of the person, the voluntary association, covenants, fiduciary responsibilities, stewardship, and the like. Many Jews and Christians have found in the corporation a suitable field in which to exercise their vocation as religious persons. Few thinkers have tried to put into words the religious content that such persons sense (rather than articulate) in their work. That is what I have tried to do. Like all things human (like the church), the corporation often fails its own ideals.

To work in a modern business corporation, no one need pass a test of faith or even reveal his or her religious convictions to others. But it would be a mistake to permit the business corporation's commendable acceptance of religious pluralism to mask the religious vocation that many see in it.

THE MULTINATIONAL CORPORATION. In speaking of the corporation, I will concentrate on those large business corporations that are found among the three hundred or so multinational corporations, two-thirds of which are American (Lea and Webley 1973, 1).

The reason one must first consider these *big* corporations is that all but very strict socialists seem to be in favor of markets, ownership, cooperatives, and *small* business. Religious socialists like John C. Cort favor the private ownership of small businesses, ownership through cooperatives, and some free-market mechanisms (Cort 1979–80, 423–34).

What are multinational corporations? They are not those that merely sell their goods in other lands, buy goods from other lands, or trade with other lands. Multinationals are corporations that build manufacturing or other facilities in other lands in order to operate there. The building of a base of operations in other lands is an important condition for qualification as a multinational corporation in the strict sense. One should not think only of factories; banks and insurance firms — important for local investment — may also establish such operations.

The training of an indigenous labor and managerial force is not a strictly necessary condition for a corporation to be considered multinational, but it is a common characteristic, particularly of American companies. Thus multinationals make four chief contributions to the host country. Of these the first two, (1) capital facilities and (2) technological transfers inherent in the training of personnel, remain forever in the host country, whatever the ultimate fate of the original company. In addition, products manufactured within the nation no longer have to be imported; thus (3) the host nation's problems with balance of payments are eased. Finally, (4) wages paid to employees remain in the country, and local citizens begin to invest in the corporation, so that most of its future capital can be generated locally. (See Muller 1973, 42.) These are important factors in any accounting of the relative wealth transferred to and from the host country and the country of the corporation's origin. Critics sometimes concentrate only on the flow of return on investment. They neglect to add up the capital investment, training, balance-of-payments relief, salaries, and stimulation of local investment.

Almost all of the two hundred American multinationals are to be found among *Fortune 500* industrial companies, though a few are among the largest banks and insurance firms. What less-developed countries want most today is manufactured goods, at prices made possible by local production, and the financial services of banks and insurance companies.

Generally speaking, only a company of the size represented by the *Fortune 500* has the capital and skills to accept the risks of operating in an unfamiliar culture. As it is, 40 percent of all foreign sales of U.S. multinationals are in Western Europe, and another 25 percent are in Japan, Canada, Taiwan, Hong Kong, South Korea, Australia, and other industrial nations. Most concerns about the multinationals, however, focus on their role in the developing nations. Only about 12 percent of the business of U.S. multinationals is to be found in Latin America, and only a tiny fraction in Africa. Vast expanses of the whole world never see an American multinational (U.S. Bureau of the Census 1979, Table 944).

The vast majority of U.S. corporations are not multinationals. Many that could be do not wish to be, believing the headaches more costly than the rewards. Some that are refuse to build operations in unstable conditions such as those that characterize most of the developing nations. That is why such a small proportion of overseas activities by U.S. multinationals is to be found in Latin America and Africa.

Other contextual matters should be noted. In most nations of the world — notably the socialist nations — private corporations are not permitted to come into existence. Only a relatively few nations of the world produce privately held corporations. Furthermore, some nations that do so (like the United States) were formerly colonies, and some others (Hong Kong) still are. Since economic development depends to a large extent upon home-based, privately held corporations, differences in moral-cultural climate are significant. Some cultures seem to develop far higher proportions of skilled inventors, builders, and managers of industry than others. In some cultures, the work force is more productive than in others.

Over time, education and training may provide new moral models and fairly swift cultural development. Simultaneously, of course, such developments may provoke intense conflicts with guardians of the earlier cultural order. It cannot be stressed too often that corporations are not merely economic agencies. They are also moral-cultural agencies. They may come into existence, survive, and prosper only under certain moral-cultural conditions.

It goes without saying that private corporations depend upon a nonsocialist, nonstatist political order. Insofar as socialist governments in Yugoslavia and elsewhere now experiment with autonomous economic enterprises, take their signals from a free market, and reward their managers and workers according to profit and loss, they are moving toward a democratic capitalist political order. As their middle class grows, so will the demand for further political rights, due process, democratic methods, a free press, freedom of worship, and the rest. Economic liberties require political liberties, and vice versa. Historically, private business enterprise has not only grown up with liberal democracy, it has also been the main engine in destroying class distinctions between aristocrats and serfs by making possible personal and social mobility on a massive scale.

The private business corporation is particularly active among Americans. The United States, with a population of just over four million in 1800, already had more corporations than all of Europe combined. Some of these corporations began to grow into large-scale organizations — roughly following the railroads — at the end of the nineteenth century.

Nearly all American corporations, and particularly those from the *Fortune 500,* originated around a novel invention. They grew in sales, size, and capital either through products never before known to the human race or through novel processes for producing them. Entire industries, like those for airplanes, automobiles, oil, gas, electricity, television, cinema, computers, copiers, office machinery, electronics, and plastics, are based on corporations initially formed by the American inventors of their products.

THEOLOGICAL BEGINNINGS. In thinking about the corporation in history and its theological significance, I begin with a general theological principle. Georges Bernanos (1962, 233) once put it this way: "Grace is everywhere." Wherever we look in the world, there are signs of God's presence: in the mountains, in a grain of sand, in a human person, in the poor and the hungry. "The earth is charged with the grandeur of God." So is human history.

If grace is everywhere, there must be signs of it even in the corporation. Let me mention seven signs of grace: creativity, liberty, birth and mortality, social motive, social character, insight, and risk of liberty and election.

Creativity. The Creator locked great riches in nature, riches to be discovered only gradually through human effort. John Locke (1947, 20) observed that the yield of the most favored field in Britain could be increased a hundredfold if human ingenuity and human agricultural science were applied to its productivity. Nature alone is not as fecund as nature under intelligent cultivation. The world, then, is immeasurably rich as it comes from the Creator, but only potentially so. This potential was hidden for thousands of years until human discovery began to release portions of it for human benefit. Yet even today we have not yet begun to imagine all the possibilities of wealth in the world the Creator designed. The limits of our present intelligence restrict the human race to the relative poverty in which it still lives.

During 1979 Atlantic Richfield ran an advertisement based on a theme first enunciated, as far as I can tell, by Father Hesburgh of Notre Dame: namely, that 40 percent of the world's energy is used by the 6 percent of the world's population residing in the United States (Hesburgh 1974, 101). This way of putting the facts cultivates guilt. A moment's thought shows that it is a preposterous formulation.

What the entire human race meant by energy until the discovery of the United States and the inventions promoted by its political economy were the natural forces of sun, wind, moving water, animals, and human muscle. History for a very long time seemed relatively static. The social order did not promote inventions and new technologies, at least to the degree lately reached. The method of scientific discovery had not been invented.

In 1809 an American outside Philadelphia figured out how to ignite anthracite coal. The ability to use anthracite, which burned hotter and more steadily than bituminous coal, made practical the seagoing steamship and the locomotive.

In 1859 the first oil well was dug outside Titusville, Pennsylvania. Oil was know in biblical times but used only for products like perfume and ink. Arabia would have been as rich then as now, if anybody had known what to do with the black stuff. The invention of the piston engine and the discovery of how to drill for oil were also achieved in the United States. The first electric light bulb was illuminated in 1879 in Edison, New Jersey.

After World War II the U.S. government dragooned the utilities into experimenting with nuclear energy. Oil and coal were cheap. The government, however, promoted the peaceful uses of the atom.

Thus 100 percent of what the modern world means by energy was invented by 6 percent of the world's population. More than 60 percent of that energy had been distributed to the rest of the world. Though the United States can, of course, do better than that, it need not feel guilty for inventing forms of energy as useful to the human race as the fire brought to earth by Prometheus.

The agency through which inventions and discoveries are made productive for the human race is the corporation. Its creativity makes available to mass markets the riches long hidden in creation. Its creativity mirrors God's. That is the standard by which its deeds and misdeeds are properly judged.

Liberty. The corporation mirrors God's presence also in its liberty, by which I mean independence from the state. That independence was the greatest achievement of the much-despised but creative 6 percent of the world's population. Advancing the works of their forebears, they invented the concept and framed the laws that for the first time in history set boundaries on the state, ruling certain activities off-limits to its interference. Rights of person and home, free speech in public, a free press, and other liberties came to be protected both by constitutional law and by powerful interests actively empowered to defend themselves under that law. Private business corporations were permitted to become agents of experimentation, of trial and error, and for good reason: to unleash economic activism. The state retained rights and obligations of regulation, and undertook the indirect promotion of industry and commerce. The state alone was prohibited from becoming the sole economic agent. A sphere of economic liberty was created.

The purpose of this liberty was to unlock greater riches than the world had ever known. Liberty was to be an experiment, for which Adam Smith and others argued, that might (or might not) prove to be in accordance with nature and with the laws of human society. Pleading for room to experiment, their practical, empirical arguments flew in the face of entrenched ideological opposition. The case for liberty prevailed.

The foundational concept of democratic capitalism, then, is not, as Marx thought, private property. It is limited government. Private property, of course, is one limitation on government, but not in the sense that *I* own something, that *I* possess (Johnson 1980, 49–58; Macpherson 1962, 263). The key is that the state is limited by being forbidden to control all rights and all goods. It cannot infringe on the privacy of one's home or on one's right to the fruit of one's labors and risks. Herbert Stein (1980) has a useful definition of capitalism: "The idea of a capitalist system has nothing to do with capital and has everything to do with freedom. I think of capitalism as a system in which ability to obtain and use income independently of other persons or organizations, including government, is widely distributed among the individuals of the population" (p. 6).

This is the distinctively American way of thinking about private property. In this framework, property is important less for its material reality than for the legal rights its ownership and use represent and for the limits it imposes on the power of the state. Such liberty was indispensable if private business corporations were to come into existence.

Such corporations give liberty economic substance over and against the state.

Birth and Mortality. In coming into being with a technological break-through, and then perishing when some new technology causes it to be replaced, a typical corporation mirrors the cycle of birth and mortality. Of the original *Fortune 500* corporations first listed in 1954, only 285 remained in 1974. Of the missing 215, 159 had merged, 50 had become too small to be listed or had gone out of business, and 6 were reclassified or had unavailable data (Martin 1975, 238). As products of human liberty, corporations rise and fall, live and die. One does not have in them a lasting home — or even an immortal enemy.

Social Motive. Corporations, as the very word suggests, are not individualistic in their conception, in their operations, or in their purposes. Adam Smith entitled his book *An Inquiry into the Nature and Causes of the Wealth of Nations.* Its social scope went beyond individuals and beyond Great Britain to include all nations. The fundamental intention of the system from the beginning has been the wealth of all humanity.

The invention of democratic capitalism, the invention of the corporation, and the liberation of the corporations from total control by state bureaucracies (although some control always, and properly, remains) was intended to be multinational. Smith foresaw an interdependent world, for the first time able to overcome immemorial famine, poverty, and misery. He imagined people of every race, every culture, and every religion adopting the new knowledge about the causes of wealth. One does not need to be Christian or Jewish, or to share the Judeo-Christian worldview, to understand the religious and economic potency of the free economy.

Social Character. The corporation is inherently and in its essence corporate. The word suggests communal, nonindividual, many acting together. Those who describe capitalism by stressing the individual entrepreneur miss the central point. Buying and selling by individual entrepreneurs occurred in biblical times. What is interesting and novel is the communal focus of the new ethos: the rise of communal risk taking, the pooling of resources, the sense of communal religious vocation in economic activism.

Corporations depend on the emergence of an infrastructure in intellectual life that makes possible new forms of communal collaboration. They depend on ideas that are powerful and clear enough to organize thousands of persons around common tasks. Moreover, these ideas must be strong enough to endure for years, so that individuals who commit themselves to them can expect to spend thirty to forty years working out their vocation. For many millions of religious persons, the daily milieu in which they work out their salvation is the communal, corporate world of the workplace. For many, the workplace is a kind of second family. Even those who hate their work often like their coworkers. Comradeship is natural to humans. Labor unions properly build on it.

Insight. The primary capital of any corporation is insight, invention, finding a better way. Insight is of many kinds and plays many roles: it is central to invention; it lies at the heart of organization; it is the vital force in strategies for innovation, production, and marketing. Corporate management works hard at communal insight. Constantly, teams of persons meet to brainstorm and work out common strategies. Insight is the chief resource of any corporation, and there cannot be too much of it. Its scarcity is called stupidity.

Karl Marx erred in thinking that capital has to do primarily with machinery, money, and other tangible instruments of production. He overlooked the extent to which the primary form of capital is an idea. (See also Roman and Loebl 1977, 22–23.) The right to patent industrial ideas is an extremely important constitutional liberty. It is indispensable to the life of corporations, as indispensable as the copyright is to writers. Money without ideas is not yet capital. Machinery is only as good as the idea it embodies. The word "capital," from the Latin *caput,* "head," points to the human spirit as the primary form of wealth. The miser sitting on his or her gold is not a capitalist. The investor with an idea is a capitalist. Insight makes the difference.

The Risk of Liberty and Election. A corporation faces liberty and election; it is part of its romance to do so. Tremendous mistakes in strategy can cripple even the largest companies. Easy Washing Machines of Syracuse once made an excellent washing machine, but Maytag's discovery of a new technology took away part of Easy's market. Easy had all its assets sunk in a plant that it could not redo quickly enough to incorporate the new technology, and the company collapsed. Thus a sudden technological breakthrough, even a relatively minor one, can cripple a company or an industry. A simple strategic mistake by a team of corporate executives about where to apply the company's energies over a year or two can end up dimming the company's outlook for many years.

The corporation operates in a world of no scientific certainty in which corporate leaders must constantly make judgments about reality when not all evidence about reality is in. Such leaders argue among themselves about strategic alternatives, each perhaps saying, "We will see who is right about this," or "The next year or two will tell." But a judgment must be made and the investment committed *before* the telling is completed. Thus decision makers often experience the risks inherent in their decisions. At the very least they always face the risk of doing considerably less well than they think they are going to do.

In these seven ways, corporations offer metaphors for grace, a kind of insight into God's ways in history. Yet corporations are of this world. They sin. They are *semper reformanda*—always in need of reform.

PROBLEMS OF BIGNESS AND OTHER ACCUSATIONS. Big corporations are despised and rejected even when the market system, small businesses, and private ownership are not. Some religious socialists do not absolutely reject certain elements in the democratic capitalist idea. But they commonly bridle at the *big* corporations. Their accusations against such corporations—many of them as true as charges made against the universities or against any large institution—are many.

One accusation is that the corporations are autocratic, that internally they are not democratic. In trying to decide how true this charge is, one could undertake a survey of the management techniques of the *Fortune 500* corporations. How are they actually managed? How does their management differ in practice from the internal management of universities, churches, government agencies, or other institutions? Let us suppose that some autocrats still function in various spheres of authority today, including business. What sanctions are available to autocrats within a corporation? Leadership in all spheres today seems to depend upon large areas of consensus; leaders seem to "manage" more than they "command." I have roughly the same impression of the chief executive officers I have met as of the American Catholic bishops I have met; namely, that out of the office they would find it hard, as Schumpeter says, to say boo to a duck. Few, as I see them, are autocrats. But empirical tests are in order to see how many autocrats are in corporations as distinct from any other sphere of life.

A second frequent accusation against big corporations is the alienation their employees experience in the workplace. To what extent is such alienation caused by the conditions of modern work under any existing system or under any imaginable system? Do laborers in auto factories in Bratislava or Poznan work under conditions any different from those faced by laborers in the United States? One ought to compare hours of work, conditions of the workplace, salaries, working procedures, and levels of pollution. There is no evidence that any real or imagined socialism can take the modernity out of modern work. Nor is boring work unique to the modern factory; it surely dominated the ancient work of European peasants and continues to dominate the fourteen-hour day of the modern potato farmer. Farming is not, in my experience, inherently less alienating than working seven hours, with time off for lunch, on an assembly line. Alienation is not a problem peculiar to capitalism or to corporations. Is work less alienating within a government bureaucracy?

A third accusation against corporations is that they represent too great a concentration of power. What is the alternative? There is indeed a circle within which small is beautiful, a relatively small and beautiful circle. But "small is beautiful" does not apply across the whole large world.

Socialist economist Robert Lekachman (1980, 40) has argued that the big corporations should be reduced in size to more manageable proportions. Maybe so. To my mind the question is a practical, experimental one. Consider the largest of all corporations, General Motors. It is already broken up into more than 200 units in more than 177 congressional districts in the United States. Its largest single facility, in Michigan, employs no more than 14,000 people. Many universities—the University of Michigan and Michigan State, to name two—comprise human communities two or three times that size. Corporations already follow the principle of subsidiarity far more thoroughly than Lekachman seems to take into account. One might argue that they should be still smaller. Yet one must note that the smaller U.S. auto companies— American Motors, Chrysler, and Ford—are apparently in danger of perishing because of inadequate capital to meet the enormous expenses of retooling for new auto technologies. The foreign auto companies competing with General Motors (even in the United States) are also very large. If small is beautiful, its beauty seems precarious indeed; big may be necessary.

In practice, I cannot imagine how human capacities and human choices of the sort needed by mass markets could be made available except through large organizations. Small organizations may suit a small country, but it seems to me absurd to imagine that a continental nation with a population of 220 million can be well served in all respects only through small organizations in small industries. If somebody can invent a system of smallness, fine; I am not, in principle, against it. I just cannot imagine that it can work in practice.

Corporations are further accused of being inherently evil because they work for a profit. Without profit no new capital is made available for research, development, and new investment. Furthermore, there is a difference between maximization of profit and optimization of profit. To aim at maximizing profit—that is, to obtain the greatest profit possible out of every opportunity—is to be greedy in the present at the expense of the future. Profit maximizers demand too much for products that can be produced more cheaply by somebody else and in the process narrow their own markets and destroy their own reputations. Inevitably, they damage themselves and, in time, destroy themselves. Adam Smith made this point a long time ago, and history is replete with examples of it. By contrast, to optimize profit is to take many other factors besides profit into account, including long-term new investment, consumer loyalty, and the sense of a fair service for a fair price.

The profit motive must necessarily operate in a socialist economy, too. Every economy that intends to progress must have as its motive the ability to get more out of the economic process than it puts in. Unless there is a return on investment, the economy simply spins its wheels. Capital accumulation is what profits are called in socialist enterprises. If the Soviets invest money in dams or in building locomotives, they must get back at least what they invest or they lose money. If they do lose money—and they often do—then they must draw on other resources. And if they do that throughout the system, economic stagnation and decline are inevitable. The same law binds both socialist and capitalist economies: economic progress, growth, and forward motion cannot occur unless the return on investment is larger than the investment itself.

It is true that under socialism profits belong to the state and are allocated to individuals by the state for the state's own purposes. Such a procedure can be institutionalized, but the costs of enforcing it are great. It tremendously affects the possibilities of liberty, of choice. It deeply affects incentives and creativity.

Objections to corporations are many. Some are clearly justified. Some are spurious. A full-dress theology of the corporation would properly evaluate each one fairly from many points of view. A convenient summary of some of them is to be found in *The Crisis of the Corporation* by Richard J. Barnet. Barnet makes three major accusations: (1) that the multinational corporations have inordinate power; (2) that they weaken the powers of the nation-state; and (3) that their actual practice destroys several "myths" about corporations.

The power of the multinational corporations, Barnet believes, springs from their ability to internationalize planning, production, finance, and marketing. In planning, each part can specialize, so that the whole pursues "profit maximization." In production, resources from various lands are integrated. In finance, computerization allows multinational corporations to take advantage of fluctuations in capital markets. In marketing, goods and consumption are standardized (Barnet 1975, 7–8).

From another point of view, these accusations seem to list advantages. Any economic organization that can work as Barnet describes would seem to be well placed to produce the maximum number of goods at the lowest cost. This efficiency should have the effect of making the most practical use of scarce capital, while increasing that capital through profitable investment. (We have already noted an important difference between "profit maximization" and "profit optimization.") The purpose of an interdependent world economic order is to match off the strengths of one region with those of another: a region with capital reserves and high labor costs is needed by a region without capital reserves but cheap labor. The cost of ignoring each other would be high for both regions. Cooperation should produce benefits for both.

Barnet argues that the powers of nation-states are weakened because multinational corporations make intracorporate transfers of funds without the knowledge of national governmental bodies. In addition, he asserts, they shift production to low-wage areas with fewer "union troubles," move productive facilities to regions where tax advantages are greatest, have no loyalty to any one country, and use dominance in one national market to achieve dominance in others because they can "out-advertise" smaller local companies (Barnet 1975, 8–11).

If all of these assertions are true, at least sometimes and in some places, not all their effects are evil. Consider, for example, the competition between Japanese, European, and U.S. automakers. The new reality is that market competition has been internationalized. Every such new development has advantages and costs. It appears that U.S. citizens benefit in quality and cost from this competition. Obviously, foreign autoworkers would seem to benefit. Unless U.S. automakers can do better, U.S. autoworkers will continue to suffer.

Would the world be a better place if each nation-state tried solely to protect its own industries? At various times, protectionism has triumphed. Nations do have the power to expel, close out, restrict, and nationalize foreign industries; often they do. This course, too, has costs as well as advantages. Barnet does not show that its costs are lower than those of the competition he opposes. No matter how Chrysler advertised during 1979, it did not seem to move the cars it tried to sell. Advertising is far less exact than he imagines.

Barnet argues, finally, that monopolization undercuts competitive free enterprise. He concedes that monopoly scarcely exists, but hastens to substitute for it oligopoly (four major firms, for example, controlling a majority of sales in several industries) whose "effects are much the same" (Barnet 1975, 14). He argues that efficiency is undercut by intracorporate transactions (as when tax laws encourage the shipment of products over long distances, when similar products could be acquired locally); that income distribution between the top 20 percent and the bottom 20 percent of income earners in the United States has "remained the same for forty-five years"; and that democracy is not enhanced by a free economy (Barnet 1975, 16–17, 20, 21–22).

Since Barnet himself is in favor of state monopolies in the socialist pattern, his objections to "oligopolies" do not have an authentic ring. Surely, four large companies in an industry are better than one state monopoly. Moreover, in the international field, the three major U.S. auto companies, for example, compete not only with each other but with Volvo, Fiat, Peugeot, Volkswagen, Toyota, and many others. In other industries, international competition is also a reality (Thurow 1980).

It is true that prices in a complex, highly technological industry are not simpleminded, but to suggest that they are "no longer a useful indicator" of cost and value falls short of sophistication. Consumers today make economic choices not only between which car to buy but whether to buy a car at all, or to invest the money, or to build an addition on the house, and so forth. In seeking the consumer's dollars, producers compete not only with others in their own industries but with other industries altogether. Pricing, however sophisticated the process through which it is reached, still affects the decisions of purchasers, as alternative marketing strategies amply demonstrate.

With respect to income distribution, most socialists today recognize that incomes are not and cannot be perfectly equal. They certainly are not in socialist countries. If persons at the top end of the income ladder receive eight times as much as those at the bottom, it follows that the total share of income of those at the top will be significantly higher than that of a similar cohort at the bottom. This relationship is strictly arithmetical. Imagine that Barnet himself earns fifty thousand dollars a year from his salary and royalties. This income would rank among the top 3 percent of all U.S. households, seven times as high as the official poverty level for a nonfarm family of four (U.S. Bureau of the Census 1980, 1). Arithmetically, his class must accumulate a disproportionate share of all U.S. income.

There is a further point. One must not compare only percentiles — snapshots of groups at one point in time. As a graduate student at Harvard Law School, Barnet's income was certainly lower than it is now; it may even have been below the poverty level. This did not, except technically, make Barnet "poor." At each decade thereafter, one would expect his income to place him in a different percentile. While percentiles may remain relatively constant, individuals in a free, mobile society rise and fall between them. Moreover, a family's relative wealth depends in the long run — say, three generations — on the sort of investment it makes with available funds. Investments in consumption at each moment preclude growth; investments in education, property, and the like make future material improvement probable. Thus, in many families, one generation works not solely for itself but for its future progeny. As it happens, families once wealthy sometimes experience economic decline, and families once poor sometimes become better off than in earlier generations. One must track not simply the statistical percentiles but the rise and fall over time of individuals and families within these percentiles. One would expect some individuals and families to be more intelligent, wiser, and luckier over time than others. Inequality of income is no more a scandal than are inequalities of looks, personality, talent, will, and luck. Inequality of income appears to be an inevitable fact in all large societies.

There is a peculiar historical link — which even Marxists recognize — between the emergence of liberal democracy in Great Britain, the United States, the Netherlands, and a few cognate lands and the emergence of a free economy. One might be satisfied to stress the *historical* character of the link. But it also seems to have a *necessary conceptual* character as well. If individuals lack fundamental economic liberties (to earn, spend, save, and invest as they see fit), they necessarily have few effective political liberties. If they are dependent upon the state for economic decisions, they must be wards of the state in other matters. Moreover, to believe that state bureaucrats are competent to make economic decisions beneficial for the common good is to make a great leap of faith, when one considers the actual economic well-being of workers in the U.S.S.R., Poland, Cuba, Yugoslavia. Even the democratic socialists of Sweden and West Germany insist upon vital economic liberties for individuals and corporations.

Socialist societies do not permit private corporations. They operate on the assumption that state officials know best what is for the common good. In reflecting on their actual practice, one may come to believe that democratic capitalism is more likely to meet the goals of socialism — plus other goals of its own — than socialism is. The social instrument invented by democratic capitalism to achieve social goals is the private corporation. Anyone can start one; those who succeed in making them work add to the common benefit. Yet corporations do not live (or die) in a vacuum. They must meet the demands of the moral-cultural system and of the political system. While corporations spring from some of our most cherished ideals about liberty, initiative, investment in the future, cooperation, and the like, they must also be judged in the light of our ideals. They are moral-cultural institutions, as well as economic institutions. Their primary task is economic. One cannot ask them to assume crushing and self-destructive burdens. Yet they are more than economic organisms alone and must be held to political and moral judgment.

THREE SYSTEMS – THREE FIELDS OF RESPONSIBILITY. The most original social invention of democratic capitalism, in sum, is the private corporation founded for economic purposes. The motivation for this invention was also social: to increase "the wealth of nations," to generate (for the first time in human history) sustained economic development. This effect was, in fact, achieved. However, the corporation – as a type of voluntary association – is not merely an economic institution. It is also a moral institution and a political institution. It depends upon and generates new political forms. In two short centuries, it has brought about an immense social revolution. It has moved the center of economic activity from the land to industry and commerce.

Beyond its economic effects, the corporation changes the ethos and the cultural forms of society. To some extent, it has undercut ancient ways of human relating, with some good effects and some bad. After the emergence of corporations, religion has had to work upon new psychological realities. The religion of peasants has given way to the religion of new forms of life: first an urban proletariat then a predominantly service and white-collar society. The productivity of the new economics has freed much human time for questions other than those of mere subsistence and survival. The workday has shrunk and after work, millions take part in voluntary activities – meetings, associations, sports, travel, politics, religion, and the like. Personal and social mobility has increased. Schooling has become mandatory. Teenagerhood has been invented. Room has been created for the emergence of the private self.

But the corporation is not only an economic institution and a moral-cultural institution. It also provides a new base for politics. Only a free political system permits the voluntary formation of private corporations. Thus, those who value private economic corporations have a strong interest in resisting both statism and socialism. It would be naive and wrong to believe that persons involved in corporations are (or should be) utterly neutral about political systems. An economic system within which private corporations play a role, in turn, alters the political horizon. It lifts the poor, creates a broad middle class, and undermines aristocracies of birth. Sources of power are created independent of the power of the state, in competition with the powers of the state, and sometimes in consort with the powers of the state. A corporation with plants and factories in, say, 120 congressional districts represents a great many employees and stockholders. On some matters, at least, these are likely to be well organized to express their special political concerns. Political jurisdictions often compete to attract corporations, but their arrival also creates political problems.

Corporations err morally, then, in many ways. They may through their advertising appeal to hedonism and escape. They may incorporate methods of governance that injure dignity, cooperation, inventiveness, and personal development. They may seek their own immediate interests at the expense of the common good. They are as capable of sins as individuals are and, in the fashion of all institutions, of grave institutional sins as well. Thus, it is a perfectly proper task of all involved within corporations and in society at large to hold them to the highest moral standards. Corporations are human institutions designed to stimulate economic polity committed to high moral-cultural ideals. When they fall short of these purposes, their failure injures all.

Private corporations are social organisms. Neither the ideology of laissez faire nor the ideology of rugged individualism suits their actual practice or their inherent ideals. They cannot come into existence, and certainly cannot function, except within political systems designed to establish and to promote the conditions of their flourishing — a sound currency, a system of laws, the regulation of competitive practices, and the construction of infrastructures like roads, harbors, airports, and certain welfare functions. The state, then, plays an indispensable role in democratic capitalism.

The ideals of democratic capitalism are not purely individualist, either, for the corporation draws upon and requires highly developed social skills like mutual trust, teamwork, compromise, cooperation, creativity, originality and inventiveness, and agreeable management and personnel relations.

Great moral responsibility, then, is inherent in the existence of corporations. Frequently enough, they err. They are properly subjected to constant criticism and reform. Some critics accept the ideals inherent in the system of private business corporations, and simply demand that corporations be faithful to these ideals. Some critics are opposed to the system *qua* system.

There is plenty of room — and plenty of evidence — for citing specific deficiencies of corporations: economic, political, and moral-cultural. To be sure, there is a difference between accusations and demonstrated error. Like individuals, corporations are innocent until proven guilty. A passionate hostility toward bigness (or even toward economic liberty), like a passionate commitment to statism, may be socially useful by providing a searching critique from the viewpoint of hostile critics. But unless it gets down to cases and sticks to a reasoned presentation of evidence, it must be recognized for what it is: an argument less against specifics than against the radical ideal of democratic capitalism and the private corporation. It is useful to distinguish these two types of criticism, and helpful when critics are self-conscious and honest about which ideals actually move them. To criticize corporations in the light of their own ideals, the ideals of democratic capitalism, is quite different from criticizing them in the name of statist or socialist ideals incompatible with their existence. Clarity about ideals is as necessary as clarity about cases.

Theologians, in particular, are likely to inherit either a precapitalist or a frankly socialist set of ideals about political economy. They are especially likely to criticize corporations from a set of ideals foreign to those of democratic capitalism. To those who do accept democratic capitalist ideals, then, their criticisms are likely to have a scent of unreality and inappropriateness. Wisdom would suggest joining argument at the appropriate level of discourse: whether the argument concerns general economic concepts, the rival ideals of democratic capitalism and socialism, or concrete cases and specific matters of fact. Each of these levels has its place. Wisdom's principal task is *distinguer.*

Managing a free society aimed at preserving the integrity of the trinitarian system — the economic system, the political system, and the moral-cultural system — is no easy task. To govern a free economy is yet more difficult than to form a free government. It is hard enough to govern a government. It is difficulty squared to govern a free economy — to establish the conditions for prosperity, keep a sound currency, promote competition, establish general rules and standards binding upon all, keep markets free, provide education to all citizens, care for public needs, and provide succor to the unfortunate. To have the virtue to do all these things is surely of some rather remarkable theological significance. It may even represent — given the inherent difficulties — a certain amazing grace. To fall short is to be liable to judgment.

Christians have not, historically, lived under only one economic system; nor are they bound in conscience to support only one. Any real or, indeed, any imaginable economic system is necessarily part of history, part of this world. None is the Kingdom of Heaven — not democratic socialism, not democratic capitalism. A theology of the corporation should not make the corporation seem to be an ultimate; it is only a means. Such a theology should attempt to show how corporations may be instruments of redemption, humane purposes and values, and God's grace. The waters of the sea are blessed, as are airplanes and plowshares and even troops making ready for just combat. Christianity , like Judaism, attempts to sanctify the real world as it is, in all its ambiguity, so as to reject the evil in it and bring the good in it to its highest possible fruition.

Most Christians do not now work for major industrial corporations. Instead, they work for the state (even in state universities), for smaller corporations, and other businesses. Still, a Christian social theology that lacks a theology of the large corporation will have no effective means of inspiring those Christians who do work within large corporations to meet the highest practicable Christian standards. It will also have no means of criticizing with realism and practicality those features of corporate life that deserve to be changed. Whether to treat big corporations as potential vessels of Christian vocation or to criticize them for their inevitable sins, Christian theology must advance much further than it has in understanding exactly and fairly every aspect of corporate life. The chief executive officer of General Electric needs such a theology. So do those critics of the corporation at the Interfaith Center for Corporate Responsibility. If we are to do better than clash like ignorant armies in the night, we must imitate Yahweh at Creation and say, "Let there be light." We have not yet done all we should in casting such light.

• • •

In this context, a final word or two about agriculture. Many family farms are incorporated; they, too, are corporations, to which the above reflections should apply. Large agribusinesses, however, are subject to the further liabilities inherent in all bigness. In some ways, large corporations are like dinosaurs; they are necessarily bureaucratic.

In premodern times, land was the basis of most wealth. And the ancient justification for private property was that families had a generational interest in improving the land. In this way, over time, private owners added to the common good. It has for centuries been the experience that, when private property is ignored and social ownership is tried, no one cares enough to stay up all night with a sick cow, as a family owner does. Family ownership—private property—has great social utility. This argument today reinforces the claim of the family farm.

On the other hand, human needs today are immense. On our own farm in Iowa, my wife and I have known our renter to use his equipment over our forty acres of corn in a single day. So it is not so hard to understand why modern technology demands larger and larger combinations of land. Will this, in the long run, be good for the land? There are families that have used the land badly; there are bound to be agri-businesses that also use it badly—and over vastly larger tracts of it. But the future exerts its exigent demands upon all present owners of land. Whether for families or for agribusiness, stewardship is pivotal. The land demands respect. Those who abuse it lose it.

No system of political economy—not democratic capitalism, not the family system, not the large corporate system, not *any* system—is fool-proof. There are characteristic limitations and evil tendencies in each; each must be disciplined by vigilance. Iowa State University is to be commended, therefore, for stimulating passionate, plural arguments about the future of America's "fields of waving grain." Our forebears, who so few generations ago were the first to cut this soil with a plow, depend on us to protect the abundance they bequeathed to us. Their creative vision—and the risks they took—brought it into being. To be good stewards in our turn is, as theirs was, at bottom a religious task.

References

Barnet, Richard J. *The Crisis of the Corporation*. Washington, D.C.: Institute for Policy Studies, 1975.

Bernanos, Georges. *Diary of a Country Priest*. Translated by Pamela Morris. New York: Macmillan, 1962.

Cort, John C. "Can Socialism Be Distinguished from Marxism?" *Cross Currents* 29 (Winter 1979–80): 423–34.

Hesburgh, Theodore. *The Humane Imperative: A Challenge for the Year 2000*. New Haven, Conn.: Yale University Press, 1974.

Johnson, Paul. "Is There a Moral Basis for Capitalism?" In *Democracy and Mediating Structures: A Theological Inquiry,* edited by Michael Novak. Washington, D.C.: American Enterprise Institute, 1980.

Lea, Sperry, and Simon Webley. *Multinational Corporations in Developed Countries: A Review of Recent Research and Policy Thinking*. Washington, D.C.: British-North American Committee, 1973.

Lekachman, Robert. "The Promise of Democratic Socialism." In *Democracy and Mediating Structures: A Theological Inquiry,* edited by Michael Novak. Washington, D.C.: American Enterprise Institute, 1980.

Locke, John. *Second Treatise of Civil Government*. New York: Macmillan, 1947.

Macpherson, C. B. *The Political Theory of Possessive Individualism: Hobbes to Locke*. New York: Oxford University Press, 1962.

Martin, Linda Grant. "The 500: A Report on Two Decades." *Fortune,* May 1975, 238.

Muller, Ronald E. "The Multinational Corporation: Asset or Impediment to World Justice?" In *Poverty, Environment and Power,* edited by Paul Hallock. New York: International Documentation on the Contemporary Church—North America, 1973.

Roman, Stephen B., and Eugen Loebl. *The Responsible Society.* New York: Regina Ryan Books/Two Continents, 1977.

Stein, Herbert. *Capitalism — If You Can Keep It.* Washington, D.C.: American Enterprise Institute, 1980.

Thurow, Lester C. "Let's Abolish the Antitrust Laws." *New York Times,* 19 October 1980.

U.S. Bureau of the Census. *Money Income and Poverty Status of Families and Persons in the United States: 1979 (Advanced Report).* Washington, D.C.: Government Printing Office, 1980.

_____. *Statistical Abstract of the United States, 1979.* Washington, D.C.: U.S. Department of Commerce, 1979.

A Catholic, Populist View

The Christian Vision of Economic Life

The basis for all that the Church believes about the moral dimensions of economic life is its vision of the transcendent worth — the sacredness — of human beings. The dignity of the human person, realized in community with others, is the criterion against which all aspects of economic life must be measured (John XXIII 1961, 219–20; see Vatican Council II 1966, 199–308). All human beings, therefore, are ends to be served by the institutions that make up the economy, not means to be exploited for more narrowly defined goals. Human personhood must be respected with a reverence that is religious. When we deal with each other we should do so with the sense of awe that arises in the presence of something holy and sacred. For that is what human beings are: we are created "in the image of God" (Gen. 1:27). Similarly, all economic institutions must support the bonds of community and solidarity that are essential to the dignity of persons. Wherever our economic arrangements fail to conform to the demands of human dignity lived in community they must be questioned and transformed. These convictions have a biblical basis. They are also supported by a long tradition of theological and philosophical reflection and through the reasoned analysis of human experience by contemporary men and women.

In presenting the Christian moral vision, we turn first to the Scriptures for guidance. Though our comments are necessarily selective, we hope that pastors and other church members will become personally engaged with the biblical texts. The Scriptures contain many passages that speak directly of economic life. We must also attend to the Bible's

Excerpts from the third draft of the U.S. Catholic bishops' pastoral letter *Economic Justice for All: Catholic Social Teaching and the U.S. Economy,* copyright 1986 by the United States Catholic Conference, Washington, D.C., are used with permission. Copies of the final letter are available from the Office of Publishing Services, U.S.C.C., 1312 Massachusetts Avenue, N.W., Washington, D.C. 20005.

deeper vision of God, of the purpose of creation, and of the dignity of human life in society. Along with other churches and ecclesial communities who are "strengthened by the grace of Baptism and the hearing of God's Word," we strive to become faithful hearers and doers of the word (Vatican Council II 1966, 341–66). We also claim the Hebrew Scriptures as common heritage with our Jewish brothers and sisters, and we join with them in the quest for an economic life worthy of the divine revelation we share.

Biblical Perspectives

The fundamental conviction of our faith is that human life is fulfilled in the knowledge and love of the living God in communion with others who, as recipients of God's love, are called to love the same God. The sacred Scriptures offer guidance so that men and women may enter into full communion with God and with each other and may witness to God's saving acts. We discover there a God who is creator of heaven and earth and of the human family. Though our first parents reject the God who created them, God does not abandon them, but from Abraham and Sarah forms a people of promise. When this people is enslaved in an alien land, God delivers them and makes a covenant with them in which they are summoned to be faithful to the *torah* or sacred teaching. The focal points of Israel's faith — creation, covenant, and saving history — provide a foundation for reflection on issues of economic and social justice.

CREATED IN GOD'S IMAGE. After the exile, when Israel combined its traditions into a written torah, it prefaced its history as a people with the story of the creation of all peoples and of the whole world by the same God who created them as a nation (Gen. 1–11). God is the creator of heaven and earth (Gen. 14:19–22; Isa. 40:28; 45:18); creation proclaims God's glory (Ps. 89:5–12) and is "very good" (Gen. 1:31). Fruitful harvests, bountiful flocks, a loving family, are God's blessings on those who heed God's word. Such is the joyful refrain that echoes throughout the Bible. One legacy of this theology of creation is the conviction that no dimension of human life lies beyond God's care and concern. God is present to creation and creative engagement with God's handiwork is itself reverence for God.

At the summit of creation stands the creation of man and woman made in God's image (Gen. 1:26–27). As such every human being possesses an inalienable dignity that stamps human existence prior to any division into races or nations and prior to human labor and human achievement (Gen. 4–11). Men and women are also to share in the creative activity of God. They are to be fruitful, to care for the earth (Gen. 2:15), and to have "dominion" over it (Gen. 1:28), which means that they are "to govern the world in holiness and justice and to render judgment in integrity of heart" (Wis. 9:3). Creation is a gift; men and women are to be faithful stewards in caring for the earth. They can justly consider that by their labor they are unfolding the Creator's work (Westermann 1974; Vawter 1977).

The narratives of Genesis 1–11 also portray the origin of the strife and suffering that mar the world. Though created to enjoy intimacy with God and the fruits of the earth, Adam and Eve disrupted God's design by trying to live independently of God through a denial of their status as creatures. They turned away from God and gave to God's creation the obedience due to God alone. For this reason the prime sin in so much of the biblical tradition is idolatry: service of the creature rather than of the creator (Rom. 1:25), and the attempt to overturn creation by making God in human likeness. The Bible castigates not only the worship of idols, but also manifestations of idolatry such as the quest for unrestrained power and the desire for great wealth (Isa. 40:12–20; 44:1–20; Wis. 13:1–14:31; Col. 3:5, "covetousness which is idolatry"). The sin of our first parents had other consequences as well. Alienation from God pits brother against brother (Gen. 4:8–16) in a cycle of war and vengeance (Gen. 5:22–23). Sin and evil abound, and the primeval history culminates with another assault on the heavens, this time ending in a babble of tongues scattered over the face of the earth (Gen. 11:1–9). Sin simultaneously alienates human beings from God and shatters the solidarity of the human community. Yet this reign of sin is not the final word. The primeval history is followed by the call of Abraham, a man of faith, who was to be the bearer of the promise to many nations (Gen. 12:1–4). Throughout the Bible we find this struggle between sin and repentance. God's judgment on evil is followed by God's seeking out a sinful people.

The biblical vision of creation has provided one of the most enduring legacies of Church teaching. To stand before God as creator is to respect God's creation, both the world of nature and of human history. From the Patristic period to the present, the Church has affirmed that misuse of the world's resources or appropriation of them by a minority of the world's population betrays the gift of creation since "whatever belongs to God belongs to all" (St. Cyprian 1958, 251; also Avila 1983).

A PEOPLE OF THE COVENANT. When the people of Israel, our forerunners in faith, gathered in thanksgiving to renew their covenant, they recalled the gracious deeds of God (Josh. 24:1–15; Deut. 6:20–25; 26:5–11). When they lived as aliens in a strange land and experienced oppression and slavery, they cried out. The Lord, the God of their ancestors, heard their cries, knew their afflictions, and came to deliver them (Exod. 3:7–8). By leading them out of Egypt, God created a people that was to be the Lord's very own (Jer. 24:7; Hos. 2:23). They were to imitate God by treating the alien and the slave in their midst as God had treated them (Exod. 22:21–22; Jer. 34:8–14).

In the midst of this saving history stands the covenant at Sinai (Exod. 19–24). It begins with an account of what God has done for the people (Exod. 19:1–6; cf. Josh. 24:1–13) and includes from God's side a promise of steadfast love (*hesed*) and faithfulness (*'emeth,* Exod. 34:5–7). The people are summoned to ratify this covenant by faithfully worshiping God alone and by directing their lives according to God's will, which was made explicit in Israel's great legal codes such as the Decalogue (Exod. 20:1–17) and the Book of the Covenant (Exod. 20:22–23:33). Far from being an arbitrary restriction of the life of the people, these codes made life in community possible (Ogletree 1983, 47–85). The specific laws of the covenant protect human life and property, demand respect for parents and the spouses and children of one's neighbor, and manifest a special concern for the vulnerable members of the community: widows, orphans, the poor, and strangers in the land. Laws such as that for the sabbath year when the land was left fallow (Exod. 23:11; Lev. 25:1–7) and for the year of release of debts (Deut. 15:1–11) summoned people to respect the land as God's gift and reminded Israel that as a people freed by God from bondage they were to be concerned for the poor and oppressed in their midst. Every fiftieth year a Jubilee was to be proclaimed as a year of "liberty throughout the land" and property was to be restored to its original owners (Lev. 25:8–17; cf. Isa. 61:1–2; Luke 4:18–19). (See North 1954; Ringe 1985.) The codes of Israel reflect the norms of the covenant: reciprocal responsibility, mercy, and truthfulness. They embody a life in freedom from oppression: worship of the One God, rejection of idolatry, mutual respect among people, care and protection for every member of the social body. Being free and being a coresponsible community are God's intent for us.

When the people turn away from the living God to serve idols and no longer heed the commands of the covenant, God sends prophets to recall his saving deeds and to summon them to return to the one who betrothed them "in right and in justice, in love and in mercy" (Hos. 2:21). The substance of prophetic faith is proclaimed by Micah: "to do justice and to love kindness, and to walk humbly with your God" (Mic. 6:8). Biblical faith in general, and prophetic faith especially, insist that fidelity to the covenant joins obedience to God with reverence and concern for the neighbor. The biblical terms that best summarize this double dimension of Israel's faith are *sedaqah,* justice (also translated as righteousness), and *mishpat* (right judgment or justice embodied in a concrete act or deed). The biblical understanding of justice gives a fundamental perspective to our reflections on social and economic justice (see Donahue 1977, 68–112; Mott 1982).

Yahweh is described as a "God of Justice" (Isa. 30:18) who loves justice (Isa. 61:8, cf. Ps. 11:7; 33:5; 37:28; 99:4) and delights in it (Jer. 9:24). God demands justice from the whole people (Deut. 16:20) and executes justice for the needy (Ps. 140:12). A distinctive aspect of the biblical presentation of justice is that the justice of a community is measured by its treatment of the powerless in society, most often described as the widow, the orphan, the poor, and the stranger (non-Israelite) in the land. The Law, the Prophets, and the Wisdom literature of the Old Testament all show deep concern for the proper treatment of such people (Exod. 22:21–27; Deut. 15:1–11; Job 29:11–17; Ps. 69:33; 72:1, 4, 12–24; 82:3–4; Prov. 14:21, 31; Isa. 3:14–15; 10:2; Jer. 22:16; Zech. 7:9–10). What these groups of people have in common is their vulnerability and lack of power. They are often alone and have no protector or advocate. Therefore it is God who hears their cries (Ps. 109:21; 113:7), and the king who is God's anointed is commanded to have special concern for them.

Justice has many nuances (Pedersen 1926, 337–40). Fundamentally it suggests a sense of what is right or should happen. For example, paths are just when they bring you to your destination (Gen. 24:48; Ps. 23:3), and laws are just when they create harmony within the community, as Isaiah says: "Justice will bring about peace; right will produce calm and security" (Isa. 32:17). God is "just" by acting as God should, coming to the people's aid and summoning them to conversion when they stray. People are summoned to be "just," that is, to be in a proper relation to God by observing God's laws, which form them into a faithful community. Biblical justice is more comprehensive than subsequent philosophical definitions. It is not concerned with a strict definition of rights and duties but with the rightness of the human condition before God and within society. Nor is justice opposed to love; rather, it is both a manifestation of love and a condition for love to grow (Alfaro 1973, 40–41; McDonagh 1982, 119). Because God loves Israel, he rescues them from oppression and summons them to be a people that "does justice" and loves kindness. The quest for justice arises from loving gratitude for the saving acts of God and manifests itself in wholehearted love of God and neighbor.

These perspectives provide the foundation for a biblical vision of economic justice. Every human person is created as an image of God, and the denial of dignity to a person is a blot on this image. Creation is a gift to all men and women, not to be appropriated for the benefit of a few; its beauty is an object of joy and reverence. The same God who came to the aid of an oppressed people and formed them into a covenant community continues to hear the cries of the oppressed and to create communities that are to hear his word. God's love and life are present when people can live in a community of faith and hope. These cardinal points of the faith of Israel also furnish the religious context for understanding the saving action of God in the life and teaching of Jesus.

THE REIGN OF GOD AND JUSTICE. Jesus enters human history as God's anointed son who announces the nearness of the reign of God. This proclamation summons us to acknowledge God as creator and covenant partner and challenges us to seek ways in which God's revelation of the dignity and destiny of all creation might become incarnate in history. It is not simply the promise of the future victory of God over sin and evil but that this victory has already begun — in the life and teaching of Jesus.

What Jesus proclaims by word he enacts in his ministry. He resists temptations of power and prestige, follows his Father's will, and teaches us to pray that it be accomplished on earth. He warns against attempts to "lay up treasures on earth" (Matt. 6:19) and exhorts his followers not to be anxious about material goods but rather to seek first God's reign and God's justice (Matt. 6:25–33). His mighty works symbolize that the reign of God is more powerful than evil, sickness, and the hardness of the human heart. He offers God's loving mercy to sinners (Mark 2:17), takes the cause of those who suffered religious and social discrimination (Luke 7:36–50; 15:1–2), and attacks the use of religion to avoid the demands of charity and justice (Mark 7:9–13; Matt. 23:23).

WEALTH AND RICHES. The life and teaching of Jesus as presented in the Gospel of Luke has special relevance today. It was written to a community composed of both poor and relatively well-off members. In Luke the first public utterance of Jesus is not (as in Matthew and Mark), "The kingdom of God is near," but "The Spirit of the Lord is upon me, because he has anointed me to preach the good news to the poor" (Luke 4:18). Jesus adds to the blessing on the poor a warning, "Woe to you who are rich, for you have received your consolation" (Luke 6:24). He warns his followers against greed and reliance on abundant possessions and underscores this by the parable of the man whose life is snatched away at that very moment he tries to secure his wealth (Luke 12:13–21).

In Luke alone Jesus tells the parable of the rich man who does not see the poor and suffering Lazarus at his gate (Luke 16:19–31). When the rich man finally "sees" Lazarus, it is from the place of torment and the opportunity for conversion has passed. Pope John Paul II has often recalled this parable to remind us of our world where great wealth and great poverty lie side by side (1979, 311–12; 1980, 139).

In Luke, Jesus lives as a poor man, and like the prophets takes the side of the poor and warns of the dangers of wealth (Dupont and George 1971; Hengel 1974; Johnson 1981; Mealand 1980; Pilgrim 1981; Stegemann 1984). The terms used for poor, while primarily describing lack of material goods, also suggest dependence and powerlessness. Throughout the Bible poverty is a misfortune and a cause of sadness. From the earliest stages of Israel's traditions throughout the later literature and into the New Testament, there is a constant refrain that the poor must be cared for and protected and that when they are exploited God hears their cries. Conversely, even though the goods of the earth are to be enjoyed and people are to thank God for material blessings, great wealth is a constant danger. The rich are wise in their own eyes (Prov. 28:11), are prone to apostasy and idolatry (Amos 5:4–13; Isa. 2:6–8), as well as to violence and oppression (Amos 4:1–3; Job 20:19; Sir. 13:4–7; James 2:6; 5:1–6; Rev. 18:11–19). Since the poor in the biblical sense are not blinded by wealth or tempted to make it into an idol, they are open to God's presence and their human powerlessness makes them a model of those who trust in God alone.

In some of the later traditions that influence the New Testament, there emerge the "righteous poor" (Sir. 12:1–4) who are related to the humble (cf. Matt. 5:3, 5) or the "little ones" to whom God reveals what was hidden from the wise (Matt. 11:25–30). When Jesus calls the poor "blessed," he is not praising their condition of poverty but their openness to God. When he summons disciples to leave all and follow him, he is calling them to share his own radical trust in the Father and his freedom from care and anxiety (cf. Matt. 6:25–34). The practice of evangelical poverty in the church has always been a living witness to the power of that trust and to the joy that comes with that freedom.

Early Christianity saw the poor as an object of God's special love, but it neither canonized material poverty nor accepted deprivation as an inevitable fact of life. Though few early Christians possessed wealth or power (1 Cor. 1:26–28; James 2:5), their communities had well-off members (Acts 16:14; 18:8). Jesus's concern for the poor was continued in different forms in the early church. The early community at Jerusalem distributes its possessions so that "there was not a needy person among them," and "holds all things in common"—a phrase that suggests not only shared material possessions but even more basically friendship and mutual concern among all its members (Acts 2:44; 4:32). While recognizing the dangers of wealth, the early church proposed the proper use of possessions to alleviate need and suffering as the ideal rather than universal dispossession. Beginning in the first century and throughout the Patristic period, Christian communities organized structures to support and sustain the weak and powerless in a society that was often brutally unconcerned about human suffering.

Such perspectives provide a basis for what today is called the "preferential option for the poor" (Eagleson and Scharper 1979, 264–67; Dorr 1983). Though in the Gospels and in the New Testament as a whole the offer of salvation is extended to all peoples, Jesus takes the side of those most in need, physically and spiritually. The example of Jesus poses a number of challenges to the contemporary church. It imposes a prophetic mandate to speak for those who have no one to speak for them, to be a defender of the defenseless, who in biblical terms are the poor. It also demands a compassionate vision that enables the church to see things from the side of the poor and powerless and to assess lifestyle, policies, and social institutions in terms of their impact on the poor. It summons the church also to be an instrument in assisting people to experience the liberating power of God in their own lives so that they may respond to the Gospel in freedom and in dignity. Finally, and most radically, it calls for an emptying of self, both individually and corporately, that allows the church to experience the power of God in the midst of poverty and powerlessness.

A COMMUNITY OF HOPE. The biblical vision of creation, covenant, and community, as well as the summons to discipleship, unfolds under the tension between promise and fulfillment. The whole Bible is spanned by the narratives of the first creation (Gen. 1–3) and the vision of a new creation at the end of history (Rev. 21:14). Just as creation tells us that God's desire was one of wholeness and unity between God and the human family and within this family itself, the images of a new creation give hope that enmity and hatred will cease and justice and peace will reign (Isa. 11:6; 25:1–8). Human life unfolds "between the times," the time of the first creation and that of a restored creation (Rom. 8:18–25). Although the ultimate realization of God's plan lies in the future, Christians in union with all people of goodwill are summoned to shape history in the image of God's creative design and in response to the reign of God proclaimed and embodied by Jesus.

A Christian is a member of a new community, "God's own people" (1 Pet. 2:9–10), who, like the people of Exodus, owes its existence to the gracious gift of God and is summoned to respond to God's will made manifest in the life and teaching of Jesus. A Christian walks in the newness of life (Rom. 6:4) and is "a new creation; the old has passed away, the new has come" (2 Cor. 5:17). This new creation in Christ proclaims that God's creative love is constantly at work, offers sinners forgiveness, and reconciles a broken world. Our action on behalf of justice in our world proceeds from the conviction that, despite the power of injustice and violence, life has been fundamentally changed by the entry of the Word made flesh into human history.

Christian communities that commit themselves to solidarity with those suffering and to confrontation with those attitudes and ways of acting that institutionalize injustice, will themselves experience the power and presence of Christ. They will embody in their lives the values of the new creation while they labor under the old. The quest for economic and social justice will always combine hope and realism and must be renewed by every generation. It involves a diagnosis of those situations that continue to alienate the world from God's creative love as well as presenting hopeful alternatives that arise from living in a renewed creation. This quest arises from faith and is sustained by hope as it seeks to speak to a broken world of God's justice and loving kindness.

Ethical Norms for Economic Life

These biblical and theological themes shape the overall Christian perspective on economic ethics. The Catholic tradition also affirms that this ethical perspective is intelligible to those who do not share Christian religious convictions. Human beings are created in God's image, and their dignity is manifest in the ability to reason and understand, in their freedom to shape their own lives and the life of their communities, and in the capacity for love and friendship. Human understanding and religious belief are complementary, not contradictory. In proposing ethical norms, therefore, we appeal both to Christians and to all in our pluralist society to show that respect and reverence owed to the dignity of every person. Intelligent reflection on the social and economic realities of today are also indispensable in the effort to respond to economic circumstances never envisioned in biblical times. Therefore, we now want to outline an ethical framework that can guide economic life today in ways that are both faithful to the Gospel and shaped by human experience and reason. First are the duties all people have to each other and to the whole community: love of neighbor, the basic requirements of justice, and the special obligation to those who are poor or vulnerable. Corresponding to these duties are the human rights of every person: the obligation to protect the dignity of all demands respect for these rights. Finally these duties and rights suggest several priorities that should guide the economic choices of individuals, communities and the nation as a whole.

THE RESPONSIBILITIES OF SOCIAL LIVING. Human life is life in community. Catholic social teaching proposes several complementary perspectives that show how moral obligations and duties in the economic sphere are rooted in this call to community.

Love and Solidarity. The commandments to love God with all one's heart and to love one's neighbor as oneself are the heart and soul of Christian morality. These commands point out the path toward true human fulfillment and happiness. They are not arbitrary restrictions on human freedom. Only active love of God and neighbor makes the fullness of community happen. Christians look forward in hope to a true communion among all persons with each other and with God. The Spirit of Christ labors in history to build up the bonds of solidarity among all persons until that day on which their union is brought to perfection in the Kingdom of God (Paul VI 1975, 24). Indeed Christian theological reflection on the very reality of God as a trinitarian unity of persons — Father, Son, and Holy Spirit — shows that being a person means being united to other persons in mutual love (Vatican Council II 1966, 199–308).

What the Bible and Christian tradition teach, human wisdom confirms. Centuries before Christ the Greeks and Romans spoke of the human person as a "social animal," made for friendship, community, and public life. These insights show that human beings cannot grow to full self-realization in isolation but need interaction with others (Vatican Council II 1966, 199–308).

The virtues of citizenship are an expression of Christian love more crucial in today's interdependent world than ever before. These virtues grow out of a lively sense of one's dependence on the commonweal and obligations to it. This civic commitment must also guide the economic institutions of society. In the absence of a vital sense of citizenship among the businesses, corporations, labor unions, and other groups that shape economic life, society as a whole is endangered. Solidarity is another name for this social friendship and civic commitment that make human moral and economic life possible.

The Christian tradition recognizes, of course, that the fullness of love and community will be achieved only when God's work in Christ comes to completion in the kingdom of God. This kingdom has been inaugurated among us, but God's redeeming and transforming work is not yet complete. Within history, knowledge of how to achieve the goal of social unity is limited. Human sin continues to wound the lives of both individuals and larger social bodies and places obstacles in the path toward greater social solidarity. If efforts to protect human dignity are to be effective, they must take these limits on knowledge and love into account. Nevertheless, sober realism should not be confused with resigned or cynical pessimism. It is a challenge to develop a courageous hope that can sustain efforts that will sometimes be arduous and protracted.

Justice and Participation. The norms of basic justice state the minimum levels of mutual care and respect that all persons and communities owe to each other. Though the rich biblical understanding of the "wholeness" (*shalom*) of a fully just society reaches beyond these levels, the criteria of basic justice establish the minimum standards for all social and economic life. Catholic social teaching, like much philosophical reflection, distinguishes three dimensions of basic justice: commutative justice, social justice, and distributive justice (Pieper 1966, 43–116; Hollenbach 1977, 207–31).

Commutative justice calls for fundamental fairness in all agreements and exchanges between individuals or private social groups. It demands respect for the equal human dignity of all persons in economic transactions, contracts, or promises. For example, workers owe their employers diligent work in exchange for their wages. Employers are obligated to treat their employees as persons, paying them fair wages in exchange for the work done and by establishing conditions and patterns of work that are truly human (Gunnemann, forthcoming).

Distributive justice requires that the allocation of income, wealth, and power in society be evaluated in light of its effects on persons whose basic material needs are unmet. The Second Vatican Council (1966, 199–308) stated: "The right to have a share of earthly goods sufficient for oneself and one's family belongs to everyone. The Fathers and Doctors of the Church held this view, teaching that we are obliged to come to the relief of the poor and to do so not merely out of our superfluous goods." Minimum material resources are an absolute necessity for human life. If persons are to be treated as members of the human community, then the community has an obligation to help fulfill these basic needs unless an absolute scarcity of resources makes this strictly impossible. No such scarcity exists in the United States today.

Justice also has implications for the way the larger social, economic, and political institutions of society are organized. Social justice implies that persons have an obligation to be active and productive participants in the life of society and that society has a duty to enable them to participate in this way. This form of justice can also be called "contributive," for it stresses the duty of all who are able to help create the goods, services, and other nonmaterial or spiritual values necessary for the welfare of the whole community. Work should enable the working person to become "more a human being," more capable of acting intelligently, freely, and in ways that lead to self-realization (John Paul II 1981, 6,9).

Economic conditions that leave large numbers of able people unemployed, underemployed, or employed in dehumanizing conditions fail to meet the converging demands of these three forms of basic justice. Work with adequate pay for all who seek it is the primary means for achieving basic justice in our society. Discrimination in job opportunities or income levels on the basis of race, sex, or other arbitrary standard can never be justified (Vatican Council II 1966, 199–308). It is a scandal that such discrimination continues in the United States today. Where the effects of past discrimination persist, society has the obligation to take positive steps to overcome the legacy of injustice. Judiciously administered affirmative action programs in education and employment can be important expressions of the drive for solidarity and participation that is at the heart of true justice. Social harm calls for social relief.

Basic justice also calls for the establishment of a floor of material well-being on which all can stand. This is a duty of the whole of society and it creates particular obligations for those with greater resources. This duty calls into question extreme inequalities of income and consumption when so many lack basic necessities. Catholic social teaching does not maintain that a flat, arithmetical equality of income and wealth is a demand of justice, but it does challenge economic arrangements that leave large numbers of people impoverished. Further, it sees extreme inequality as a threat to the solidarity of the human community, for great disparities lead to deep social divisions and conflict (Vatican Council II 1966, 199–308).

This means that all of us must examine our way of living in light of others' needs. Christian faith and the norms of justice impose distinct limits on what we consume and how we view material goods. The great wealth of the United States can easily blind us to the poverty that exists in this nation and the destitution of hundreds of millions of people in other parts of the world. Americans are challenged today as never before to develop the inner freedom to resist the temptation constantly to seek more. Only in this way will the nation avoid what Paul VI (1967, 19) called "the most evident form of moral underdevelopment," namely greed.

These duties call not only for individual charitable giving but also for a more systematic approach by businesses, labor unions, and the many other groups that shape economic life—as well as government. The concentration of privilege that exists today results far more from institutional relationships that destribute power and wealth inequitably than from differences in talent or lack of desire to work. These institutional patterns must be examined and revised if we are to meet the demands of basic justice. For example, a system of taxation based on "assessment according to ability to pay" (John XXIII 1961) is a prime necessity for the fulfillment of these social obligations.

Overcoming Marginalization and Powerlessness. These fundamental duties can be summarized this way: basic justice demands the establishment of minimum levels of participation in the life of the human community for all persons. The ultimate injustice is for a person or group to be actively treated or passively abandoned as if they were nonmembers of the human race. To treat people in this way is effectively to say that they simply do not count as human beings. This can take many forms, all of which can be described as varieties of marginalization, or exclusion from social life (Synod of Bishops Second General Assembly 1971, 10, 16; Paul VI 1971, 15). This exclusion can occur in the political sphere: restriction of free speech, concentration of power in the hands of a few, or outright repression by the state. It can also take economic forms that are equally harmful. Within the United States, individuals, families, and local communities can fall victim to a downward cycle of poverty generated by economic forces they are powerless to influence. The poor, the disabled, and the unemployed too often are simply left behind. This pattern is even more severe beyond our borders in the least-developed countries. Whole nations are prevented from participating in the international economic order because they lack both the resources to do so and the power to change their disadvantaged position. Many people within the less-developed countries are excluded from sharing in the meager resources available in their homelands by unjust elites and unjust governments. These patterns of exclusion are created by free human beings. In this sense they can be called forms of social sin (Vatican Council II 1966, 199–308; Synod of Bishops Second General Assembly 1971, 51; John Paul II 1984b). Acquiescence in them or failure to correct them when it is possible to do so is a sinful dereliction of Christian duty.

MORAL PRIORITIES FOR THE NATION. The common good demands justice for all, the protection of the human rights of all (John XXIII 1961, 65). Dedication to the common good, therefore, implies special duties toward those who are economically vulnerable or needy. Many in the lower middle class in this country are barely getting by and fear becoming the victims of economic forces over which they have no control. If the common good is to be truly common, greater economic freedom, power, and security for these vulnerable middle-class members of the community is an important national goal.

The obligation to provide justice for all means that the poor have the single most urgent claim on the conscience of the nation. The fulfillment of the basic needs of the poor is of the highest priority. Increasing active participation in economic life by those who are presently excluded or vulnerable is a high social priority. The investment of wealth, talent, and human energy should be specially directed to benefit those who are poor or economically insecure.

Economic and social policies as well as the organization of the work world should be continually evaluated in light of their impact on the strength and stability of family life. The long-range future of this nation is intimately linked with the well-being of families for the family is the most basic form of human community (Vatican Council II 1966, 199–308). Efficiency and competition in the marketplace must be moderated by greater concern for the way work schedules and compensation support or threaten the bonds between spouses and between parents and children. Health, education, and social service programs should be scrutinized in light of how well they ensure both individual dignity and family integrity.

These priorities are not policies. They are norms that should guide the economic choices of all and shape economic institutions. They can help the United States move forward to fulfill the duties of justice and protect economic rights. They were strongly affirmed as implications of Catholic social teaching by Pope John Paul II during his visit to Canada in 1984: "The needs of the poor take priority over the desires of the rich; the rights of workers over the maximization of profits; the preservation of the environment over uncontrolled industrial expansion; production to meet social needs over production for military purposes" (p. 248). There will undoubtedly be disputes about the concrete applications of these priorities in our complex world. We do not seek to foreclose discussion about them. However, we believe that an effort to move in the direction they indicate is urgently needed.

The economic challenge of today has many parallels with the political challenge that confronted the founders of our nation. In order to create a new form of political democracy they were compelled to develop ways of thinking and political institutions that had never existed before. Their efforts were arduous and their goals imperfectly realized, but they launched an experiment in the protection of civil and political rights that has prospered through the efforts of those who came after them. We believe the time has come for a similar experiment in securing economic rights: the creation of an order that guarantees the minimum conditions of human dignity in the economic sphere for every person. By drawing on the resources of the Catholic moral-religious tradition, we hope to make a contribution through this letter to such a new "American Experiment": a new venture to secure economic justice for all.

•   •   •

Food and Agriculture

The fundamental test of an economy is its ability to meet the essential human needs of this generation and future generations in an equitable fashion. Food, water, and energy are essential to life; and their abundance in the United States has tended to make us complacent. But these goods—the foundation of God's gift of life—are too crucial to be taken for granted. God reminded the people of Israel that "the land is mine; for you are strangers and guests with me" (Lev. 25:23). Our Christian faith calls us to contemplate God's creative and sustaining action and to evaluate our own collaboration with the Creator in using the earth's resources to meet human needs. While Catholic social teaching on the care of the environment and the management of natural resources is still in the process of development, a Christian moral perspective clearly gives weight and urgency to their use in meeting human needs.

No aspect of this concern is more pressing than the nation's food system. Just as food is a unique, life-sustaining commodity, farming is a special vocation. As pastors we feel the distress of those farm families threatened with bankruptcy or facing foreclosure (U.S. Department of Agriculture 1985, viii–x). We are also conscious of the damage to natural resources associated with many modern agricultural practices: the overconsumption of water, the depletion of topsoil, and the pollution of land and water. Though our farms continue to produce food surpluses, millions of people here and abroad suffer the consequences of hunger. Our food system is urgently in need of rethinking and reform.

Two aspects of the current situation especially concern us: first, landownership is becoming further concentrated as units now facing bankruptcy are added to existing farms. Diversity of ownership and widespread participation are declining in this sector of the economy as they have in others. Since differing scales of operation and the investment of family labor have been important for American farm productivity, this increasing concentration of ownership in almost all sectors of agriculture points to an important change in that system.

Second, diversity and richness in American society are lost as farm people leave the land and rural communities decay. It is not just a question of coping with additional unemployment and a need for retraining and relocation. It is also a matter of maintaining opportunities for employment and human development in a variety of economic sectors and cultural contexts.

The situation of racial minorities in the U.S. food system is a matter of special pastoral concern. They are largely excluded from significant participation in the farm economy. Despite the agrarian heritage of so many Hispanics, for example, they operate only a minute fraction of America's farms (U.S. Department of Commerce 1982). Black-owned farms, at one time a significant resource for black participation in the economy, have been disappearing at a dramatic rate in recent years (U.S. Commission on Civil Rights 1982), a trend that the U.S. Commission on Civil Rights has warned "can only serve to further diminish the stake of blacks in the social order and reinforce their skepticism regarding the concept of equality under the law" (p. 8).

It is largely as hired farm laborers rather than farm owners that minorities participate in the farm economy. Along with many white farmworkers, they are, by and large, the poorest paid and least benefited of any laboring group in the country. Moreover, they are not as well protected by law and public policy as other groups of workers; and their efforts to organize and bargain collectively have been systematically and vehemently resisted, usually by farmers themselves. Migratory field workers are particularly susceptible to exploitation. This is reflected not only in their characteristically low wages but in the low standards of housing, health care, and education made available to these workers and their families (U.S. Department of Labor 1984).

Farm owners and farmworkers are the immediate stewards of the natural resources required to produce the food that is necessary to sustain life. These resources must be understood as gifts of a generous God. When they are seen in that light and when the human race is perceived as a single moral community, we gain a sense of the substantial responsibility we bear as a nation for the world food system.

The United States has set a world standard for food production, but not without cost to our natural resource base (*Soil Conservation* 1984; LeVeen 1984; U.S. Department of Agriculture 1981a). The continuation of current practices, reflecting short-term investment interests or immediate income needs of farmers and other land owners, constitutes a danger to future food production. Meeting human needs today and in the future demands an increased sense of stewardship and conservation from owners, managers, and regulators of all resources, especially those required for the production of food.

GUIDELINES FOR ACTION. We are convinced that current trends in the food sector are not in the best interests of the United States or of the global community. The decline in the number of moderate-sized farms and the mounting evidence of poor resource conservation raise serious questions of morality and public policy. As pastors, we cannot remain silent while thousands of farm families caught in the present crisis lose their homes, their land, and their way of life. We approach this situation, however, aware that it reflects longer-term conditions that carry consequences for the food system as a whole and for the resources essential for food production.

While much of the change needed must come from the cooperative efforts of farmers themselves, we strongly believe that there is an important role for public policy in the protection of family farms, as well as in the preservation of natural resources. We suggest three guidelines for both public policy and private efforts aimed at shaping the future of American agriculture.

First, moderate-sized farms operated by families on a full-time basis should be preserved and their economic viability protected. Similarly, small farms and part-time farming, particularly in areas close to cities, should be encouraged. There is a genuine social and economic value in maintaining a wide distribution in the ownership of productive property. (For further thoughts on this see the full text of this pastoral letter, Chapter II, no. 112.). The democratization of decision making and control of the land resulting from wide distribution of farm ownership are protections against concentration of power and a consequent possible loss of responsiveness to public need in this crucial sector of the economy (U.S. Department of Agriculture 1981b, 148). Moreover, when those who work in an enterprise also share in its ownership, their active commitment to the purpose of the endeavor and their participation in it are enhanced. Ownership provides incentives for diligence and is a source of an increased sense that the work being done is one's own. This is particularly significant in a sector as vital to human well-being as agriculture.

Furthermore, diversity in farm ownership tends to prevent excessive consumer dependence on business decisions that seek maximum return on invested capital, thereby making the food system overly susceptible to fluctuations in the capital markets. This is particularly relevant in the case of nonfarm corporations that enter agriculture in search of high profits. If the return drops substantially or if it appears that better profits can be obtained by investing elsewhere, the corporation may cut back or even close down operations without regard to the impact on the community or on the food system in general. In similar circumstances, full-time farmers with a heavy personal investment in their farms and strong ties to the community are likely to persevere in the hope of better times. Family farms also make significant economic and social contributions to the life of rural communities (Tweeten 1983, 1041). They support farm suppliers and other local merchants, and their farms support the tax base needed to pay for roads, schools, and other vital services.

This rural interdependence has value beyond the rural community itself. Both Catholic social teaching and the traditions of our country have emphasized the importance of maintaining the rich plurality of social institutions that enhances personal freedom and increases the opportunity for participation in community life. Movement toward a smaller number of very large farms employing wage workers would be a movement away from this institutional pluralism. By contributing to the vitality of rural communities, full-time residential farmers enrich the social and political life of the nation as a whole. Cities, too, benefit soundly and economically from a vibrant rural economy based on family farms. Because of outmigration of farm and rural people, too much of this enriching diversity has been lost already.

Second, the opportunity to engage in farming should be protected as a valuable form of work. At a time when unemployment in the country is already too high, any unnecessary increase in unemployed people, however small, should be avoided. Farm unemployment leads to further rural unemployment as rural businesses lose their customers and close. The loss of people from the land also entails the loss of expertise in farm and land management and creates a need for retraining and relocating another group of displaced workers.

Losing any job is painful, but losing one's farm and having to leave the land can be tragic. It often means the sacrifice of a family heritage and a way of life. Once farmers sell their land and their equipment, their move is practically irreversible. The costs of returning are so great that few who leave ever come back. Even the small current influx into agriculture attracted by lower land values will not balance this loss. Society should help those who would and could continue effectively in farming.

Third, effective stewardship of our natural resources should be a central consideration in any measures regarding U.S. agriculture. Such stewardship is a contribution to the common good that is difficult to assess in purely economic terms, because it involves the care of resources entrusted to us by our Creator for the benefit of all. Responsibility for the stewardship of these resources rests on society as a whole. Since farmers make their living from the use of this endowment, however, they bear a particular obligation to be caring stewards of soil and water. They fulfill this obligation by participating in soil and water conservation programs, using farm practices that enhance the quality of the resources, and maintaining prime farmland in food production rather than letting it be converted to nonfarm uses.

POLICIES AND ACTIONS. The human suffering involved in the present situation and the long-term structural changes occurring in this sector call for responsible action by the whole of society. A half-century of federal farm-price supports, subsidized credit, production-oriented research and extension services, and special tax policies for farmers have made the federal government a central factor in almost every aspect of American agriculture (U.S. Department of Agriculture 1984). No redirection of current trends can occur without giving close attention to these programs.

A prime consideration in all agricultural trade and food assistance policies should be the contribution our nation can make to global food security. This means continuing and increasing food aid without depressing Third World markets or using food as a weapon in international politics. It also means not subsidizing exports in ways that lead to trade wars and instability in international food markets.

We therefore offer the following six suggestions for governmental action with regard to the farm and food sector of the economy.

1. The current crisis calls for special measures to assist otherwise viable family farms that are threatened with bankruptcy or foreclosure. Operators of such farms should have access to emergency credit and programs of debt restructuring. Rural lending institutions facing problems because of nonpayment or slow payment of large farm loans should also have access to temporary assistance. Farmers, their families, and their communities will gain immediately from these and other short-term measures aimed at keeping these people on the land.

2. Established federal farm programs, whose benefits now go disproportionately to the largest farmers (*The Distribution of Benefits* 1984), should be reassessed for their long-term effects on the structure of agriculture. Income-support programs that help farmers according to the amount of food they produce or the number of acres they farm should be subject to limits that ensure a fair income to farm families and restrict participation to producers who genuinely need such income assistance. There should also be a strict ceiling on price-support payments that assist farmers in times of falling prices so that benefits go to farms

of moderate or small size. To succeed in redirecting the benefits of these programs while holding down costs to the public, consideration should be given to a broader application of mandatory production control programs ("The Great Debate" 1984).

3. We favor reform of tax policies that now encourage the growth of large farms, attract investments into agriculture by nonfarmers seeking tax shelters, and inequitably benefit large and well-financed farming operations (U.S. Department of Agriculture 1981b). Offsetting nonfarm income with farm "losses" has encouraged high-income investors to acquire farm assets with no intention of depending on them for a living as family farmers must. The investment tax credit and the ability to depreciate capital equipment faster than its actual decline in value have benefited wealthy investors and farmers. Lower tax rates on capital gains have stimulated farm expansion and larger investments in energy-intensive equipment and technologies as substitutes for labor. Changes in estate tax laws have consistently favored the largest estates. All of these results demonstrate that reassessment of these and similar tax provisions is needed (Dunford 1984). We continue, moreover, to support a progressive land tax on farm acreage to discourage the accumulation of excessively large holdings. (See United States Catholic Conference 1972.)

4. Although it is often assumed that farms must grow in size in order to make the most efficient and productive use of sophisticated and costly technologies, numerous studies have shown that medium-sized commercial farms achieve most of the technical cost efficiencies available in agriculture today. We therefore recommend that the research and extension resources of the federal government and the nation's land grant colleges and universities be redirected toward improving the productivity of small and medium-sized farms (Miller 1981).

5. Since soil and water conservation, like other efforts to protect the environment, are contributions to the good of the whole society, it is appropriate for the public to bear a share of the cost of these practices and to set standards for environmental protection. Government should therefore encourage farmers to adopt more conserving practices and distribute the costs of this conservation more broadly.

6. Justice demands that worker guarantees and protections such as minimum wages and benefits and unemployment compensation be extended to hired farmworkers on the same basis as all other workers. There is also an urgent need for additional farmworker housing, health care, and educational assistance.

FARM COMMUNITY. While there is much that government can and should do to change the direction of farm and food policy in this country, that change in direction also depends upon the cooperation and good will of farmers. The incentives in our farm system to take risks, to expand farm size, and to speculate in farmland values are great. Hence farmers and ranchers must weigh these incentives against the values of family, rural community, care of the soil, and a food system responsive to long-term as well as short-term food needs of the nation and the world. The ever-present temptation to individualism and greed must be countered by a determined movement toward solidarity in the farm community. It is possible to approach farming in a cooperative way, working with other farmers in the purchase of supplies and equipment and in the marketing of produce. It is not necessary for every farmer to be in competition against every other farmer. Such cooperation can be extended to the role farmers play through their various general and community organizations in shaping and implementing governmental farm and food policies. (For further ideas on this subject see Chapter IV of the full text of this pastoral letter.) Likewise, it is possible to seek out and adopt technologies that reduce costs and enhance productivity without demanding increases in farm size. New technologies are not forced on farmers; they are chosen by farmers themselves.

Farmers also must end their opposition to farmworker unionization efforts. Farmworkers have a legitimate right to belong to unions of their choice and to bargain collectively for just wages and working conditions. In pursuing that right they are protecting the value of labor in agriculture, a protection that also applies to farmers who devote their own labor to their farm operations.

CONCLUSION. The U.S. food system is an integral part of the larger economy of the nation and the world. As such this integral role necessitates the cooperation of rural and urban interests in resolving the challenges and problems facing agriculture. The very nature of agricultural enterprise and the family farm traditions of this country have kept it a highly competitive sector with a widely dispersed ownership of the most fundamental input to production, the land. That competitive, diverse structure, proven to be a dependable source of nutritious and affordable food for this country and millions of people in other parts of the world, is now threatened. The food necessary for life, the land and water resources needed to produce that food, and the way of life of the people who make the land productive are at risk. Catholic social and ethical traditions attribute moral significance to each of these. Our response to the present situation should reflect a sensitivity to that moral significance, a determination that the United States will play its appropriate role in meeting global food needs, and a commitment to bequeath to future generations an enhanced natural environment and the same ready access to the necessities of life that most of us enjoy today.

References

Alfaro, J. *Theology of Justice in the World.* Rome: Pontifical Commission on Justice and Peace, 1973.

Avila, C. *Ownership: Early Christian Teaching.* Maryknoll, N.Y.: Orbis Books, 1983.

The Distribution of Benefits from the 1982 Federal Crop Programs. Washington, D.C.: U.S. Senate Committee on the Budget, 1984.

Donahue, J. R. "Biblical Perspectives on Justice." In *The Faith That Does Justice,* edited by John C. Haughey, 68–112. New York: Paulist Press, 1977.

Dorr, Donald. *Option for the Poor: A Hundred Years of Vatican Social Teaching.* Maryknoll, N.Y.: Orbis Books, 1983.

Dunford, Richard. *The Effects of Federal Income Tax Policy on U.S. Agriculture.* Washington, D.C.: Subcommittee on Agriculture and Transportation of the Joint Economic Committee of the Congress of the United States, 1984.

Dupont, J., and A. George, eds. *La pauvrete evangelique.* Paris: Cerf, 1971.

Eagleson, J., and P. Scharper, eds. *Puebla and Beyond.* Maryknoll, N.Y.: Orbis Books, 1979.

"The Great Debate on Mandatory Production Controls." In *Farm Policy Perspectives: Setting the Stage for 1985 Agricultural Legislation.* Washington, D.C.: United States Senate Committee on Agriculture, Nutrition, and Forestry, 1984.

Gunnemann, Jon P. "Capitalism and Commutative Justice." In *The Annual of the Society of Christian Ethics.* Forthcoming.

Hengel, M. *Property and Riches in the Early Church.* Philadelphia: Fortress Press, 1974.

Hollenbach, David. "Modern Catholic Teachings Concerning Justice." In *The Faith That Does Justice,* edited by John C. Haughey. New York: Paulist Press, 1977.

John Paul II. Address at Yankee Stadium. *Origins* 9, no. 19(October 1979): 311–12.

———. Address on Christian Unity in a Technological Age, Toronto, 14 September 1984. *Origins* 14, no. 16 (October 1984a): 248.

———. Address to Workers at Sao Paulo, no. 8. *Origins* 10, no. 9 (July 1980): 139.

———. *Apostolic Exhortation on Reconciliation and Penance. Origins* 14, no. 27 (December 1984b): 441–42.

———. *On Human Work.* 1981.

Johnson, L. *Sharing Possessions: Mandate and Symbol of Faith.* Philadelphia: Fortress Press, 1981.

John XXIII. *Mater et Magistra* ("Christianity and Social Progress"). 1961.

LeVeen, E. Philip. "Domestic Food Security and Increasing Competition for

Water." In *Food Security in the United States,* edited by Lawrence Busch and William B. Lacy. Boulder, Colo.: Westview Press, 1984.

McDonagh, E. *The Making of Disciples.* Wilmington, Del.: Michael Glazier, 1982.

Mealand, D. L. *Poverty and Expectation in the Gospels.* London: SPCK, 1980.

Miller, Thomas E., et al. *Economies of Size in U.S. Field Crop Farming.* Washington, D.C.: ERS, USDA, 1981.

Mott, S. C. *Biblical Ethics and Social Change.* New York: Oxford University Press, 1982.

North, R. *Sociology of the Biblical Jubilee.* Rome: Biblical Institute, 1954.

Ogletree, T. *The Use of the Bible in Christian Ethics.* Philadelphia: Fortress Press, 1983.

Paul VI. *Evangelization in the Modern World* ("Evangelii Nuntiandi"). 1975.

_____. Octogesima Adveniuns ("The Eightieth Anniversary of 'Rerum Novarum"). 1971.

_____. *On the Development of Peoples.* 1967.

Pedersen, J. *Israel: Its Life and Culture.* Vols. 1 and 2. London: Oxford University Press, 1926.

Pieper, Josef. *The Four Cardinal Virtues.* Notre Dame, Ind.: University of Notre Dame Press, 1966.

Pilgrim, W. *Good News to the Poor: Wealth and Poverty in Luke-Acts.* Minneapolis: Augsburg, 1981.

Ringe, S. *Jesus, Liberation and the Biblical Jubilee: Images for Ethics and Christology.* Philadelphia: Fortress Press, 1985.

St. Cyprian. "On Works and Almsgiving." In *St. Cyprian: Treatises,* translated by R. J. Defarrari. Fathers of the Church, 36. New York: Fathers of the Church, 1958.

Soil Conservation in America: What Do We Have to Lose? Washington, D.C.: American Farmland Trust, 1984.

Stegemann, W. *The Gospel and the Poor.* Philadelphia: Fortress Press, 1984.

Synod of Bishops Second General Assembly. *Justice in the World* ("Justitia in Mundo"). 1971.

Tweeten, Luther. "The Economics of Small Farms." *Science* 219 (March 1983): 1041.

U.S. Commission on Civil Rights. *The Decline of Black Farming in America.* Washington, D.C.: U.S. Commission on Civil Rights, 1982.

U.S. Department of Agriculture. *America's Soil and Water: Condition and Trends.* Washington, D.C.: U.S. Department of Agriculture Soil Conservation Service, 1981a.

_____. *The Current Financial Condition of Farmers and Farm Lenders.* Ag. Info. Bull. No. 490. Washington, D.C.: ERS, USDA, 1985.

_____. *History of Agricultural Price-Support and Adjustment Programs, 1933–84.* U.S. Ag. Info. Bull. No. 485. Washington, D.C.: ERS, USDA, 1984.

_____. *A Time to Choose: Summary Report on the Structure of Agriculture.* Washington, D.C.: U.S. Department of Agriculture, 1981b.

U.S. Department of Commerce, Bureau of the Census. *Census of Agriculture, 1982.* Washington, D.C.: Government Printing Office.

U.S. Department of Labor. *Hearings Concerning Proposed Full Sanitation Standards.* Document No. H 308. Washington, D.C.: Government Printing Office, 1984.

United States Catholic Conference. *Where Shall the People Live?* Washington, D.C.: U.S. Catholic Conference, 1972.

Vatican Council II. *Decree on Ecumenism,* 341–66; *Pastoral Constitution on the Church in the Modern World* ("Gaudium et Spes"), 199–308. In *The Documents of Vatican II,* edited by Walter M. Abbott, S.J. New York: The Guild Press, 1966.

Vawter, B. *On Genesis: A New Reading.* Garden City, N.Y.: Doubleday, 1977.

Westermann, C. *Creation.* Philadelphia: Fortress Press, 1974.

MERRILL J. OSTER

A Christian Businessman's View

IN A FREE SOCIETY the old axiom still holds, "Where there is no risk, there is no reward." Farming is a business of risk management. But today, in agriculture's hour of crisis, loud cries rise from the meeting hall as well as from some pulpits defending "socialized agriculture" as the answer to the farm crisis. It is time for a serious look at the trade-offs that go along with government-mandated farm size, price and profit "guarantees," and land reform — ideas held up as answers to the agricultural crisis of the 1980s.

Let's face it, this is a risky business. Risk exposure is masked during periods of inflation. During deflationary times, the mask falls off and we find which businesses had risk management techniques in place and which ones simply had the "pedal to the metal" and were headed into the economic turn full throttle. Holding the pedal to the metal has an economic description: profit maximization. Such a stance carries maximum risk in hope of maximum reward. That is the smart way to run a business during inflationary times. And, it is a formula for trouble in deflationary periods. Most trouble today stems from debts incurred in the late 1970s.

We have a choice in this economic race. We can blame the sharp, unmarked turn. It came at an inconvenient time in the race. Blame Paul Volcker, bad monetary policy. Blame the factors that caused the architect of the turn to act: a public demanding more from a government that already had a bloated appetite for credit. Bad fiscal policy. Blame factors associated with the turn in our economic world such as the value of the dollar or bumper crops that created low prices. Blame others on the track, the "big" farmers.

But in our hearts we all know that we must also check the position

---

MERRILL J. OSTER is president of Professional Farmers of America.

of our own foot on the leverage pedal. Fast growing operations fed by borrowed capital speed down the straightaway during inflation. But not a one of us can say we were unaware of the flip side: how high economic speeds threaten our safety when a turn of events suddenly shows up.

In my mind, the family farm or the commercial family farm or the farm business are all one and the same. Twenty-eight percent of farmers gross forty thousand dollars plus. And they produce 82 percent of the farm output. The consolidation of family farms, which so many fear, is already a fact. Look at the farms over one hundred thousand dollars. About 12 percent of all farms, they produce nearly 70 percent of the nation's output. To suggest that small farms, grossing between forty thousand dollars and two hundred thousand dollars are more desirable than the larger ones takes a real stretch of the imagination. It certainly has no biblical base.

A "family farmer" in this range is probably not providing enough goods and services to generate anything near a full-time living. This is because he or she has part-time employment. For example, if a corn grower produced 130 bushels of corn per acre at $2.50 per bushel, the gross would be $325.00 per acre. This farmer would gross over $40,000.00 on just 150 acres of corn. Since it takes about three person-hours to grow an acre of corn or soybeans, this 150 acres would employ the farmer only 450 hours. But an Iowa State University study shows that three thousand person-hours per year is the average employed on the farm. This means that a corn-soybean farmer needs one thousand acres to have a full-time job. Iowa State's data further suggest that one family providing the work unit equivalents of one full-time person (three thousand hours) would gross $325,000.00 (assuming 130 bushels yield selling at $2.50).

Pushed by this economic reality and striving to fully employ a father and a son, many family farming operations in Iowa set their sights on two thousand acres or more. The two-person family farm either moves to this level, finds income from livestock or some other source, or is saddled with the higher overhead cost per acre associated with the failure to employ labor full-time.

Conferring the title "family farmer" to those over forty thousand dollars but under two hundred thousand dollars doesn't come close to fitting the economic reality of full-time family farmers. Corn/soybean farms that small just cannot produce enough productive employment to support a family. Those who work on "family farms," as defined by Comstock in the Introduction, are grossly underemployed.

I would rather define the family farm as one whose size can be expected to produce a full-time income to the family running the business. That allows for variability of enterprises and price fluctuations. And it encompasses the intent of the individual farmer who wants to get one or more children started farming.

A Value Base for Decisions

I believe that "religious values" come from the Bible and that the Bible lays out principles that relate to life in general as well as economic life in particular:

1. Humans: Created by God. Walked in harmony until the fall caused by humans' elevation of their will over God's. Result: The ground was cursed so it wouldn't yield good alone (Gen. 3:17). Physical labor was required to subsist (Gen. 3:19), and physical and spiritual death (Gen. 3:19, 3:2, Rom. 5:12) entered the picture.

2. Purpose: The major purpose in life is to serve, worship, and emulate our Creator and Redeemer (I Cor. 10:24).

3. Government: The Bible teaches that unless changed by Christ's salvation, mankind is fallen and by and large self-interested and blinded by that self-interest. Governments, made up of fallen people, are as corrupt and self-seeking as the people who form them (Rom. 3:23, I John 2:11).

4. Responsibility and freedom: Christ's teaching was one of freeing man and woman from the burden of sin. In word and example he pointed out that serving our fellow beings is the way to live responsible lives resulting from that freedom. Freedom and responsibility are usually coupled in the teaching of the apostles.

5. Ownership: God owns everything. We are his stewards. As trustees, we are responsible to him for how we invest our knowledge, skills, and time.

6. Institutional responsibility: We are created to live in families, gather in churches, communities, workplaces where we carry out specific stewardship responsibilities.

7. Right to private property: This right is protected by half of the Ten Commandments, one of many indications that the right to private property is the individual's not the state's.

8. Work: God gives the power to get wealth, and he gives it to diligent workers in his kingdom.

9. Profit: Profits are possible because of a biblical principle of dominion over the earth through service.

10. Investment: Efficiency of investment of capital is encouraged by the parables of the talents and the pounds.

11. Prices: The Bible does not teach a just pricing system, but creates conditions for the free market. In biblical capitalism, the seller does not cheat the buyer, but produces something the buyer needs so both benefit.

12. Charity: Biblical law commands it. Never is charity enforced by the state in scripture. Voluntary giving of one-tenth to the local church is a fundamental provision for destitute widows and poor.

13. Rich and poor: The Lord makes them both and places no particular esteem or curse on either. In New Testament times there were rich and poor in the church. The rich are never condemned but are warned not to trust in riches and not to forget the poor.

14. Big and small: Economic equality, a central theme of socialism, which requires the redistribution of property and the prohibition of economic freedoms by the state, is not condoned in scripture.

15. Scarcity: God's imposed scarcity (Gen. 3:17–19) induces people to become less wasteful, more efficient. It forms the base for the "law of supply and demand," which is always operable and can't be repealed by government action.

The argument that the so-called family farm has some special biblical merit is without foundation. In the Bible, the rich are indicted for exploiting the poor, yet the prophets never suggest that the remedy is economic redistribution. The prophets consistently pinpointed the root cause of economic trouble as spiritual: the nation had departed from God, and economic injustice was one result. In that framework, the priority for us in this nation is not socioeconomic reform or land reform but individual and national spiritual repentance.

In terms of biblical stewardship of resources it could be argued that the larger family farm is more desirable than the smaller one. The votes cast by a free society over the past two hundred years would certainly support that view. One combine can serve one thousand acres as well as it serves two hundred. By spreading the capital cost over more acres, there is a lower cost of production. Lower production costs mean cheaper food to the consumer. That is what the free market is supposed to do: bring the most benefits to the most people.

There is no inconsistency between the Bible and the idea of the free market. Biblical capitalism allows the resources of an industry to drift to the most efficient producers. This makes for efficient food production and low food prices, both of which help the needy most.

When we look at various methods of improving our lot in life, there are two fundamental views. One is to redistribute wealth by using government to rob one group and support another. Demands for equal income for all farmers and for land reform are consistent with Marxist principles. This requires a strong central decision-making power.

The other is to emphasize wealth creation, taking a "carrot and stick" approach to the advancement of our welfare. This free-market approach is associated with personal freedom of choice.

One choice available to all who live under capitalist systems is whether or not to be a risk taker. Thus, we have the opportunity for high profit (if the risk is successful) or big losses (if the venture fails). The principles of wealth creation are totally consistent with biblical principles. Christ deals harshly with those who misuse people to get wealth, as well as those who worship wealth. But he expected inequality; he knew that there would always be both the rich and the poor.

Opposition to Wealth Creation is Growing

Proponents of centrally planned economies have for centuries undermined public confidence in the concept of profit, the right to private property, and the free market. They do this by trying to contrast a commercial society with a Christian society. For example, T. S. Eliot argued, "The organization of society on the principle of private profit, . . . is leading both to the deformation of humanity by unregulated industrialization and the exhaustion of natural resources" (Griffiths 1984, 114). In recent years the trickle of clergy aligned with such a philosophy has grown into a stream. That stream is a flood in rural America today. Ideological opportunists play every day with the emotions of distraught farmers.

Some clergy even imply that owners of large farming operations are greater sinners than smaller farmers. For example, Bishop Maurice J. Dingman of Iowa says that "landownership is being restructured, agricultural production is becoming more heavily industrialized and concentrated in fewer hands. The earth all too frequently is being subjected to harmful farming, mining, and development practices. A corporate takeover is imminent. Family farms are disappearing. A national policy should be adopted stating clearly and forceably that the ideal type of farm is that of family farming" (Dingman 1985). Here is a Catholic bishop speaking, one who claims the support of other denominational leaders. His comments are worthy of careful scrutiny.

First, there is neither empirical data nor directives in the Bible indicating that ideal soil conservation is best practiced by farmers of any particular size. Soil loss is caused by violations of certain commonsense principles by farmers big and small. Rather than using this emotional issue to attack larger family farms, religious leaders could make a greater contribution by teaching specific conservation principles. They should not urge wholesale social and economic reform.

Second, emotionally charged statements like Bishop Dingman's do not hold up in the face of factual data. According to a USDA study in one recent year, more than 85 percent of the land purchased in America was by a farmer. Thus, a takeover by corporations does not seem to be "imminent," as Bishop Dingman suggested.

Third, land redistribution is not a biblical principle nor is it to be practiced today. If such a statement were to be accepted, one would have to wonder what God was thinking when he blessed Abraham, a "large farmer," again and again. The evil the Bible warns us against is the love of money, not money itself.

Some ministers, like David Ostendorf (Chap. 25), claim that the Jubilee land law applies today. Under this law, every fifty years the land was to return to its original owner. It is this aspect of the law that is usually seized upon by those who would use the Bible to justify socialism or land reform by government edict. But Jubilee land law doesn't apply to today's landowners.

David Chilton (1981) says the Jubilee law "referred specifically to the land of Israel, which God had divided among the tribes. By divine fiat, the Israelites became the original 'owners.' No other landowner could make this claim. I may buy or sell property, but I cannot claim a 'divine right' to anything in the sense that the Israelites could. We cannot establish the Jubilee anywhere outside Palestine, for we have no starting point. Who is the original owner of your property? The Indians?" [But compare Bishop Dingman's use of the Jubilee idea (Dingman 1985).]

Moreover, consider the parable of the laborers in the vineyard. Here, a farmer hired workers at various times of day but paid them the same wages. The implication is that total discretion about the payment of wages is a right associated with ownership. Jesus was not concerned with income or land redistribution. He respected the *freedom* that property ownership implied. Of course, he also urged that ownership be exercised with the interests of others in mind.

A final biblical example: in the case of Abram and Lot, God in no way questioned the legitimacy of wealth. Those who like Ostendorf and the Catholic bishops want to establish a national land reform or a policy favoring the "family farm" are advocating a policy of nationalized theft; steal property from large commercial farms and give it to smaller farms. This would be a great affront to God, to biblical principles, and to the rights of free people. It amounts to people playing God through a central government. It is on this point that the same clergymen play into the hands of the socialists.

Socialism, a product of humanism, is state-worship, viewing the state as Lord and Savior (Lindsell 1982). The fact is that nations that practice this are demonstrably antihuman. From the Soviet Union's gulag to Sweden and Finland's alcoholism and decline of family and moral values, those states reveal themselves to be far less friendly entities than is my local neighbor. Socialism fails to treat people as people created in God's image. It is a power theory that empowers the state to enforce human goals.

Ostendorf supports land reform as a goal for America. Speaking at a seminar in Cedar Falls, Iowa, he explained his efforts as, "hammering, hollering and politicking for three years until we got a moratorium on farm foreclosures. People gathered together organized the power to get something done. People with power can make changes in their institutions" (Ostendorf 1985).

Some of the changes he advocates include the equivalent of an economic fence around Iowa so capital in certain forms is not allowed in. Money, like water, seeks its own level through a path of least resistance. Actions such as the foreclosure moratorium and the so-called family farm protection act make Iowa a less-healthy climate for capital, one of the commodities needed to spark an economic recovery, an upward bounce in land prices and a healing of our balance sheets.

These efforts to limit the ownership of farms to certain "preferred groups" adds to economic problems. Illinois has no family farm protection laws and the balance sheets of their family farms look substantially better. Ostendorf espouses the view that, "community can't be built where people aren't equal. This nation is headed for land reform." But a policy of economic equality enforced by the state is simply legal inequality. That is not justice. It is tyranny: legislated lawlessness. The biblical principle "Thou shalt not steal" is violated. Equality, mandated by the majority, is not a biblical command or model.

Ultimately, as Chairman Mao said, political power grows out of a gun barrel. That is what happens when the state is exalted over God.

Entitlement to Profit – Not in the Bible

The suggestion by Bishop Dingman that all farmers have an inherent "God-given" entitlement to a profit every year has no biblical support. The landowner's right to a profit is no more a biblical right than that of the family druggist, doctor, or grocer. Government doesn't have enough money to guarantee us all an equal living. If it tried, it could only guarantee us a very low standard of living. Communist nations have implemented the philosophy that every working person has an entitlement to a wage roughly equal to everyone else. In nations under Communist flags, "justice" has been guaranteed. But what sort of justice is it? With "just pricing" determined by a government acting in behalf of its people, everyone is equally poor and, frequently, equally hungry.

There is a body of biblical knowledge that, understood and applied, can give us Christian answers to these questions.

1. Individual and state greed. The Bible speaks out strongly against wanting something someone else has. Because so many in our society want so much more than the nation can afford, we use the power of the state to rob (tax) each other in the interest of "fairness." This underlying sin of envy or covetousness is a root problem that this nation has traditionally overcome. But with the nation and individuals drowning in a sea of debt, national and individual repentance is needed.

2. Government or individual overspending. The Bible speaks in negative terms of indebtedness. "The borrower becomes the lender's slave." Living beyond our means as a family or as a nation is immoral because of the sin of greed and the impact it has on the currency. The Bible supports a fair currency, not one that government can change from year to year to finance its promises to a gullible electorate that wants its "fair share."

3. Inflation. One of the results of greed and overspending is gradual inflation — or a debasing of the currency or robbing from one generation to pay another's bills.

4. People playing God. Whether it be in the marketplace or in the Garden of Eden, it is forbidden. To the extent that governments attempt to interfere with the marketplace to force "fair prices," it is an attempt to "play God."

5. Freedom. The Bible points us to a world in which the individual makes choices and assumes responsibility for his or her choices. The extent to which we try through government programs to make everyone economically equal, we increase their dependence on and trust in big government. Policies should be evaluated on the basis of whether they would move a nation toward a free-market economy or toward a controlled-market economy. In the controlled economy with strong central power and authority, freedoms of all kinds are suppressed. The rhetoric we hear about the farm crisis should be judged within this framework. When we do, it becomes easier to classify an idea as potentially socialist, Marxist, or free market.

6. Responsibility. The Bible clearly speaks of individual responsibility. Certainly each farmer must ask, "To what extent am I responsible for my own dilemma?" To suggest that we are victims of circumstances

beyond our control and therefore all the problems and solutions lie outside of agriculture would be an improper stance.

These biblical principles will not allow us to reach the conclusion that Christianity favors small farms over large farms or that we should institute policies giving preferential treatment to farm businesses over nonfarm businesses. They do, however, contain many references to freedom of choice. I fear a loss of this precious freedom if we mandate farm size. I have a moral obligation to save my own family farm but not everyone else's. I only wish to continue to have the right to, and responsibility for, my own choices and my own successes or failures.

References

Chilton, David. *Productive Christians in an Age of Guilt Manipulators.* Tyler, Texas: Institute for Christian Economics, 1981.

Dingman, Maurice J. "Our Moral Responsibility to Soil Conservation." Presented at Marion County Care Facility, Knoxville, Iowa, 8 August 1985.

Griffiths, Brian. *The Creation of Wealth.* London: Hodder and Stoughton, 1984.

Lindsell, Harold. *Free Enterprise: A Judeo-Christian Defense.* Wheaton, Ill.: Tyndale, 1982.

Ostendorf, David. Speech at L'Chaim's Family Farm Workshop. Cedar Falls, Iowa. 18 November 1985.

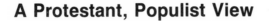

DAVID L. OSTENDORF

A Protestant, Populist View

THE LAST TWO DECADES of the twentieth century may mark the virtual completion of a process begun by Europeans upon their arrival and settlement in North America: the continuous displacement of people from land and the concentration of control over land by an ever-diminishing number of owners and interests. From the uprooting of native Americans to forced living on reservations, to the massive displacement of blacks from the South to their forced resettlement in northern cities following the Civil War; from the eradication of black-owned farms in the rural South to the contemporary farm crisis resulting in the liquidation of thousands of white farmers in the Midwest and Plains, land control in America has *not* been rooted in the religious values of the Judeo-Christian tradition. It has been rooted in the fundamental values of unbridled free-market economics and raw political power. The economic and political displacements I have in mind were (and, as Merrill Oster's essay shows, still are) justified by talk about Christian values and biblical principles. But this talk comes from those who have one foot in the Christian community and one in the corporate boardroom. This second foot is busy kicking out people who stand in the way of "efficiency," "freedom," and "progress."

From the time the early Europeans began settling North America, there existed a conflicting and often ambiguous view of the continent. On the one hand, it was seen as a lush and glorious garden where God's will would be carried out by the "chosen people." On the other, it appeared as a threatening wilderness that had to be cleared and conquered to assure the prosperity of the chosen. As early as the 1670s, worldly ways and the accumulation of land had firmed their grip on pietistic New

---

DAVID L. OSTENDORF is an ordained minister of the United Church of Christ and is director of Prairiefire Rural Action. In slightly different form, this essay appeared as "Displacement, Vision and Hope—The Contemporary American Journey to Naboth's Vineyard," copyright Prairiefire Rural Action, Inc., Des Moines, Iowa, 1986. Reprinted by permission.

England. The fears of the founders themselves—that land-hungry Englishmen might soon see nothing but vast tracts of land to be cleared and exploited—were being affirmed. In 1676 Increase Mather could write, "Land! Land! hath been the Idol of many in New England," and that whereas the first settlers were satisfied with small plots, "how many Men since coveted after the earth, that many hundreds, nay thousands of Acres, have been engrossed by one man, and they that profess themselves Christians, have forsaken Churches, and Ordinances, and all for land and elbow-room enough in the World" (Mather 1676, 7).

In ensuing decades, Mather's warnings went unheeded. By the middle of the nineteenth century, religion and technology had formed something of an unholy alliance. As greater value was placed on practical science and technique, the need to use God to account for the forces of history and the natural environment diminished. In fact, God no longer accounted for the society and its progress; the society and its progress accounted for God and was identified with God's plan, which included the conquering of the land through westward expansion, settlement, industrialization, and control.

By the 1880s Increase Mather's concern about the coveting and control of land, with God moving out of the picture, would be even more serious. As historian Roy Robbins (1942) wrote: "The actual settler by the eighties was not in a favored position; in fact, corporations and speculators were definitely in the ascendancy. This was not only the situation with regard to the agricultural lands of the country, but with regard to other lands as well. . . . Legal regulations were evaded, the honest settler was thwarted; in fact, a system of landlord-tenant and land concentration was growing up on American soil" (p. 268).

To this day, Americans have, for the most part, looked upon land as simply another commodity to be used and accumulated for the primary benefit of self. The individual and his or her right to make free choices always came first. The rights and interests of the broader community always came second, if at all. If residual benefits accrued to others as a result of the use of land, all the better. But it was and is the principle of individual ownership that has dictated public policy and personal decisions about the use and dispersal of land.

For the most part, the lingering wholesome values associated with life on the farm and in rural America have helped this society cover up the destructive, operative values in the countryside. For example, we glorify the family, its hard work and good life on the farm, while we ignore the policies that have been destroying that life for years. We see and believe television commercials depicting the simple beauty of family farming but do not recognize that they are paid for by multinational corporations intent on dominating agriculture. We are concerned about what we see happening to family farmers in the 1980s, but we really do not want their incomes to go up if it means that our food bill will also go up. We speak of farming as a way of life, while industrial agriculture makes it a straight-out business. We want to keep untarnished our pastoral images of rural America, while the unleashed mentality of efficiency and good management pushes farmers toward all-out production and maximization of profits at the expense of their families and their land. Get bigger, get better, or get out!

The Gift of Land

In recent years, the Judeo-Christian understanding of land as a gift from God has been recovered. The books of the Pentateuch (especially Leviticus and Deuteronomy) and of the prophets have provided us renewed understandings of the land as a base for the covenant relationship between God and God's people. The fundamental understanding of land as gift, with inherent responsibilities of stewardship, care, and control, has challenged us to rethink our individualistic values and profit-oriented attitudes. Perhaps land is not just a commodity to be owned, bought, sold, liquidated, inventoried, traded, recorded, and otherwise treated as a disposable item. For one does not treat a gift—especially a gift of God's—as one's property.

It was through the distribution of the land among the Israelites that God and the people shaped the covenantal community of faith. The land was apportioned by lots in order to meet the needs of families (Numbers 26) and was to be passed on by inheritance through generations to come. God had promised land to the people. And it was the land, justly distributed among families and passed on to succeeding generations, that would be the focal point of Israel's relationship to the liberating God.

Land and Power

Walter Brueggemann reminds us in *The Land* (1977) about Increase Mather's warning of 1676: "The central temptation of the land is coveting." Merrill Oster is correct in pointing this out, for it is exactly what occurred in Israel's later history; the kings assumed—as many rulers and aspiring owners are prone to do—that they were above the laws of God and nation. To control land is to control power, and the monarchs of Israel understood that as clearly as any other rulers before or since. King David and King Solomon understood it as clearly as Tenneco and Prudential and the local wealthy farmer today understand it. Land is power.

The relentless accumulation of power, or the worship of land as one's own property, is sin.

But now consider King Ahab, who sought to acquire Naboth's vineyard to use for a garden. Perceiving the vineyard as property, Ahab offered a direct exchange for a better vineyard somewhere else or its purchase value in silver. Naboth wanted nothing to do with the exchange, saying to the king, "The Lord forbid that I should let you have land which has always been in my family."

Ahab went home mad, lay down in his bed, covered his face, pouted, and refused to eat. At that point, his wife Jezebel entered and said: "Are you or are you not king in Israel?" In other words, she told him, "Get up and exercise your power as king. *Take* the property you want!" Ahab, however, was content to let Jezebel do his dirty work. She conspired to have Naboth stoned, trumping up charges of blasphemy and disloyalty. Ahab moved in and took possession of the vineyard he coveted.

In his rush to take land, however, Ahab forgot that God was still in charge. The prophet Elijah was sent to Ahab with the word of the Lord: "Have you killed Naboth and taken his land as well?" Before Ahab could answer Elijah delivered God's judgment: "Where dogs licked the blood of Naboth, there dogs shall lick your blood."

God avenges the people victimized by those seeking wealth and power through accumulation of land. This is contrary to the gift relationship established in the covenant. God does not overlook the sins of the people. This is the message repeated again and again by the prophets. Isaiah, Micah, Amos, and Jeremiah came boldly into the midst of the people of Israel and admonished them to return to the covenant relationship with God. They knew that God's judgment was imminent if they did not.

Land as the Base of Community

When the covenant was broken, so was the community and so were families. In the New Testament the message does not change: our relationship to land is fundamental to the viability of the community of faith. Consider the fourth and fifth chapters of Acts. Here we see a young community of Christians united in heart and soul, struggling for survival and growth against the odds. As a community of God's people, no one among them claimed any possession "but everything was held in common. . . . They had never a needy person among them, because all who had property in land or houses sold it, brought the proceeds of the sale, and laid the money at the feet of the apostles; it was then distributed to any who stood in need" (Acts 4:32–35).

As is often the case, there were those who sought to maximize their own self-interest at the expense of the community. Ananias and his wife Sapphira sold a property but kept back a portion of the purchase money for themselves. The rest they brought to Peter and the apostles. Peter was wise to the scheme and challenged the thief: "Ananias, how was it that Satan so possessed your mind that you lied to the Holy Spirit, and kept back part of the price of the land? While it remained, did it not remain yours? When it was turned into money, was it not still at your own disposal? What made you think of doing this thing? You have lied not to men but to God" (Acts 5:4).

When Ananias heard these scathing words of Peter, "he dropped dead, and all the others who heard were awestruck." Awestruck, indeed! One does not often witness God's wrath in so dramatic a fashion. But as if God's striking down Ananias were not enough, Sapphira later repeated her husband's story and was also challenged by Peter: "Why did you conspire to put the Spirit of the Lord to the test? Hark! there at the door are the footsteps of those who buried your husband; and they will carry you away. And suddenly she dropped dead at his feet" (Acts 5:9).

We can imagine the shock of the community at these happenings. Following the burials, "a great awe fell upon the whole church, and upon all who heard of these events; and many remarkable and wonderful things took place among the people at the hands of the apostles." Once again, God had intervened when the values of the Christian community had been threatened by individualists wanting to pursue only their own free choices.

The point of all this is that some of the most fundamental religious values of the Judeo-Christian heritage mandate a radically different approach to land distribution, control, and use than our current American approach. These values do, indeed, justify the view that family farms are more desirable than large corporate farms. What sort of different approach? The first thing we need to do is *remember* who we are.

Remembrance

Success in life often breeds contempt for origins. People of faith are admonished to remember that it is by God that they are given the opportunity to prosper and they need to exercise responsibility for the God-given resources that enabled them to flourish. The eighth chapter of Deuteronomy lays this out in no uncertain terms:

> When you have plenty to eat and live in fine houses of your own building, when your herds and flocks increase, and your silver and gold and all your possessions increase too, do not become proud and forget the LORD your God who brought you out of Egypt, out of the land of slavery. . . . Nor must you say to yourselves, "My own strength and energy have gained me this wealth," but remember the LORD your God; it is he that gives you strength to become prosperous, so fulfilling the covenant guaranteed by oath with your forefathers, as he is doing now.

Remembrance of God's gift of land compels us to continually reexamine our patterns of land tenure, our attitude toward accumulation and control of land, and our ways of using it. Remembrance is a value that compels us to look upon land as gift, not as property, as a trust and inheritance to be used for the benefit of others and to the glory of God, as a locus of vocation and stewardship.

This is not to say that family farmers themselves inherently hold this value or inevitably act according to its mandates. In fact, the lack of remembrance is a major problem across all American agriculture. However, I am convinced that family-owned and -operated farms, bound together in socioeconomic communities, and often tied by generations to the past and future, provide greater possibilities for fulfillment of this value. Agribusiness operations whose sole purpose is maximization of profit have no incentive to remember the past or plan for the long-term future. The only thing that counts for a manager is how efficient he or she was *last* year and how large the profit margin will be *next* year. That is not conducive to the development of memory or to planning for future generations.

Community

In my view, Christian values and biblical principles dictate that the good of the community should take precedence over the good of the individual. The continuing deterioration of American family farm agriculture is taking an extraordinary toll on the larger community of people associated with it and is edging the nation toward a deepening economic crisis. This is a condition that Christians everywhere should seek to rectify.

Over the course of 1986 and 1987, at least 10 percent of the nation's farmers will be liquidated (as Neil Harl affirms in Chap. 11). They will default on some twenty-five billion dollars in farm loans. Wharton Econometrics Forecasting Associates, located in Philadelphia, estimates that such a massive default would lead to a cumulative reduction in the GNP of thirty to fifty billion dollars over the ensuing eight years, a twenty-five billion dollar decline in business investment, a twelve billion dollar loss in personal income, and by 1989, some three hundred thousand fewer jobs in construction, manufacturing, and retailing.

In the same *Wall Street Journal* article outlining this grim scenario, several Midwest bankers were reported to have said that one-fourth of their downtown businesses will close in the next year. An economist at Data Resource, Inc., stated flatly that "a large sector of rural America will simply go out of business." With falling land and property values, school districts are feeling the pinch. City, county, and state governments are faced with continuing cutbacks as income and sales tax revenues decline. Municipal bond ratings for farm-dependent cities are dropping, resulting in higher interest payments. Real estate values have plummeted in towns and cities hit by high rates of farm-related unemployment.

When 21 percent of the entire workforce in the United States is linked to food production and distribution, it is obvious that the vitality of family farm agriculture is critical to the well-being of those far beyond the farm gate. Only through the economic revitalization of family farming — and the just distribution of farmland — can rural communities and values be saved.

Justice

Of all values we claim as a people of faith, none is more incumbent on us than that of justice. As Joseph Sittler (1985) wrote, "Justice is not an invented and imposed virtue; it is a precondition for human life" (p. 245). As we know from the experience of Ahab and the people of Israel, where justice does not prevail, judgment does.

The current condition of American family farm agriculture and the accelerating concentration of control over land is a travesty of justice, especially in a society that prides itself on democratic principles rooted in a religious heritage. The movement of this nation toward a two-tier structure of agriculture is contrary to the values inherent in our faith and democracy.

The spectre of a landed elite — of aspiring Ahabs — in this country may not be that far from reality. We are building a framework of land control similar to that which we abhor in Central America. We all know the figures; farms with annual sales over one hundred thousand dollars account for at least 54 percent of all agricultural production. By the year 2000, some fifty-thousand farms may account for as much as 96 percent of all production.

When it comes to the issue of land control, distributive justice must be put into play. We must have laws preventing the accumulation of land by the few. There must be adequate economic and social mechanisms providing for access to land. Distributive justice mandates that even in America land reform be a key piece of our emerging social agenda. It demands that we wrestle with the "preposterous" notion of the Jubilee year as a mechanism for redistribution of land. Unlike Merrill Oster, I do not think that that is a notion God has discarded. Our Christian values should compel us to understand anew the profound implications of the Magnificat, in which God's intervention in history on behalf of the poor, the displaced, and the dispossessed is reaffirmed.

Ours is a vision of covenant with one another and with God, realized through our care and support for the displaced and dispossessed. Through our prophetic actions we seek to challenge the values, policies, practices, and norms of rampant individualism that have brought us to this crisis. Through our commitment to memory, community, and justice, we intend to maintain and strengthen the system of family farm agriculture that effectively distributes landownership in a just way.

This vision enables us to have the audacity to proclaim that the prevailing values of this society toward farm people and farmland are intolerable. It enables us to challenge the manner in which many of us now farm. We can question high-tech, chemical-dependent, capital-intensive agriculture and its economic, social, and environmental costs. The vision empowers us to raise other, broader moral issues. Who farms? How is land controlled? It enables us to suggest that we should come out of this period of crisis having made significant progress on the development of a sustainable, regenerative agriculture.

Our contemporary journey to Naboth's vineyard puts Americans in a precarious position. We can choose to be the firm-standing, God-revering Naboth. Or we can choose to be the covetous Ahab, who is worried only about himself and the accumulation of property. Our responsibility is clear. It is to fulfill the word of God, given to Amos and the people of Israel in an earlier period of crisis:

> I will restore the fortunes of my people Israel;
> they shall rebuild cities and live in them,
> they shall plant vineyards and drink their wine,
> make gardens and eat the fruit.
> Once more I will plant them on their own soil,
> and they shall never again be uprooted
> from the soil I have given them.
> It is the word of the Lord your God.
>   (Amos 9:14–15)

References

Brueggemann, Walter. *The Land.* Philadelphia: Fortress Press, 1977.

Mather, Increase. *An Earnest Exhortation to the Inhabitants of New England.* Boston, 1676. Cited in *Nature's Nation* by Perry Miller, 29. Cambridge, Mass: Harvard University Press, 1967.

Robbins, Roy M. *Our Landed Heritage.* Princeton, N.J.: Princeton University Press, 1942.

Sittler, Joseph. "Moral Discourse in the Nuclear Age." *Christian Century,* 6 March 1985, 245.

CHARLES P. LUTZ

A Lutheran View

WHAT IS THE DISTINCTIVE OFFERING that the Christian tradition can bring to the present trauma in our country's heartland? Let me begin by stating two themes that I will not develop. First, I will not claim that Christian theology or its ethical tradition has any specific solutions in terms of public policy decisions. I do not think one can draw from that tradition the outlines of particular farm legislation other than in very broad terms, such as a bias toward nurturing the earth and toward decentralization of landownership or control. Second, I do not consider the church's call to minister amid crisis to be a unique call to minister in *this* crisis. We are asked by God to be in ministry with people who are hurting, displaced, perplexed, anxious, or oppressed wherever that is true: in urban or rural communities, in this country, or in other parts of the world.

Yet I think there is something special about the present crisis in rural America. It is special for the churches because the theology about the land and the people who tend it is central to the biblical witness. It is special for the churches also because, in a profound sense, the future of the churches in rural society is itself at stake. If that be institutional self-interest—and I think it is—let us make the most of it.

Christian Theology Has a Word of Compassion

"Compassion" means to suffer with. That is more than feeling sorrow for. It means to identify with another in the other's suffering. It is what our Lord did and does with the suffering of us all. God in Christ enters into our passion, our suffering, and takes it upon himself. That is also what the people who follow Christ are called to do.

CHARLES P. LUTZ is director of the Office of Church in Society of The American Lutheran Church, one of the predecessor bodies of the newly formed Evangelical Lutheran Church in America.

It is the first and most important thing for churches and church people to do in any time of crisis: to be with those who are suffering, who are hurting, who are disjointed and bewildered because of things they cannot control. We are called by our faith, I believe, to minister to all the people who suffer because of the rural crisis: farm families, certainly. But also those who work in rural lending institutions, in the farm machinery business, in other parts of the rural economy. I think there is special need today for the larger church to be in particular ministry with and for its rural pastors, many of whom are themselves showing compassion fatigue, who are burned out, who are in various forms of despair.

Part of compassion — of suffering with — is to understand. Christians in our cities need to try to understand what is happening, to resist the temptation to blame the victims, and to learn to grieve with those who are experiencing loss. Another part of compassion is simply the ministry of presence. We should not think that we must always say something or do something. When the word is one of compassion, at times it will be a silent word spoken by presence, by standing in solidarity with those who suffer.

St. Paul writes in 1 Cor. 12:26, "If one member suffers, all suffer together." That is a distinctive contribution that the family of God can make. In the community of faith, we all rejoice together, and we all suffer pain together. If one part of the body hurts, the whole body experiences distress.

I recall in my church, The American Lutheran Church (ALC), how rural congregations responded in the late sixties, when the crisis we talked about was an urban one. The cities were in great distress every summer for several years straight. They were hurting all year round, but seemed to erupt in the heat of summertime. Our church, along with many others, developed special programs of response, with education and financial support for some creative new approaches to racial justice, to economic development, to advocacy for the poor. There was a ready response from our congregations in rural America. I would like to see a similar depth of commitment now from nonrural congregations. I am not sure it's there. But we are called in the churches to reciprocate. That's what the word of compassion means: when you suffer I am with you; when I suffer you are with me. If we can make that basic affirmation at the beginning, all else that we want to talk about will flow therefrom.

The word of compassion is also a word about worth. Most farm people I have known take great pride in their work. It is their success at farming that gives them a sense of self-worth. In that, farmers are not different from other Americans. We all tend to value ourselves according to our achievement in our work. Biblically, that may be inadequate, but it is a statement of the way we are; in the United States probably more than anywhere else we equate worth with economic achievement. What word does our theology have, then, for those who are not achieving in economic terms, maybe for the first time in decades? Listen to Bishop Darold Beekmann of the ALC Southwestern Minnesota District:

> Farmers are accustomed to measuring worth on the basis of how straight the crop rows are, how many bushels of corn are produced per acre, how clean the fields are kept, and what condition machinery is in. These symbols of worth are all very visible to the whole community, and they are easily measured by everyone. Suddenly the word is out, "They lost the farm." The flip side of the close-knit, supportive community becomes evident. And the sense of failure can be overwhelming, not only for the farmer but also for the spouse and children. These people need to hear the Gospel word that their identity and worth are not determined by such a turn of events alone, but by an act of a loving and merciful God who assesses a person's worth in totally different ways.

We have in our theology a profound word for the one whose life is a "failure," as viewed in standard human terms. But the words we speak in the church are often contradicted by our actions, which express another theology. Iowa's Dave Ostendorf has spoken elsewhere of the struggling farmer who was asked to resign his position as church council president because his Chapter 11 bankruptcy did not reflect well on the congregation.

Can our churches learn to live and act in grace, as well as to mouth the word of grace? Can we behave in a way that mirrors the way God looks upon people — or are we doomed to mirror the common culture? E. W. Mueller, another Iowan and longtime leader in rural ministry, has observed that the church "lacks a doctrine of adjustment to social decline . . . and finds it difficult to minister to people when economic life, family life and community life fail" (Beekmann 1986). Perhaps the current rural crisis will help call us again to the heart of Christian theology — that God in pure grace is at work seeking to bring wholeness to broken people and broken communities.

Finally, a compassionate word sometimes includes the harder word that some of our farming families will not be able to continue in the work they love. It is compassionate, I believe, to say to those who have no way of continuing, "There is life after farming" — not just say it but be available to help those friends with the transition to a new life.

Christian Theology Has a Word about Creation and Its Care

The biblical faith offers a vision of the land and the people God places on the land as its caretakers. Walter Brueggemann and others have helped us to see the central place that the land occupies in our theology. This is not the time to detail that land theology, but I do wish to suggest something about the biblical warning against concentration in land control. I will approach that issue by recalling a policy decision in our own country just over a century ago: the Homestead Act. In the Homestead Act we were implementing a national decision about land tenure. By that legislation we decided essentially two things: first, that agricultural land in this country would be held privately, not publicly; second, and perhaps more important, that U.S. farmland would be held by the many and not by the few. I argue, therefore, that the Homestead Act was a decision of public policy in favor of the smaller landholder and against having a landed gentry, in favor of many modest-sized operations and against a plantation system, or what is called *latifundia* in Latin America. Latin America is a good place to look, if we want to see the results of another kind of system of land tenure (a bad system, in my judgment). Indeed, one could argue that every revolution in Latin America today, if not in the entire planet since World War II, has as a central cause the inequitable control of the land by the few.

By some measures, the U.S. has already surpassed many Latin American societies in concentration of farmland. Ours today is held by less than 3 percent of the population. We are not particularly upset about it because we are no longer a basically agrarian society. It doesn't seem a matter of basic justice to us, since most Americans don't need to or care to own farmland.

But I am concerned that the concentration trend not go any further. My reasons are ones of broad social good — not just the goals of preserving in farming those who are now there and wish to remain, or of letting new operators enter farming. Those are both good reasons. But I want to mention three others, and these are arguments about the self-interest of us all, including the vast majority of our people who will never own farmland. We all, I contend, have a stake in keeping the control of agricultural land as dispersed as is possible.

1. Dispersed ownership and management of farmland offers the best chance that the land will be preserved for future generations, stewarded carefully, built up over the long term, not used up in the short term. When farmland stays in the same family over several genera-

tions—and when there is a family expectation that such will be the case—there is simply more incentive to care gently for the land as good stewards than there is under other types of land tenure.

2. Dispersed ownership of the land is a more efficient way of making the land productive. At least, moderate-sized operations in the Midwest—gross sales, say, of $100 thousand to $250 thousand annually—have proven to be more efficient than hired-labor operations. In part that's because owner-operator family members are willing to give more of themselves for less economic return. To state it negatively, as the title of a publication from the Center for Rural Affairs in Walthill, Nebraska, says it: "Who Will Sit up With the Corporate Sow?" Most of our family farms in the Upper Midwest are big enough to be efficient. They are *not* in trouble because of inefficiencies in the production of food but because of high credit costs, low commodity prices, and some public tax and farm policies that artificially reward bigness or expansion.

3. Without a dispersed ownership of farmland and a resulting population base on the land, rural community is difficult to maintain; and rural institutions of all sorts—schools, libraries, health care, churches—are eroded away and starve to death. It is good for the United States that nonurban communities, lots of them, survive and prosper.

The policy implications of all this are that we should remove those policies that have negative impact on dispersion of land control. We should support those with positive impact. One of these is a graduated land tax. It was championed by Jefferson. It was in place for a short time early in the republic's life, but the larger holders got rid of it.

So I return to the biblical message. The writer of Isaiah, chapter five, was right: joining of field to field until you alone are left in the land is a prescription for woe upon any people. Let us heed the biblical warning. It's good for the creation. It's good for the people.

Christian Theology Has a Word about Community

This third word is the toughest word, I think because there are strong counterwords in our culture, especially in the culture of agriculture as we have known it. One reason for rural America's distress is that rural America is not together and never has been. We have a Gospel word about community, togetherness. It is not a word in support of individualism or independence. Yet individualism and independence are part of the traditional fabric and mythology of rural America. Harold Breimyer, a keen observer of the family farm's evolution, says in "Farm Policy: 13 Essays":

> The problem of the family farmer's survival begins with his own self-image. At the drop of a hint the farmer will recount all the heroic independence and self-reliance he displays and what virtues they are in our flimsy, character-less society. These qualities are not to be decried. They are splendid. They are particularly well suited to the agrarian portion of agriculture. But those qualities omit group consciousness. They have no social content—no schooling for joint action. They contribute nothing toward developing a sense of what is required not for *a* farmer but for *farmers* to survive in an increasingly industrial agriculture. . . . Often the individual farmer, lacking group consciousness, judges and responds to laws and rules only according to how they relate to him directly. He can be oblivious to how they affect him indirectly via consequences upon all his fellow farmers (pp. 70–71).

Farmers, by definition, are independent entrepreneurs. They tend to view other farmers less as allies than as competitors. Farmers have always had a deep suspicion of collective activity, especially if government is the vehicle for it.

Most farmland sold today in the Midwest is bought by other farmers getting larger, not by off-farm investors or by corporations or by new people entering farming. So in some ways, that trend coupled with a get-big-or-get-out mentality among farmers has helped to undermine the structure of family farming. What is good for a single farm family may be very bad for the institution of family farming. Two other examples could help to illustrate the point. The ending of a rigorous estate tax may help to keep farmland together in one family from generation to generation. But if it lets land-holding in agriculture drift into a hereditary pattern, it also works against the American tradition of providing access to new farmers. Selling off parcels of farm estates in order to pay taxes is not, therefore, objectionable. It may be exactly what is needed to keep open the door of opportunity for new farmers. The other example relates to controls on land use. It is entirely possible that within fifty miles of metropolitan areas family farming will survive only under the provision of land-use laws that, while repugnant to some individual farmers, offer farming as a structure an essential protection.

Let me say something more positive about the farming tradition and this matter of community consciousness. I often think of the threshing ring of the 1940s in northern Iowa where I was reared. I think of the working together in times of crisis—severe medical problems or barn fires. Can farmers in today's crisis join together in order to save one another for farming? Or does the system require cannibalism, so only a few will make it? Can farmers learn to purchase collectively? Can they learn to market their product mutually and agree mutually not to overproduce? Can farmers help young families get started in farming and not grab every bit of land that enters the market for their own expansion? Can farmers overcome their differences on specific political questions and come together on what is in the best interest of the future of family farming itself? Rural community may be no more divided than urban America — but it hurts more in rural settings.

I am suggesting that Christian theology has a vision of community that speaks especially to the rural crisis. The churches can lift up the principle of cooperation instead of competition, of communitarian rather than individualist approaches to our problems. It is tough to do while also retaining a basic neutrality on the specific policy issues. The churches have a great opportunity to perform the role of convenors, but that becomes closed to the degree that we are identified with the Farm Bureau alone or the American Agriculture Movement alone, or the NFO alone.

What I am calling for, to put it another way, is that our churches add something to compassion and a commitment to the creation — a willingness to be political. I use it in the broadest sense: having to do with the well-being of the group, with power relationships, with societal change, and structuring for justice. At the same time, I think the churches are called to avoid becoming partisan — partisan in the sense of identifying with one ideology about the rural economy, or one organizational or party approach to it, or one legislative strategy. The churches, on the basis of their biblical theology, can be powerfully committed to the need for community building over individual values. At the same time, they can retain their bridging role, their convenor opportunity, provided they are scrupulous in avoiding partisanship. Is that possible? I don't know. I think we must try.

References

Beekmann, Darold. "Ministry Among the People of the Land in the 1980s." *Word and World* 6 (Winter 1986): 5–17.

Breimyer, Harold. *Farm Policy: 13 Essays.* Ames: Iowa State University Press, 1977.

SUGGESTED READINGS

Brueggemann, Walter. *The Land: Place as Gift, Promise, and Challenge in Biblical Faith.* Philadelphia: Fortress, 1977.

Fanfani, Amintore. *Catholicism, Protestantism and Capitalism.* Notre Dame: University of Notre Dame Press, 1986.

Hart, John. *The Spirit of the Earth: A Theology of the Land.* Mahwah, N.J.: Paulist, 1984.

Lutz, Charles P., ed. *Farming the Lord's Land: Christian Perspectives on American Agriculture.* Minneapolis: Augsburg, 1980.

Novak, Michael. *The Spirit of Democratic Capitalism.* New York: Simon & Schuster, 1982.

Raines, John C., and Donna C. Day-Lower. *Modern Work and Human Meaning.* Philadelphia: Westminster, 1986.

Wogaman, J. Philip. *Economics and Ethics: A Christian Inquiry.* Philadelphia: Fortress, 1985.

Do Family Farms Support a Better Way of Life?

VII

BOTH THE Catholic bishops and David Ostendorf argue that family farms have unique spiritual value. They give theological justifications for their view; land is a gift of God to all God's people. Many families then—and not a few corporations—ought to own the land and control its resources. We—all of us—are the intended recipients of God's gift.

Are family farms sacred? Do they necessarily represent a better way of life than other agricultural arrangements? In Chapter 13, Glenn Johnson calls for "more and better . . . positive knowledge" about this sort of question. The essays in Part Seven respond by addressing the quality and spirit of life in rural America. Wendell Berry offers a defense of the family farm that is at once lyrical and hard-nosed. He recalls the simple virtues farming engenders while insisting that the power of property ownership must be democratically distributed. Michael Boehlje contributes an economic analysis in which family farm life does not appear clearly superior to life on corporate farms.

It may be useful at this point to think about our definition of the family farm. As used in this volume, the phrase refers to an agricultural operation loved, worked, and owned by a family or nonfamily corporation, with gross annual sales of forty to two hundred thousand dollars, hiring less than 1.5 person-years of labor. I noted in the Introduction that some writers were unhappy with a preliminary definition that characterized family farms in strictly economic terms. Wendell Berry is one of the most outspoken of such critics. In the essay printed here, he defines the family farm as a farm "small enough to be farmed by a family." Similarly, he characterizes good farming as "farming that does not destroy either farmland or farm people." Defending this way of life is easy, says Berry, since it represents work that has "a quality and dignity that it is dangerous for human work to go without." Ending on a hopeful note, he suggests that we can recover the social, economic, and spiritual values found in agrarian culture to remake the fabric of rural America.

Boehlje presents a social-scientific analysis of family and corporate farms, comparing them at nine different points: economic efficiency, financial stability, standard of living, resource conservation, employment, entrepreneurial prerogatives, flexibility, contributions to community, and independence. In none of these areas does he find the family farm with a clear-cut advantage over corporate farming. He concludes that these and other attributes traditionally suggested as representing the superiority of family farms are of dubious value. It is not clear to him that family farms represent a way of life better than that possible on corporate farms.

Berry responds with a comment on Boehlje's own commitment to scientific procedure, charging that that commitment blinds Boehlje to our most important cultural values.

27 WENDELL BERRY

A Defense of the Family Farm

TO BE ASKED TO DEFEND the family farm is like being asked to defend the Bill of Rights or the Sermon on the Mount or Shakespeare's plays. One is amazed at the necessity for defense, and yet one agrees gladly, knowing that the family farm is both eminently defensible and a part of the definition of one's own humanity.

And yet, having agreed to this defense, one remembers uneasily that there has been a public clamor in defense of the family farm throughout all the years of its decline — that, in fact, during all the years of its decline the family farm has had virtually no professed enemies, but that some of its worst enemies have been its professed friends. That is to say that "the family farm" has become a political catchword, like democracy and Christianity, and that, as with democracy and Christianity, much evil has been done in its name.

Before taking up this defense, then, it is necessary to make several careful distinctions. What I shall mean by the term "family farm" is a farm small enough to be farmed by a family, and one that is farmed by a family — perhaps with a small amount of hired help. I shall not mean a farm that is owned by a family and worked by other people. The family farm is both the home and the workplace of the family that owns it.

By the verb "farm" I do not mean just the production of marketable crops, but also the responsible maintenance of the health and usability of the place while it is in production. A family farm is one that is properly cared for by its family.

Furthermore, the term "family farm" implies longevity in the connection between family and farm. A family farm is not a farm that a family has bought on speculation and is only occupying and using until it

---

can be profitably sold. And neither, strictly speaking, is it a farm that a family has newly bought, though, depending on the intentions of the family, we may be able to say that such a farm is potentially a family farm. This suggests that we may have to think in terms of ranks or degrees of family farms. A farm that has been in the same family for three generations may rank higher as a family farm than a farm that has been in a family only one generation; it may have a higher degree of familiness or familiarity than the one-generation farm. Such distinctions have a practical usefulness to the understanding of agriculture, and, as I hope to show, there are rewards of longevity that do not accrue only to the farm family.

I mentioned the possibility that a family farm might use a small amount of hired help. This greatly complicates matters, and I wish it were possible to say, simply, that a family farm is farmed with family labor. But it seems important to allow for the possibility of supplementing family labor with wage-work or some form of sharecropping. For one reason, family labor may become insufficient as a result, say, of age or debility. But another and probably more important reason is that an equitable system of wage-earning or sharecropping would permit unpropertied families to earn their way to farm ownership. The critical points in defining "family farm" are that the amount of nonfamily labor should be small, and that it should supplement, not replace, family labor. It is understood that on a family farm the family members are workers not overseers. If a family on a family farm does require supplementary labor, it seems desirable that the hired help should live on the place and work year-round. The idea of a family farm is jeopardized by supposing that the farm family might be simply the guardians or maintainers of crops planted and harvested by seasonal workers. This, of course, implies both small scale and diversity.

Finally, I think that one other possibility must be allowed for: that a family farm might be one that is very small or marginal, or one that does not entirely support its family. In such cases, though the economic return and dependency might be reduced, the values of the family-owned and family-worked small farm are still available both to the family and to the nation.

The idea of the family farm, as I have just defined it, is conformable in every way to the idea of good farming: farming that does not destroy either farmland or farm people. The two ideas may in fact be inseparable. If family farming and good farming are as nearly synonymous as I suspect they are, that is because of a law that is well understood, still, by most farmers but that has been ignored in the colleges and offices and corporations of agriculture for thirty-five or forty years. The law reads something like this: land that is in human use must be lovingly used; it requires intimate knowledge, attention, and care.

The practical meaning of this law is that (to borrow an insight from Wes Jackson of the Land Institute in Salina, Kansas) there is a ratio between eyes and acres, between farm size and farmhands, that is correct. We know that this law is unrelenting — that, for example, one of the meanings of our current high rates of soil erosion is that we do not have enough farmers; we have enough farmers to use the land, but not enough to use it and protect it at the same time.

In this law, which is not subject to human repeal, is the justification of the small, family-owned, family-worked farm. It is this law that gives a preeminent and irrevocable value to familiarity: the family life that alone can properly connect a people to a land.

This is a connection, admittedly, that is easy to sentimentalize, and we must be careful not to do so. There is no guarantee that family farming will be good farming. We all know that small family farms can be abused because we know that sometimes they have been. It is nevertheless true that familiarity tends to mitigate and to correct abuse. A family that has farmed a farm through two or three generations will possess not just the land but a remembered history of its mistakes and of the remedies of those mistakes. It will know, not just what it can do, what is technologically possible, but also what it must do, and what it must not do. The family will have understood the ways in which it and the farm empower and limit one another. That is the value of longevity in landholding. In the long term, knowledge and affection accumulate, and in the long term, knowledge and affection pay. They do not just pay the family in goods and money, but they pay the family and the whole country in health and satisfaction.

But the justifications of the family farm are not merely agricultural. They are political and cultural as well. The question of the survival of the family farm and the farm family is one version of the question of who will own the country — which is, ultimately, the question of who will own the people. Shall the usable property of our country be democratically divided or not? Shall the power of property be a democratic power or not? If many people do not own the usable property, then they must submit to the few who do own it. They cannot eat or be sheltered or clothed except in submission. They will find themselves entirely dependent on money; they will find costs always higher and money always harder to get. To renounce the principle of democratic property, which is the only basis of democratic liberty, in exchange for specious notions of efficiency or the economics of the so-called free market is a tragic folly.

There is one more justification, among many, that I want to talk about: namely, that the small farm of a good farmer, like the small shop of a good craftsman, gives work a quality and a dignity that it is dangerous for human work to go without. Work without quality and without dignity is a danger both to the worker and to the nation. If using ten workers to make one pin results in the production of many more pins than the ten workers could produce individually, that is undeniably an improvement in production, and perhaps uniformity is a virtue in pins. But in the process ten workers have been demeaned; they have been denied the economic use of their minds; their work has become thoughtless and skill-less. Robert Heilbroner (1984, 20) says that such "division of labor reduces the activity of labor to dismembered gestures."

Eric Gill (1983, 61) sees in this industrial dismemberment of labor a crucial distinction between making and doing: "Skill in making . . . degenerates into mere dexterity, i.e. skill in doing, when the workman . . . ceases to be concerned for the thing made or . . . has no longer any responsibility for the thing made and has therefore lost the knowledge of what it is that he is making. . . . The trouble is simply the degradation of the mind."

The degradation of the mind, of course, cannot be without consequences. One obvious consequence is the degradation of products. When workers' minds are degraded by loss of responsibility for what is being made, they cannot use judgment; they have no use for their critical faculties; they have no occasions for the exercise of workmanship, or workmanly pride. And the consumer is likewise degraded by loss of the opportunity for qualitative choice. This is why we must now buy our clothes and immediately resew the buttons. It is why our expensive purchases quickly become junk.

But there is an even more important consequence: by the dismemberment of work, by the degradation of our minds as workers, we are denied our highest calling. Writing as a Christian artist and thinker, Gill (1983,65) says that "every man is called to give love to the work of his hands. Every man is called to be an artist."

The small family farm is one of the last places—they are getting rarer every day—where men and women, girls and boys, can answer that call to be an artist, to learn to give love to the work of their hands. It is one of the last places where the maker—and some farmers still do talk about "making the crops"—is responsible from start to finish for the thing made. This will perhaps be thought a spiritual value, but it is not for that reason an impractical or uneconomic one. In fact, from the exercise of this responsibility, this giving of love to the work of hands, the farmer, the farm, the consumer, and the nation stand to gain in the most practical ways: they gain the means of life; they gain the goodness of food; they gain longevity and dependability of the sources of food, both natural and cultural. The proper answer to the spiritual calling becomes, in turn, the proper fulfillment of physical need.

The family farm, then, is good. To show that it is good is easy. To find evil in it and to argue against it would be extremely difficult. Those who have done most to destroy it have, I think, found no evil in it. They have not argued against it. With them, perhaps, the family farm and family farmer have had the misfortune only of being in the way.

If a good thing is failing among us, pretty much without being argued against and pretty much without professed enemies, then it is necessary to ask why it should fail. I have spent years trying to answer this question, and I am sure of some answers, but I am also sure that the complete answer will be hard to come by. That is because the complete answer has to do with who and what we are as a people; the fault lies in our identity, and therefore will be hard for us to see.

But we must try to see, and the best place to begin, maybe, is with the fact that the family farm is not the only good thing that is failing among us. The family farm is failing because it belongs to an order of values and a kind of life that are failing. We can only find it wonderful, when we put our minds to it, that many people now seem willing to mount an emergency effort to "save the family farm" who have not yet thought to save the family or the community or the neighborhood schools or the small local businesses or the domestic arts of household and homestead or cultural and moral tradition — all of which are also failing and on all of which the survival of the family farm depends.

The family farm is failing because the pattern it belongs to is failing. And the principle reason for this failure is the universal adoption, by our people and our leaders alike, of industrial values, which are based on three assumptions:

1. That value equals price — that the value of a farm, for example, is whatever it would bring on sale, because both a place and its price are "assets." There is no essential difference between farming and selling a farm.

2. That all relations are mechanical — that a farm, for example, can be used like a factory, because there is no essential difference between a farm and a factory.

3. That the sufficient and definitive human motive is competitive-

ness—that a community, for example, can be treated like a resource or a market, because there is no difference between a community and a resource or a market.

The industrial mind is a mind without compunction. It simply accepts that people, ultimately, will be treated as things, and that things, ultimately, will be treated as garbage.

With industrialization has come a general depreciation of physical work. As the price of work has gone up, the value of it has tended to go down. The value of work has now been so depressed that people simply do not want to do it any more. We can say without exaggeration that the present national ambition of the United States is unemployment. People live for quitting time, for weekends, for vacations, and for retirement; and this ambition seems to be classless, as true in the executive suites as on the assembly lines. One works, not because the work is necessary or valuable or useful to a desirable end or because one loves to do it, but only to be able to quit—a condition that a saner time would regard as infernal, a condemnation. This is explained, of course, by the dullness of the work, by the loss of responsibility for, or credit for, or knowledge of, the thing made. But what can be the status of the working small farmer in a nation whose motto is a sign of relief: "Thank God it's Friday"?

The industrial mind is indifferent to the connections, necessarily both practical and cultural, between people and land, which is to say that it is indifferent to the fundamental economy and economics of human life. Our economy is increasingly abstract, increasingly a thing of paper, unable either to describe or to serve the real economy that determines whether or not people will eat and be clothed and sheltered. And it is this increasingly false or fantastical economy that is invoked as a standard of national health and happiness by our political leaders.

That this so-called economy can be used as a universal standard can only mean that it is itself without standards. Industrial economists cannot measure the economy by the health of nature, for they regard nature as a source of "raw materials." Industrial economists cannot measure it by the health of people, for they regard people as "labor" (that is, as tools or machine parts) or as "consumers." The health of the economy, then, can be measured only in sums of money.

And here we come to the heart of the matter: the absolute divorce that the industrial economy has achieved between itself and all ideals and standards outside itself. It does this of course by arrogating to itself the status of primary reality. Once that is established, all its ties to principles of morality or religion or government necessarily fall slack.

But a culture disintegrates when its economy disconnects from its principles of government and religion. If we are dismembered in our economic life, how can we be members in our communal and spiritual life? We assume that we can have an exploitive, ruthlessly competitive, profit-for-profit's-sake economy, and yet remain a God-fearing and a democratic nation, as we still apparently wish to think ourselves. This simply means that our highest principles and standards have no practical force or influence, and are reduced merely to talk.

That this is true was acknowledged by William Safire in a column (1986, A7) in which he declared that our economy is driven by greed — which, therefore, should no longer count as one of the seven deadly sins. "Greed," he said, "is finally being recognized as a virtue . . . the best engine of betterment known to man." It is, moreover, an agricultural virtue: "the cure for world hunger is the driving force of Greed." Such statements would be possible only to someone who sees the industrial economy as the ultimate reality. Safire attempts a disclaimer, perhaps to maintain his status as a conservative: "I hold no brief for Anger, Envy, Lust, Gluttony, Pride, Envy or Sloth." But this is not a cat that can be let only partly out of the bag. In fact, all seven of the deadly sins are "driving forces" of this economy as its advertisements and commercials plainly show.

As a nation, then, we are not very religious and not very democratic, and that is why we have been destroying the family farm for the last forty years — along with other small local economic enterprises of all kinds. We have been willing for millions of these people to be condemned to failure and dispossession by the workings of an economy utterly indifferent to any claims they may have had either as children of God or as citizens of a democracy. "That's the way a dynamic economy works," we have said. We have said, "Get big or get out." We have said, "Adapt or die." And we have washed our hands of them.

Throughout this period of drastic attrition on the farm we have supposedly been "subsidizing agriculture," but, as Wes Jackson says, this is a misstatement. What we have actually been doing is using the farmers to launder money for the agribusiness corporations, which have controlled both their supplies and their markets, while the farmers have overproduced and been at the mercy of the markets. The result has been that the farmers have failed by the millions and the agribusiness corporations have prospered — or they prospered until the present farm depression, when some of them finally realized that, after all, they were dependent on their customers, the farmers.

And throughout this period, such a desperate one for farmers, the

colleges of agriculture, the experiment stations, and the extension services have been working under their old mandate (the Hatch Act sec.
361b) to promote "a sound and prosperous agriculture and rural life," to
"aid in maintaining an equitable balance between agriculture and other
segments of the economy," to contribute "to the establishment and maintenance of a permanent and effective agricultural industry," and to help
"the development and improvement of the rural home and rural life."

That the land grant system has failed this commission is by now
simply obvious. I am aware that there are many individual professors,
scientists, and extension workers whose lives have been dedicated to the
fulfillment of this commission and whose work has genuinely served the
rural home and rural life. But, in general, it can no longer be denied that
the system has failed its trust. One hundred and twenty-four years after
the Morrill Act, ninety-nine years after the Hatch Act, seventy-two years
after the Smith-Lever Act, "the industrial classes" are not liberally educated, agriculture and rural life are not sound or prosperous or permanent, and there is no equitable balance between agriculture and other
segments of the economy. Anybody's statistics on the reduction of the
farm population, on the decay of rural communities, on soil erosion, soil
and water pollution, water shortages, and farm bankruptcies, tell indis-
putably a story of failure.

This failure cannot be understood apart from the complex allegiances between the land grant system and the aims, ambitions, and
values of the agribusiness corporations. The willingness of land grant
professors, scientists, and extension experts to serve as state-paid researchers and traveling salesmen for those corporations has been well
documented and is widely known.

The reasons for this state of affairs, again, are complex. I have
already given some of them; I don't pretend to know them all. But I
would like to mention one that I think is probably the most telling: that
the offices of the land grant complex—like the offices of the agricultural
bureaucracy—have been looked upon by their aspirants and their occupants as a means, not to serve farmers, but to escape farming. Over and
over again, one hears the specialists and experts of agriculture introduced as "old farm boys" who have gone on (as it is invariably implied)
to better things. And the reason for this is plain enough. The life of a
farmer has characteristically been a fairly hard one, and the life of a
college professor or professional expert has characteristically been fairly
easy. Farmers—working family farmers—do not have tenure, business
hours, free weekends, paid vacations, sabbaticals, and retirement funds.
They do not have professional status. Like other classes that have been
oppressed, they tend to have a low opinion of themselves; too often they

introduce themselves as "just a farmer" or "just an old country boy" or "just an ignorant farmer." Both the farmers and the land grant professional seem to have assumed that those who need advice are therefore inferior to those who give it, both of them disregarding the fact that the land grand professional has rarely been in a position to take his or her own advice: that is, the land grant professional has seldom been a working farmer, and a working farmer is the last thing he or she has wished to be.

The direction of the career of agricultural professionals is typically not toward farming or association with farmers. It is "upward" through the hierarchy of a university or a bureau or an agribusiness corporation. They do not, like Cincinnatus, leave the plow to serve their people and return to the plow. They leave the plow, simply, for the sake of leaving the plow.

This means that there has been for several decades a radical disconnection between the land grant institutions and the farms. And this disconnection has left the land grant professionals free to give bad advice; indeed, if they can get this advice published in the right place, from the standpoint of their careers it does not matter whether their advice is good or not.

For example, after years of milk glut, when today's dairy farmers are everywhere threatened by their surplus production, university experts are still working to increase milk production and still advising farmers to cull their least productive cows. They are apparently oblivious both to the possible existence of other standards of judgment and to the fact that this culling of the least productive cows is ultimately the culling of the smaller farmers.

Perhaps this could be dismissed as human frailty or inevitable bureaucratic blundering — except that the result is damage caused by people who probably would not have given such advice if they were themselves in a position to suffer from it. Serious responsibilities are undertaken by public givers of advice, and serious wrong is done when the advice is bad. And surely a kind of monstrosity is involved when tenured professors with protected incomes recommend or even tolerate Darwinian economic policies for farmers or announce (as one university economist after another has done) that the failure of so-called inefficient farmers is good for agriculture and good for the country. They see no inconsistency, apparently, between their own protectionist economy and the free-market economy that they recommend to their supposed constituents to whom the free market has proved, time and again, to be fatal.

Nor do they see any inconsistency, apparently, between the economy of a university, whose sources like those of any tax-supported institution are highly diversified, and the extremely specialized economies that they have recommended to their farmer-constituents. These inconsistencies nevertheless exist, and they explain why so far there has been no epidemic of bankruptcies among professors of agricultural economics.

These, of course, are simply instances of the notorious discrepancy between theory and practice. But this need not exist, or it need not be so extreme, in the colleges of agriculture. The answer to the problem is simply that those who profess should practice — or at least a significant percentage of them should. This is, in fact, the rule in other colleges and departments of the university. A professor of medicine who was no doctor would be readily seen as an oddity; so would a law professor who could not try a case; so would a professor of architecture who could not design a building. What, then, would be so strange about an agriculture professor who would be, and who would be expected to be, a proven farmer?

But it would be wrong, I think, to imply that the farmers are merely the victims of their predicament and share none of the blame. In fact, they along with all the rest of us do share the blame, and their first hope of survival is in understanding that they do.

Farmers as much as any other group have subscribed to the industrial fantasies that I listed earlier: that value equals price, that all relations are mechanical, and that competitiveness is a proper and sufficient motive. Farmers like the rest of us have assumed, under the tutelage of people with things to sell, that selfishness and extravagance were merely normal. Like the rest of us, farmers have believed that they might safely live a life prescribed by the advertisers of products rather than the life required by fundamental human necessities and responsibilities.

It could be argued that the great breakthrough of industrial agriculture occurred when most farmers became convinced that it would be better to own a neighbor's farm than to have a neighbor, and when they became willing, necessarily at the same time, to borrow extravagant amounts of money. They thus violated the two fundamental laws of domestic or community economics: you must be thrifty and you must be generous; or to put it a more practical way: you must be (within reason) independent, and you must be neighborly. With that violation, farmers became vulnerable to everything that intended their ruin.

An economic program that encourages the unlimited growth of individual holdings not only anticipates but actively proposes the failure of many people. Indeed, as our antimonoploy laws testify, it proposes the failure, ultimately, of all but one. It is a fact, I believe, that many people have now lost their farms and are out of farming who would still be in place had they been willing for their neighbors to survive along with themselves. In light of this we see that the machines, chemicals, and credit that farmers have been persuaded to use as "labor-savers" have, in fact, performed as neighbor-replacers. And whereas neighborhood tends to work as a service free to members, the machines, chemicals, and credit have come at a cost set by people who were not neighbors.

That is a description of the problem of the family farm, as I see it. It is a dangerous problem. I do not think it is hopeless. On the contrary, a number of solutions to the problem are implied in my description of it.

What, then can be done?

The most obvious, the most desirable solution would be to secure that "equitable balance between agriculture and other segments of the economy," which is one of the stated goals of the Hatch Act. To avoid the intricacies of the idea of parity, which we inevitably think of here, let us just say that the price of farm products as they leave the farm should be on a par with the price of those products that the farmer must buy.

In order to achieve this with minimal public expense, it will be necessary to control agricultural production. Supply must be adjusted to demand. Obviously this is something that individual farmers or individual states cannot do for themselves. It is a job that belongs appropriately to the federal government. As a governmental function, it is perfectly in keeping with the ideal everywhere implicit in the originating documents of our government that the small have a right to certain protections from the great. We have within limits that are obvious and reasonable the right to be small farmers or small businessmen or businesswomen just as, or perhaps insofar as, we have a right to life, liberty, and property. The individual citizen is not to be victimized by the rich any more than by the powerful. When Marty Strange (1984, 118) writes, "To the extent that only the exceptional succeed, the system fails," he is biologically, economically, and agriculturally sound, but he is also speaking directly from American political tradition.

The plight of the family farm would be improved also by other governmental changes, for example, in policies having to do with taxation and credit.

Our political problem, of course, is that small farmers are neither numerous enough nor rich enough to be optimistic about government help. The government tends rather to find their surplus production useful and their economic failure ideologically desirable. And so it seems to me that we must concentrate on those things that farmers and farming communities can do for themselves — striving in the meantime for policies that would be desirable.

It may be that the gravest danger to farmers is their relatively new inclination to look to the government for help — after the agribusiness corporations and the universities to which they have already looked have failed them. In the process they have forgotten how to look to themselves, to their farms, to their families, to their neighbors, and to their tradition. This, as I have already said, has been their great mistake.

Marty Strange (1984, 116) has written also of his belief "that commercial agriculture can survive within pluralistic American society, as we know it — *if* [my emphasis] the farm is rebuilt on some of the values with which it is popularly associated: conservation, independence, self-reliance, family, and community. To sustain itself, commercial agriculture will have to reorganize its social and economic structure as well as its technological base and production methods in a way that reinforces these values." I agree. Those are the values that offer us survival not just as farmers but as human beings. And I would point out that the transformation that Marty is proposing cannot be accomplished by the governments or the corporations or the universities. If it is to be done, the farmers themselves, their families, and their neighbors will have to do it.

What I am proposing, in short, is that farmers find their way out of the gyp-joint known as the industrial fantasy.

The first item on the agenda, I suggest, is the remaking of the rural neighborhoods and communities. The decay or loss of these has demonstrated their value; we find as we try to get along without them that they are worth something to us — spiritually, socially, and economically. And we hear again the voices out of our cultural tradition telling us that to have community, people don't need a "community center" or "recreational facilities" or any of the rest of the paraphernalia of "community improvement" that is always for sale. Instead, they need to love each other, trust each other, and help each other. That is hard. All of us know that no community is going to do those things easily or perfectly. And yet we know there is more hope in that difficulty and imperfection than in all the neat instructions for getting big and getting rich that have come out of the universities and the agribusiness corporations in the past fifty years.

Second, the farmers must look to their farms and consider the losses, human and economic, that may be implicit in the way those farms are structured and used. If they do that, many of them will understand how they have been cheated by the industrial orthodoxy of competition: how specialization has thrown them into competition with other farmer-specialists, how bigness of scale has thrown them into competition with neighbors and friends and family, how the consumer economy has thrown them into competition with themselves.

If it is a fact that for any given farm there is a ratio between people and acres that is correct, there are also correct ratios between dependence and independence and between consumption and production. For a farm family, a certain degree of independence is possible and is desirable. But no farmer and no family can be entirely independent. A certain degree of dependence is inescapable. Whether or not the dependency is desirable is a question of who is helped by it. If a family removes its dependency from its neighbors — if, indeed, farmers remove their dependency from their families — and gives it to the agribusiness corporations (and to moneylenders), the chances are, as we have seen, that the farmers and their families will not be greatly helped. This suggests that dependence on family and neighbors may constitute a very desirable kind of independence.

It is clear in the same way that a farm and its family cannot be *only* productive. There must be some degree of consumption. This also is inescapable; whether or not it is desirable depends on the ratio. If the farm consumes too much in relation to what it produces, then the farm family is at the mercy of its suppliers and is exposed to dangers that it need not be exposed to. When, for instance, farmers farm on so large a scale that they cannot sell their labor without enormous consumption of equipment and supplies, then they are vulnerable. I talked to an Ohio farmer recently who cultivated his corn crop with a team of horses. He explained that when he was plowing his corn he was *selling* his labor and that of his team (labor fueled by the farm itself and therefore very cheap) rather than *buying* herbicides. His point was simply that there is a critical difference between buying and selling and that the name of this difference at the years's end ought to be 'net gain.'

Similarly, when farmers let themselves be persuaded to buy their food instead of grow it, they became consumers instead of producers and lost a considerable income from their farms. This is simply to say that there is a domestic economy that is proper to the farming life, and that it is different from the domestic economy of the industrial suburbs.

Finally, I want to say that I have not been talking from speculation but from proof. I have had in mind throughout the one example known to me of an American community of small family farmers who have not only survived but thrived during some very difficult years: I mean the Amish. I do not recommend, of course, that all farmers become Amish. Nor do I want to suggest that the Amish are perfect people or that their way of life is perfect. What I want to recommend are some Amish principles:

1. They have preserved their families and communities.
2. They have maintained the practices of neighborhood.
3. They have maintained the domestic arts of kitchen and garden, household and homestead.
4. They have limited their use of technology so as not to displace or alienate available human labor or available free sources of power (the sun, wind, water, etc.).
5. They have limited their farms to a scale that is compatible both with the practice of neighborhood and with optimum use of low-power technology.
6. By the practices and limits already mentioned, they have limited their costs.
7. They have educated their children to live at home and serve their communities.
8. They have esteemed farming as both a practical art and a spiritual discipline.

These principles define a world to be lived in by human beings, not a world to be exploited by managers, stockholders, and experts.

References

Gill, Eric. *A Holy Tradition of Working,* edited by Brian Keeble. West Stockbridge, Mass.: Lindisfarne Press, 1983.

Heilbroner, Robert. *The Act of Work.* Occasional Papers of the Council of Scholars, Library of Congress, 1984.

Safire, William. "Make That Six Deadly Sins; A Re-examination Shows Greed to Be A Virtue." *Courier-Journal* (Louisville, Ky.), 7 January 1986, A7.

Strange, Marty. "The Economic Structure of a Sustainable Agriculture." In *Meeting the Expectations of the Land,* edited by Wes Jackson, Wendell Berry, and Bruce Colman. Berkeley, Calif.: North Point Press, 1984.

MICHAEL BOEHLJE

Costs and Benefits
of Family Farming

THE FAMILY FARM has been a revered dimension of U.S. agriculture. Maintenance of family-based farming is the espoused objective of much of U.S. public policy, whether it be price and income supports, credit programs, rural development and industrialization, or tax policy. Farmers, rural businesspeople, and urban consumers alike attribute unique, almost mystical, virtues to family farming. Anyone who would question the legitimacy of family farming and family-based agriculture must be a heretic. But that is precisely what my research leads me to do.

My purpose is not to deride the virtues of the family farm but only to attempt as a social scientist to identify its costs and benefits, both economic and noneconomic. Lest I promise more than I can deliver, my objective is modest: to attempt to structure our thoughts as to what the key dimensions of family-based agriculture are, and what economic, social, and moral attributes the family farm actually possesses.

A Structural Framework

Two aspects of the family farm structure are often confused. It is important that we identify these right from the start and keep them distinct throughout the following analysis. The first aspect concerns the relationship of the owner of production assets (particularly land) to the user of those assets (the operator or farmer). One way this relationship can be structured is for the owner to be the operator. Here, the operator of the farm owns the land as well as the machinery, equipment, and working capital. A second way this relationship can be structured is for the owner to act as a landlord renting out the land to a tenant. This is called a tenant-landlord arrangement. Some farmers use a mixed strat-

---

MICHAEL BOEHLJE is head of the Department of Agricultural and Applied Economics, University of Minnesota, St. Paul.

egy for control of the land resource; they own and farm a portion of land while renting and farming land owned by someone else.

The second aspect of the family farm structure of agriculture concerns the type of ownership, the economic control of the farm. Who owns or operates it? Here, a family or an extended family may be the locus of control, a situation in which the linkages between the individuals involved are dominated by personal ties rather than economic considerations. Alternatively, a nonfamily or "corporate" structure may control the farm. Here, the linkages between the individuals are primarily economic with only limited personal ties. The differences between these two ways of controlling a farm are numerous, but one of the more obvious ones centers on the degree of interpersonal commitment; within the family structure there is typically (but not always) more commitment to the relationship than there is in the corporate structure. Often in the corporate structure, the primary bonding is economic; interpersonal commitments are much less permanent or important.

Delineation of these two dimensions—the structure and control of farm ownership—is useful in evaluating the attributes of family farms. The distinction assists us in assessing the actual attributes of owner-operator versus tenant agriculture as well as of family versus nonfamily control. These two dimensions are frequently intertwined in discussions of the family farm, but family-controlled agriculture does not necessarily imply owner-operator agriculture.

The structural dimensions of size and financing method are not insignificant to discussions of the family farm. Some commentators imply that family farms are small and use primarily internal sources of financing (i.e., equity rather than debt or leased assets). This is not necessarily the case as evidenced by large farm and nonfarm businesses owned by the same family (for example, the Cargill family owns substantial farm as well as nonfarm business enterprises) as well as the substantial use of debt and even external sources of equity by owner-operator, family-controlled farms. Clouding the family farm issue with issues of size and financing again makes the debate confusing.

A Structural Overview of U.S. Production Agriculture

Four important structural characteristics of U.S. agriculture are size of firm, financial structure, tenancy arrangement, and business form, which implies the family nature of the industry.

As the analyses by Lasley and Harl in this volume make clear, U.S. agriculture seems to be headed for a bimodal structure. In 1982, for example, 28.4 percent of the farms in the United States had acreages of fifty acres or less (U.S. Bureau of the Census 1959, 1969, 1974, 1982). The acreage included in these farms was 1.5 percent of the total acreage in all farms. In contrast, 7.2 percent of the farms had acreages in excess of one thousand acres; these farms accounted for 40.2 percent of the total acreage. Average farm size in the United States continues to rise slowly, although evidence of the movement to a bimodal size distribution makes the average increasingly meaningless as a description of U.S. agriculture. In essence, farms with lower gross sales comprise a large proportion of total farms and a relatively small portion of total sales, whereas large farms exhibit the opposite distributional characteristics.

The fundamental financial structure of U.S. agriculture is summarized in Table 28.1. Note that approximately 15 percent of the farms in the United States had rates of return on equity of -20 percent or less as of 1 January 1985; in contrast over 15 percent had rates of return of 20 percent or greater. Clearly, the financial condition of agriculture is again characterized by a distribution; some farmers are suffering very severe financial stress, while others are exhibiting very strong financial performance.

As further data from the U.S. Bureau of Census show, the percentage of full-owner farms remained relatively constant from 1959 to 1982 at slightly less than 60 percent, whereas the part-owner farms increased from 21.9 percent to 29.3 percent of the total and the full-tenant farms decreased from 20.0 percent to 11.5 percent of the total during this same time period. In 1982 approximately 35 percent of the farmland was contained in full-owner farms, 54 percent in the part-owner farms, and slightly less than 12 percent in tenant farms.

Finally, the form of business organization used by farmers is also important. Census bureau data show that in 1982 approximately 87 percent of the farms were organized as individual or family units, 10 percent as partnerships, and less than 3 percent as corporations. These three forms of business organization accounted for 69 percent, 16 percent, and 14 percent, respectively, of the value of land and buildings and 59 percent, 16 percent, and 24 percent, respectively, of the value of farm products sold. Furthermore, of the approximately sixty thousand corporate farms in the United States in 1982, almost fifty-three thousand, or 90 percent, were family-held corporations.

Attributes of Family Farms

The following discussion will attempt to evaluate selected economic, social, and ethical attributes of family farms. The discussion should be viewed more accurately as a set of testable hypotheses rather than defini-

Table 28.1. Financial structure of U.S. agriculture: distribution of farm operators, debts, and assets by return to equity and region, 1 January 1985

Region	Insolvent farms %	Less than -20% %	-20 to -10% %	-10 to -5% %	-5 to 5% %	5 to 10% %	10 to 20% %	Greater than 20% %	All farms %
Northeast									
Operators (farms)	0.22	0.97	0.90	0.82	2.25	0.63	0.64	0.79	7.21
Debt[a]	0.76	0.68	0.47	0.43	1.17	0.47	0.60	0.79	5.38
Assets[b]	0.10	0.39	0.63	0.69	2.86	0.66	0.66	0.63	6.62
Lake state									
Operators (farms)	0.42	1.19	1.08	0.94	3.65	1.74	1.83	1.88	12.72
Debt	1.55	2.09	1.21	0.78	3.70	2.18	2.64	1.58	15.74
Assets	0.25	0.76	0.79	0.68	3.95	2.16	2.26	0.98	11.84
Corn Belt									
Operators (farms)	0.82	2.19	1.44	1.82	5.54	2.67	2.70	4.09	21.27
Debt	3.02	3.66	1.15	1.56	4.74	2.55	2.32	4.74	23.74
Assets	0.49	1.39	0.92	1.44	6.43	3.29	2.47	2.64	19.05
Northern plains									
Operators (farms)	0.43	1.45	0.76	0.71	3.10	1.44	1.33	1.49	10.72
Debt	1.39	2.14	1.14	1.44	3.45	1.57	1.52	1.96	14.61
Assets	0.23	0.96	0.76	0.88	4.65	2.09	1.69	1.02	12.28
Appalachia									
Operators (farms)	0.13	1.85	1.27	1.31	3.65	1.36	1.91	2.23	13.71
Debt	0.24	0.60	0.30	0.25	1.14	0.49	0.78	1.17	4.97
Assets	0.03	0.42	0.45	0.64	3.16	0.97	1.19	1.03	7.88
Southeast									
Operators (farms)	0.18	0.77	0.68	0.51	1.49	0.53	0.87	0.94	5.98
Debt	0.97	0.59	0.47	0.19	0.65	0.35	0.32	0.60	4.15
Assets	0.15	0.40	0.41	0.46	2.48	0.60	0.56	0.72	5.76

364

Table 28.1. *(continued)*

Region	Insolvent farms %	Less than −20% %	−20 to −10% %	−10 to −5% %	−5 to 5% %	5 to 10% %	10 to 20% %	Greater than 20% %	All farms %
Delta									
Operators (farms)	0.22	0.80	0.66	0.66	1.45	0.38	0.43	0.76	5.36
Debt	0.77	0.70	0.35	0.49	1.13	0.22	0.21	0.64	4.51
Assets	0.11	0.38	0.45	0.54	1.81	0.41	0.41	0.44	4.54
Southern plains									
Operators (farms)	0.25	1.83	1.30	0.97	3.27	1.35	1.17	1.81	11.96
Debt	1.16	1.60	0.98	0.55	1.40	0.64	0.58	0.96	7.86
Assets	0.19	1.01	0.99	1.24	6.20	1.36	1.12	1.09	13.19
Mountain									
Operators (farms)	0.13	0.62	0.32	0.41	2.10	0.58	0.52	0.68	5.37
Debt	1.03	1.21	0.68	0.71	2.70	0.81	0.89	1.08	9.12
Assets	0.16	0.65	0.42	0.74	5.27	1.28	0.73	0.62	9.87
Pacific									
Operators (farms)	0.18	0.43	0.47	0.51	1.77	0.74	0.81	0.81	5.71
Debt	2.38	1.10	0.49	0.65	2.06	1.24	0.78	1.21	9.91
Assets	0.28	0.47	0.53	0.63	3.99	1.38	0.88	0.79	8.96
United States									
Operators (farms)	2.99	12.10	8.87	8.65	28.27	11.43	12.21	15.48	100.00
Debt	13.27	14.37	7.25	7.05	22.15	10.52	10.66	14.72	100.00
Assets	1.99	6.82	6.34	7.93	40.79	14.21	11.97	9.95	100.00

Source: USDA, *1984 Farm Costs and Returns Survey* in Jolly 1985.

Note: Return to equity is net cash income from the farming operation plus nonfarm income minus estimated living allowance divided by operator farm equity.

[a]Operator debt
[b]Operator assets

tive conclusions. In a number of cases, concrete empirical studies have not been completed. In those cases, my arguments are based on logic and theory rather than verifiable observation.

ECONOMIC EFFICIENCY. The relative efficiency of owner-operator versus tenant agriculture depends to a significant degree on the tenancy arrangement. Ignoring issues of risk, which will be discussed later, theoretical arguments would suggest that assuming proportionate sharing of income and expenses by the tenant and landlord, resource utilization should be equally efficient with a rental arrangement as with an owner-operator arrangement. Differences in efficiency may be a result of different financial structures and risk characteristics of tenant versus owner-operator agriculture.

As to differences in efficiency between family and nonfamily agriculture, the key questions relate to compensation of management and incentives for superior performance. If owner-operators do not charge full cost for their management services (instead management receives a residual return or alternatively compensation is combined with the labor return) while the nonfamily operator does charge the full cost, then costs for managerial services will be higher and efficiency lower with the nonfamily structure. Furthermore, if salaried management in a nonfamily entity is not rewarded for performance, then the owner-operator who receives full benefits of improved performance will have more economic incentive to reduce costs and improve efficiency. Again, the result would be a more efficient agriculture if structured with family compared to nonfamily operators.

FINANCIAL STABILITY AND RISK BEARING. During periods of rising land values, owner-operators have the financial cushion of equity accumulated through appreciation that can be used to refinance operating losses and cash-flow shortfalls. Consequently, during such periods, owner-operators have more financial reserves and exhibit more financial stability than tenants. However, in periods of declining asset values, these reserves disappear. Furthermore, if substantial debt was incurred to attain owner-operator status, the large fixed charges of debt servicing for an owner-operator with high leverage results in increased financial instability. This is particularly the case because land (as with all real estate in general) does not generate a particularly competitive *cash* rate of return. Thus the financial stability of owner-operator versus tenant agriculture, given the current institutional structure, would seem to be indeterminate.

The type of rental arrangement also has an impact on financial stability. Those who cash rent have fixed charges similar (in the short run) to those of an owner-operator, even though these charges are not necessary long-run commitments since most rural land leases are contracted for just one season. With a crop-share lease, the risk is more evenly shared between tenant and landlord thus resulting in increased financial stability for the firm. One risk the tenant faces that is not faced by the owner-operator is the potential loss of the use of resources because of termination of the lease. However, such a loss may not be much different from the loss of property due to foreclosure by a highly leveraged owner-operator.

The financial stability of family versus nonfamily farms also seems to be indeterminate. It might be argued that family farms have more financial resiliency because the family is contributing labor and management resources and may be willing to forego compensation for these resources to offset losses, whereas managers and employees in nonfamily operations expect to be compensated irrespective of the financial performance of the firm. However, if financial reversals are sufficiently large that cash infusions are required, the nonfamily unit may have access to a larger pool of funds to maintain solvency than the family-controlled unit. Nonfamily units may also have a more diversified financial structure (debt, owner equity, outside equity, leased assets), which may give it increased financial stability.

STANDARD OF LIVING. Concerns about the economic standard of living and income generation capacity of farming operations with various characteristics is an important issue in the family farm debate. Undoubtedly, farms with different financial characteristics will generate different levels of net or cash income and consequently support different standards of living. With the cost-price pressure of recent years, concern has been expressed about the capability of the moderate-sized farm (forty to one hundred thousand dollars in gross sales) to generate levels of income that will support acceptable living standards.

From an economic perspective, the differences between family- and nonfamily-controlled agriculture and tenant and owner-operator based agriculture with respect to standard of living are not clear. As has been noted earlier, depending upon the tenancy arrangements and terms, tenants may generate higher or lower levels of income than owner-operators on the same resource base. It does appear that tenants receive a higher proportion of their total return in a cash form whereas landlords receive a larger proportion of their return in the form of noncash capital gains, particularly capital gains on farmland. Consequently, given the same resource base, it is possible that tenants have more cash income that could be allocated to family living and other consumption purposes.

Employee or managerial compensation policies are a significant determinant of differences between family- and corporate-controlled agriculture in terms of standard of living. If corporations compensate employees and managers at a lower level than the income generated by similar resources within family farms, the standard of living will be lower in corporate-based agriculture. However, it is frequently argued that family farmers have been willing to accept lower compensation for their labor and management resources than might be obtained elsewhere because of the nonmonetary benefits of family farming. If this is indeed the case, the economic standard of living may actually be lower with family- compared to corporate-structured agriculture.

RESOURCE CONSERVATION. Although the traditional argument has been that owner-operators are more concerned about soil erosion and more likely to adopt conservation practices, various studies have refuted this argument (Lee 1980; Lee and Stewart 1983; Bills 1985). In essence, it appears that a combination of tax rules, cost-sharing arrangements, and legal constraints have provided adequate incentives for landlords and investors to adopt conservation practices so that differences between owner-operators and landlord-tenant operated property in the adoption of conservation practices do not exist.

As to differences between family and nonfamily operations in resource conservation, one might argue that a family operation may have more incentive to adopt and utilize conservation techniques because of the expectation that succeeding generations will be involved in farming and the desire to pass on a highly productive and well-maintained operation to the next generation. The incentive to conserve resources for future generations may not be as large in the nonfamily situation since the future generation quite likely will not be direct operators of the farm. However, empirical evidence does not necessarily support this argument (Lee 1980; Lee and Stewart 1983).

EMPLOYMENT. Differences in total employment opportunities in agriculture between a family and a nonfamily structured industry and between tenant and owner-operator agriculture are not expected to be significant. Differences in total employment opportunities would appear to be more fundamentally a function of the efficiency of one's workers and the extent to which one substitutes capital for labor. If owner-operator structured agriculture or family-owned farming operations are more efficient in labor utilization, total employment opportunities may actually be lower than would be the case with tenant-based or nonfamily structured agriculture. Clearly, the type of employment opportunities and who can take advantage of those opportunities may be significantly impacted by a family compared to nonfamily structured agriculture. Likewise, as noted in the next section, the combination of labor and management skills may also be impacted by the owner-operator compared to tenant structure or the family versus nonfamily organization of the agricultural sector.

Entrepreneurial Prerogatives

The opportunity to exercise entrepreneurial authority and who has the prerogative to do so will be impacted by the tenancy and ownership structure of agriculture. With family compared to nonfamily ownership of agricultural assets, the entrepreneurial authority is vested within a family unit or in the corporate unit, respectively. Which entity — the family or a corporate unit — is the most efficient and equitable in exercising this entrepreneurial authority is not clear from the evidence.

The difference between owner-operator and tenant structures of ownership also bears on the issue of the exercise of entrepreneurial rights. The owner-operator has complete managerial authority to exercise entrepreneurial responsibilities, but is, of course, at the mercy of outside constraints such as credit and other legal, institutional, and business relationships. In the tenant-landlord situation, this entrepreneurial responsibility is shared. In cash rent tenancy arrangements, most of the entrepreneurial authority will typically be exercised by the tenant, whereas a crop share or custom farming operation will frequently enable the landlord to exercise more entrepreneurial decision making. Again, the differences in overall efficiency and equity consequences of having entrepreneurial authority exercised by owner-operators or tenants and landlords is indeterminate.

ABILITY TO ADOPT NEW TECHNOLOGY. One of the hallmarks of U.S. agriculture has been constant technological advances and the concomitant willingness of farmers to innovate and adopt new production techniques. A frequent assertion is that this attribute is a result of the family structure of agriculture, and that corporate farms do not have the same incentive to innovate and adopt new technology. Similarly, the argument is that owner-operators receive more benefits from technological innovation than tenants and so would be expected to have more economic incentive to try new production techniques and management practices.

The evidence supporting these arguments appears to be more anecdotal than empirical. As to tenant versus owner-operator agriculture, equitable sharing of the costs and benefits of new techniques and innovation, much like the sharing of other cost and revenues, provides the same incentive for adaption of new technology on the part of tenants and landlords as owner-operators. Likewise, there do not appear to be inherent differences in family- versus corporate-controlled agriculture in innovativeness, if the proper incentive mechanisms are used.

It would appear that other factors, including capital constraints, size of firm, access to information on new technology, and so forth, would be more important in explaining innovation and adoption of new technology than differences in the tenancy or family versus nonfamily structure of agriculture. Furthermore, work by Ruttan (1982) and others indicates that technical advance in the agricultural sector is to a significant degree induced by the competitive structure of the industry, the relative availability of capital, labor, and land resources, the sizes of the firms in the industry, and the demand characteristics of the product. None of these factors is uniquely associated with the tenancy or locus of control characteristics of agriculture.

COMMUNITY CONTRIBUTIONS. Differences in community contributions (sometimes called "connectedness") are frequently espoused for family-based compared to corporate-controlled agriculture (Goldschmidt 1978). Concerns about whether large-scale farming operations will bypass local input supply and product purchasing firms for regional and wholesale outlets are legitimate. But size, not ownership structure, is most likely to be the determining factor here (Heady and Sonka 1974; Krenz 1974; Brown 1979).

Because of the connectedness inherent in family-controlled units, particularly those controlled by extended families that include a number of generations, there may also be more commitment to the economic and social institutions in the local community compared to corporate-owned and managed farms. However, some of the earlier studies in California that documented significant differences in communities as a function of differences in farm structure (Goldschmidt 1978) have been challenged as to their scientific methodology in recent years (Emerson and Vertrees 1979; Drache 1978) and similar recent studies in the Midwest focusing on trade patterns are not supportive of the conclusions of the California work (Korsching 1984; Heffernan and Campbell 1986). I would suggest that this issue merits more explicit scientific exploration and documentation.

Differences between owner-operator and tenant structured agriculture with respect to connectedness and community contributions may exist because of the differences in permanency of tenants versus owner-operators. Current law gives very limited property rights to tenants, which may result in their becoming more transient members of local communities compared to owner-operators. Because of this potential transient nature, tenants may become less committed to and connected within the social and economic structure of rural communities. If these differences in community contribution and connectedness do in fact occur between tenants and landlords, such differences might be altered by giving tenants more property rights, thus increasing their security of tenure, reducing their transient nature, and increasing their commitment to community economic and social structures.

Independence

A significant potential difference between tenant and owner-operator structured agriculture is the control the farmer has over his or her own future or destiny. As suggested earlier, owner-operators have more control over their resource base and consequently the potential to exercise independent decision making in business as well as some personal/family living matters. In contrast, tenants must respond to the conditions and decisions of landlords, particularly given the balance of property rights between tenants and landlords in U.S. agriculture. Consequently, it can logically be argued that owner-operators exercise significantly more independence and control over their own future than a tenant-based agriculture.

Similar arguments can be made for the independence and control of the family member in family-based agriculture compared to the employee in corporate-based agriculture. If the family members comprise the decision-making unit, they exercise control subject to the typical business constraints associated with credit transactions, government regulations, and so on. In a corporate-controlled agriculture with the farmer operating as an employee or manager, there may be less opportunity to exercise unconstrained control and independence in decision making.

It may be appropriate at this point to emphasize the critical importance of property rights, particularly relative property rights of tenants compared to landlords with respect to the issue of exercising control. Currently institutional structure and law give farm tenants very few property rights, typically only one-year leases and no compensation for improvements. Thus tenants have little control over a large part of their resource base; it follows that they have a strong economic incentive to become owner-operators. Changing the balance of property rights of tenants versus landlords, including the potential for longer-term leases and compensation to the tenant for improvements made, may have a significant impact on the economic and social attractiveness of renting land. If reasonable terms of trade are maintained between owners and users, the perceived negative social consequences of renting may be partially offset, and increased tenancy may in fact improve the financial resiliency of the agricultural sector.

An additional dimension of the issue of control is that of political and economic power, the ability to influence the decision making of those who are not direct members of the firm. Because of their relatively stronger property rights, landlords would appear to have more potential political and economic power than tenants. As for corporate compared to noncorporate agriculture, it is not clear that there would be significantly different economic and political power in each structure, except for the greater power held over those individuals directly employed or financially responsible to the firm. Again, political and economic power may be related to farm size, but that is a separate issue from the question of family versus corporate control or owner-operator versus tenancy arrangements.

Conclusion

A number of noneconomic moral issues might be identified that bear on the value of the two different structures of agriculture. Such issues as respect for life, human dignity, and honesty suggest themselves. My cursory assessment of these attributes suggests that definitive differences as a function of these dimensions of the structure of agriculture are not obvious.

A brief observation on the attribute of honesty or commitment may be appropriate at this point. There appear to be changing standards in rural communities compared to earlier years. The "your word is your bond" attitude is no longer standard. Rural people are not necessarily becoming blatantly dishonest, but they are more willing to accept the grey area between right and wrong and accept less than pure business decisions. The reasons for this change in standards of honesty may be twofold: one, that people's standards sometimes are adjusted when financial survival is at stake, and two, farmers feel that their current financial problems are not all their own fault and that others (including lenders, business firms, and the government) are partly to blame. This leads them to feel that it is justified to transfer part of the loss to others through defaulting on commitments. A fundamental question is whether these perceived changes in the moral fiber of rural communities are transitory or permanent. We should also ask what they imply about future business arrangements and even personal and social commitments of those who live in rural communities.

In conclusion, the title of Part Seven, "Do Family Farms Support a Better Way of Life?" hints that family farming may lead to a better way of life. According to my analyses, that claim is only clearly true with respect to one of the ten issues I have examined—independence and control of one's future. With respect to all of the other nine criteria, there seems to be no clear-cut advantage for the owner-operator farm structure over the landlord-tenant arrangement. Nor, in nine out of ten cases, does there appear to be a distinct advantage for family-owned farms over corporate-owned farms.

Yet unanswered questions are, what are the specific attributes of the family farming way of life that are desirable? And are these attributes a function of tenancy, family versus corporate locus of control, financial structure, or farm size? Definitive analyses of the hypotheses presented here would sharpen the debate concerning the costs and benefits of family farming.

References

Bills, Nelson L. *Cropland Rental and Soil Conservation in the United States.* Agricultural Economics Report No. 529. Washington, D.C.: U.S. Department of Agriculture, March 1985.

Brown, David L. "Farm Structure and the Rural Community." In *Structure Issues of American Agriculture,* 283–88. AER-438. Washington, D.C.: U.S. Department of Agriculture, November 1979.

Community Services Task Force Report. *The Family Farm in California: Report of the Small Farm Viability Project.* Sacramento: State of California, November 1977.

Drache, Hiram M. *Tomorrow's Harvest: Thoughts and Opinions of Successful Farmers.* Danville, Ill.: Interstate Publishers, 1978.

Emerson, Peter M., and James G. Vertrees. Review of Walter Goldschmidt, *As You Sow: Three Studies in the Social Consequences of Agribusiness* and Hiram M. Drache, *Tomorrow's Harvest: Thoughts and Opinions of Successful Farmers.* In *American Journal of Agricultural Economics* 61, no. 4 (November 1979): 712–14.

Goldschmidt, Walter. *As You Sow: Three Studies in the Social Consequences of Agribusiness.* 2nd ed. Montclair, N.J.: Allanheld, Osmun & Co., Inc., 1978.

Heady, Earl O., and Steven T. Sonka. "Farm Size, Rural Community Income, and Consumer Welfare." *American Journal of Agricultural Economics* 56, no. 3 (August 1974): 534–42.

Heffernan, William D., and Rex R. Campbell. "Agriculture and the Community: The Sociological Perspective." In *Interdependencies of Agriculture and Rural Communities in the Twenty-first Century: The North Central Region,* edited by Peter F. Korsching and Judith Gildner. Ames: Iowa State University Press, 1986.

Jolly, Robert W., et al. "Incidence, Intensity, and Duration of Financial Stress among Farm Firms." Iowa State University, Ames, Iowa, 1985.

Korsching, Peter F. "Farm Structural Characteristics and Proximity of Purchase Location of Goods and Services." In *Research in Rural Sociology and Development,* Vol. 1, edited by Harry K. Schwarzweller. Greenwich, Conn.: Jai Press, 1984.

Krenz, Ronald D., et al. *Economics of Large Wheat Farms in the Great Plains.* AER-264. Washington, D.C.: U.S. Department of Agriculture, July 1974.

Lee, Linda K. "The Impact of Landownership Factors on Soil Conservation." *American Journal of Agricultural Economics* 62, no. 5 (December 1980).

_____, and William H. Stewart. "Landownership and the Adoption of Minimum Tillage." *American Journal of Agricultural Economics* 65, no. 2 (May 1983): 265–64.

Ruttan, Vernon R. *Agricultural Research Policy.* Minneapolis: University of Minnesota Press, 1982.

U.S. Bureau of the Census. *Census of Agriculture: 1982.* Vol. 1. Washington, D.C.: Government Printing Office, 1982.

_____. *Census of Agriculture: 1959.* Washington, D.C.: Government Printing Office, 1959.

_____. *Census of Agriculture: 1974.* Vols. 1 and 2. Washington, D.C.: Government Printing Office, 1974.

_____. *Census of Agriculture: 1969.* Vol. 2. Washington, D.C.: Government Printing Office, 1969.

U.S. Department of Agriculture. *1984 Farm Costs and Returns Survey.* Washington, D.C.: Government Printing Office, 1984.

Responses of Berry and Boehlje

A Reply to Michael Boehlje

The first thing I notice about Michael Boehlje's essay is that he has no commitment and that the avoidance of commitment is apparently his operating principle. Thus he sees both of the possibilities he is talking about — that farmers should own their farms or rent them from landlords — merely as "alternatives" to each other.

Professor Boehlje is driven to this extreme by the very unscientific language that adheres to the idea of "the family farm," which, he says, has "a revered dimension"; virtues that are "almost mystical" are attributed to it. So tenacious are these feelings that "anyone who would question the legitimacy of family farming and family-based agriculture must be a heretic." What is necessary, he feels, is an objective comparison; we should step aside from emotion and cultural attribution and examine the issue scientifically.

Thus one might object — as indeed some have — to the "almost mystical" virtues attributed to family life, or to human life. Why, if Professor Boehlje's assumptions are correct, should we not have a cost-benefit analysis of family life? Why should we not dispassionately question the advisability of family-based housing, when more efficient resource utilization could obviously be obtained by keeping the whole citizenry in concentration camps?

My point is that when one allows scientific procedure to rule out cultural value as "almost mystical," then one is reduced to considering monstrosity "on its merits" — and inevitably, by the same academic parody of fair-mindedness, finding merit in it.

Professor Boehlje assumes too complacently that avoidance of commitment assures him the status of a neutral observer. In fact, his neutrality makes him highly serviceable to others who belittle "almost mystical" values, not for the sake of scientific objectivity but for power and money. His essay, for example, is of no use whatever to defenders of family-based agriculture, but it *is* useful, and it lends the prestige of his professorship and his university, to the enemies of family farming.

I notice, secondly, that the promised scientific analysis is not scientific at all, but a highly speculative comparison between two very tentatively described suppositions, too often requiring qualification by "if" and "may" and confessing at the end that "definitive differences . . . are not obvious." Professor Boehlje does not produce, on either side, a single example of a working farm. This is of critical importance, for the stated issues are those of efficiency of resource utilization and the economic stability of farm families. To make sense of these issues, we need painstaking descriptions of the workings and the economies of many individual farms. We must ask, and answer with responsible particularity, what it is that makes a farm good, that is, able to support its people and preserve its fertility. All the determining questions, so far as I can tell, have to do with structure, proportion, and size or scale—all of which Professor Boehlje entirely ignores.

He concerns himself instead with the question of the relative merits of tenancy and ownership for American farm families. But this is a question that was answered by American history. Our people did not come here to *rent* land, they came here to *own* land or to own usable properties of other kinds. They might as well have stayed in Europe or in Asia if their aim had been the tenant-landlord arrangement. But their experience had been that the tenant-landlord arrangement was invariably arranged in favor of the landlord. They wanted the liberty, political and economic, that they understood to be dependent upon the democratic (that is, small) ownership of usable property. Thomas Jefferson spoke for these people and their hope. There is a connection, as he saw, between economy and politics—a connection that can be made only by the "almost mystical" language that Professor Boehlje would discard. The Declaration of Independence and the Constitution are full of it.

A Response to Wendell Berry

My response to Berry's comments will be brief and succinct.

1. I am disappointed that Berry confuses the concept of commitment with attempts to be objective and analytical. I am committed, committed to a better understanding of what is good and bad about family farming.

2. Berry has convoluted my analysis of family farming to one of challenging the attributes of family life and human life. The dialogue and issues in the paper are focused on the family approach to farming

versus other ways of organizing farming activity. It is not concerned with the attributes of the family or family-based housing.

3. There was nothing in my discussion that suggested that scientific procedure should rule out cultural values. In fact, my whole objective was to propose that we use scientific procedures to obtain a better understanding of cultural values.

4. My attempts to be analytical, given very little data or research to draw upon, forced me to develop some arguments based on logic rather than empirical analysis. The appropriate response in such a situation is to counter the logic and undermine the "speculative results" in that fashion rather than challenge the results with no analytical framework.

5. It is not clear to me how various descriptions of "a working farm" would have significantly contributed to the arguments. With respect to the assertion that structure, proportion, and size were ignored, I ask Berry and the reader to reread my discussion of a structural framework and the sections on economic efficiency and financial stability and risk bearing.

6. In spite of the strong cultural focus in this country on institutions that encourage ownership while discouraging rental arrangements — particularly in rural real estate markets — the fastest growing group of farmers in the past three decades has been part owners/part renters. Rental is a legitimate method of resource control and management, and to suggest otherwise ignores the economic realities in both rural and urban markets.

7. I would hope that we have a more definitive base for supporting family farming than the Declaration of Independence and the Constitution. I do not find much in either document that talks definitively or prescriptively about this issue.

I am concerned that the proponents of family farming continue to espouse emotional arguments that are not well substantiated except by selective references to history. Instead, why not specifically focus on the ten attributes I have identified (and add ten more if they seem appropriate) and numerically document or logically argue the contribution of family farming to those attributes? My discussion was unable to focus Berry's attention on specific attributes and whether such attributes were possessed by family farms. I only hope to have more success with other readers.

SUGGESTED READINGS

Allan, George. *The Importances of the Past: A Meditation on the Authority of Tradition.* Albany: State University of New York Press, 1986.

Berry, Wendell. *The Unsettling of America: Culture and Agriculture.* New York: Avon, 1977.

Jackson, Wes, Wendell Berry, and Bruce Colman, eds. *Meeting the Expectations of the Land.* Berkeley, Calif.: North Point Press, 1984.

Korsching, Peter R., and Judith Gildner, eds. *Interdependencies of Agriculture and Rural Communities in the Twenty-first Century: The North Central Region.* Ames: Iowa State University Press, 1986.

Schertz, Lyle P., et al. *Another Revolution in U.S. Farming?* Washington, D.C.: U.S. Department of Agriculture, 1979.

Shi, David E. *The Simple Life: Plain Living and High Thinking in American Culture.* New York: Oxford, 1985.

Structure Issues of American Agriculture. Agricultural Economic Report 438. Washington, D.C.: ESCS, USDA, Nov. 1979.

VIII

What Ought Congress To Do?

THE STUDY of economics may give us the feeling that the future is determined by inexorable market forces. The truth, of course, is that human decisions greatly affect trade, and if political choices do not exactly determine our economic future, they dramatically influence it. This fact can sometimes benefit the farmer, sometimes harm the farmer. Believing that inflation would continue to add value to their acres, farmers bought land at record prices in the late 1970s and early 1980s. In 1979 the Federal Reserve Board acted to tighten monetary supply. Inflation slowed, land values stabilized and then began a precipitous decline. The current crisis was the result. There are probably many lessons in this tale, not the least of which is that we can change economic conditions if we want to do so. In fact, we have great power to intervene in market forces and tremendous capacity to influence societal structures.

The next two articles address this issue. What should we do at the national level to help agriculture? Tom Harkin, Democratic senator from Iowa, suggests that we empower local farmers by forming a national referendum. In it, each farmer would be given one vote to determine how much corn and beans would be raised. By limiting production, prices would go up. Harkin's plan, needless to say, is consistent with the agrarianizing tradition Kirkendall identified in Chapter 9.

Jesse Helms, Republican senator from North Carolina, defends a very different approach. In line with those Kirkendall called "modernizers," he advocates turning agriculture toward the free market, allowing "market forces," not farmers, to set prices. Helms's plan has already turned out to be the most successful politically. As Helms tells us, President Reagan backed it and signed it into law in February 1986.

But Harkin's plan is not dead. Many are dissatisfied with Helms's 1985 farm bill; in 1986 alone it has cost taxpayers over twenty-six billion dollars. There promises to be an ongoing debate in the Senate and House over agricultural policy. Harkin's bill, which surprised all political observers by getting thirty-six out of one hundred votes in the Senate in 1985, may be reconsidered. For that reason, both the 1985 farm bill (formerly the Food Security Act of 1985) and Harkin's Save the Family Farm Act (formerly the Farm Policy Reform Act of 1985) are included here.

JESSE HELMS

The 1985 Farm Bill

THE 1985 FARM BILL, approved unanimously by a joint Senate/House conference committee on 14 December 1985 and signed into law by President Reagan on 23 December 1985, is the beginning of a transition to market-oriented farm policy. It will restore the United States as a strong competitor in world markets and thereby increase exports of U.S. farm commodities. No longer will our nation encourage farmers to produce commodities just to be sold to the government for storage. We are heading back to the free-market concept, which should never have been abandoned in the first place.

The present five-year bill will reduce loan rates for wheat, feed grains, cotton, and rice, thereby reducing the cost of agricultural programs to the federal government. It retains target prices for many program commodities, reauthorizes other commodity programs, and extends and expands export programs of the U.S. Department of Agriculture. Furthermore, the bill contains strong, new, soil conservation provisions and improves the effectiveness of the credit programs of the Farmers Home Administration.

The bill is the beginning of a slow but decisive transition to market-oriented farm policy. Its true significance is that it shows that Congress has begun the process of correcting the failures of past farm policies. A new era of hope is in store for American farmers. The bill sends a clear and unmistakable message to countries that have been using export subsidies to unfairly increase their share of world markets: the American farmer is back as the major competitor in world markets. The loan rate reductions in the bill are supplemented by a strong export title that expands current export programs and creates new tools for the secretary of agriculture to use to promote U.S. exports. At the same time, farm

---

JESSE HELMS is a U.S. senator from North Carolina and chairs the Senate Committee on Agriculture, Nutrition, and Forestry.

income protection will be maintained at record levels in order to aid farmers financially during the transition to market-oriented pricing of their products. The problem has been, and still is, that those income protection mechanisms were set at high levels and in a way that directed billions of dollars in taxpayer subsidies to support the income of the largest and wealthiest farmers. In a time of massive budget deficits and the need to reduce those deficits, this kind of policy just doesn't make sense. These high support levels will also induce large surplus production, which can only delay a solid recovery in the farm economy.

The bill also makes historic reforms in conservation policy. It makes U.S. agriculture more efficient by protecting soil and water resources and ensuring that the most productive land will be used, and not abused. Important reforms are even made in the dairy program. The price supports will be reduced over the life of the bill, reducing government costs, helping to balance supply and demand. I am extremely disappointed, though, at the imposition of a milk tax in the bill.

Despite its shortcomings, the bottom line is that this bill makes many significant and effective reforms in farm policy. It signals the intention of Congress to make a decisive transition to market-oriented farm policy. At the same time, there are important reforms yet to be made. Congress must continue to try to reduce the cost of farm programs by devising a way to more effectively limit large subsidies to those producers least in need of assistance.

Some of the major provisions of the bill can be found in the following nine categories:

1. Wheat and Feed Grains. Basic initial loan rates for grains would start at $3.00 a bushel for 1986 wheat and $2.40 for 1986 corn. From 1987 on, loans for grains would be set each year at between 75 and 85 percent of the average price received by producers during the immediate preceding five marketing years, excluding the high and low years, with annual reductions limited to 5 percent.

After calculating the basic initial rates, the secretary of agriculture could then further reduce the rate for any year by up to 20 percent if (1) market prices in the previous season failed to top 110 percent of the previous year's basic loan rate, or (2) the secretary determines that a further cut is needed to compete on world markets.

For the 1986 crop only, the secretary would be required to use this authority to drop the loan at least 10 percent.

The secretary would have discretionary authority to allow repayment of price-support loans at levels that could be set as low as 70 percent of the original loan rate. The secretary would have additional discretionary authority to give producers one of two alternate types of marketing certificates to help promote exports.

Target-price income protection would operate alongside whichever loan system was used, and any reduction below the basic initial loan rates would be offset by increased target-price deficiency payments, which would not be subject to payment limits.

Target prices, which provide direct payments to farmers when market prices are below the target rate, would be frozen at current levels ($4.38 a bushel for wheat and $3.03 a bushel for corn) through 1987. For the 1988 crop, the secretary would have authority to set the target price level at not less than 98 percent of the 1986 level ($4.38 and $3.03); at not less than 95 percent for the 1989 crop; and at not less than 90 percent for the 1990 crop, but not lower than $4.00 for wheat and $2.75 for corn.

To qualify for benefits in any years in which carryover wheat stocks exceed one billion bushels, wheat producers would be required to reduce acreage as follows: in 1986, a maximum diversion of 25 percent (including a mandatory minimum reduction of 15 percent, a mandatory in-kind paid diversion of 2.5 percent, and further discretion for reductions of 7.5 percent); in 1987, a maximum reduction of 27.5 percent (including a mandatory minimum reduction of 20 percent and a further 7.5 percent at the secretary's discretion); and in 1988–90, a maximum of 30 percent (including a mandatory minimum of 20 percent and a further 10 percent at the secretary's discretion). For the 1986 crop only, the secretary would be required to offer growers who planted before announcement of the program a chance to idle 10 percent of their base in return for payments.

The secretary is required to allow haying and grazing on diverted acres in 1986, permit grazing in 1987–90 if state agricultural stabilization committees request it. Current law gives the secretary discretionary authority to permit haying in 1987–90.

2. Soybeans. The bill continues the basic price-support loan rate through 1986 and 1987 at the current minimum rate of $5.02 a bushel. For the next three years, rates would be based on an average market price formula with reductions limited to no more than 5 percent a year and with a floor of $4.50 per bushel. In all of the five years, the secretary would have authority to reduce loan levels by an additional 5 percent if necessary to keep the crop competitive in world markets. The secretary would also have discretionary power to allow repayment of loans under a "marketing loan" system.

3. Payment Limitations and General Commodity Provisions. The bill continues the present fifty thousand dollar annual per producer ceiling for program payments. Exempt from the ceiling would be (1) target-price payments required to offset support loan cuts below basic loan

levels, (2) payments made under several program cost-reduction items, which the bill provides for discretionary use by the secretary, and (3) any gains farmers realize when paying off support loans at less than the initial loan level.

Advance deficiency payments would be required for the 1986 crops of wheat, feed grains, cotton, and rice, and would be discretionary in future years for the life of the bill. The secretary also would be given authority to make up to 5 percent of the total deficiency payment with surplus commodities.

4. Export Title. The secretary is required to offer surplus commodities owned by the agriculture department to exporters, processors, or foreign buyers to encourage the development and expansion of overseas markets for American crops including processed farm products. The secretary would be directed to use one billion dollars worth of commodities for the program over the next three years to counter unfair foreign trade practices and generally to make American products more competitive; and at least 15 percent of the program could be devoted to exports of poultry, meat, and meat products. To the extent practical, the secretary must use the commodities in equal amounts in each of the three years.

The secretary is required to make available not less than five billion dollars annually in short-term export credit guarantees for the life of the bill where such guarantees would improve the competitive position of American exports.

The secretary is required to make not less than $325 million available annually in fiscal years 1986–88 in cash or in surplus commodities for direct export credits to counter the subsidies, import quotas, or unfair trade practices of foreign countries. (The direct credits, whether in cash or in kind, could be used in so-called blended credit export programs.)

The Food for Peace program is extended, and at least 75 percent of the foods shipped under the donation phase of the program must be in the form of processed, fortified, or bagged products. For fiscal 1986, shipments under the donation phase would be maintained at the 1985 level of 1.8 million tons. The bill also authorizes expanded operations under a related program that provides surplus commodities for needy people abroad, and it provides that not less than one-tenth of 1 percent of Food for Peace (P.L. 480) funds for fiscal 1986 and 1987 must be used for a farmer-to-farmer technical assistance program.

The president is authorized to donate up to at least seventy-five thousand and up to five hundred thousand metric tons of eligible section 416 or P.L. 480 commodities or any combination thereof, in each of the fiscal years 1989–90 under a new Food for Progress program to promote private free enterprise policy and development. At least seventy-five thousand tons of such commodities must be from section 416. In return for the commodities, the recipient country must promote economic freedom in the production of food for domestic consumption and must be able to use the donated commodities without disrupting its own agricultural markets.

5. Conservation. For highly erodible land that has not been cultivated since 1980, the bill provides a sodbuster program to discourage plowing up fragile soils. If a farmer planted a crop on fragile land in violation of the terms of the bill, he or she would lose price supports and other farm benefits for all of his or her crops in the year of the violation. Highly erodible land that was used for crops (or idled under a government acreage control program) between 1981 and 1985 would initially be exempt from the sodbuster penalties, but this "grandfather clause" exemption would disappear for any affected producer who fails to begin applying a conservation plan by 1990 or two years after completion of a soil survey of his or her land, whichever is later. Producers would have until 1995 to complete application of the conservation plan. A companion "swampbuster" provision would deny farm benefits to producers who convert wetlands to crop use in the future except in cases where the impact of the action is found to be minimal.

For highly erodible soils that are already in crop use, the bill provides a long-term conservation reserve program under which farmers would contract for periods of ten to fifteen years to return forty to forty-five million of such acres to less-intensive uses such as grass or trees, which in some cases may be used to establish shelterbelts. In return for compliance with the contracts, growers would get cash or in-kind land rental payments (established on a bid basis) plus payments covering a part of the cost of needed land treatment measures. No more than 25 percent of the land in any county could be enrolled in the reserve except in counties where the secretary of agriculture decides that higher levels would not hurt the county economy. There would be a fifty-thousand dollar limit on annual payments to farmers under reserve contracts.

6. Credit and Rural Development. The bill reauthorizes and in some cases revises federal farm credit and rural development programs. It requires the secretary, through September of 1988, to operate a $490 million program under which the agriculture department and private lenders would share equally in the cost of reduction in interest rates for hard-pressed farmers who hold loans guaranteed by the Farmers Home Administration. The government could pay for 2 percent of the "buy-down," or one-half of the total, whichever was less.

The agriculture department would be ordered to observe a number of new restrictions on the way it handles farmland acquired by the government in future foreclosures of FmHA loans. Among other provisions, USDA would be forbidden to sell such land if the sales would depress local farmland values, and first priority in any sales or leases would have to be given to operators of farms that are not larger than family size. In leases, the bill calls for giving priority to former owners of the land and sales prices must reflect the probable income the land can produce. Where FmHA-owned land is administered under management contracts, contracts must be let under competitive bids with preference to local small businesspeople, and the use of conservation practices may be required on land that has been classed as highly erodible as a condition of sale or lease.

7. Research. The bill authorizes a three-year program of special grants for educational and counseling programs to develop income alternatives for producers who have been forced out of farming by economic stress. It also directs the secretary of agriculture to develop appropriate controls on the development and use of biotechnology in agriculture.

A provision is included urging USDA to emphasize, in its research and teaching programs, new technology suitable for small- and moderate-sized farms. The secretary of agriculture would be directed to report on the feasibility of more complete studies of the relationship between diet and blood cholesterol in humans and dietary calcium and its importance in human health and nutrition. The bill allows the secretary to make cooperative, cost-sharing agreements with private agencies, organizations, or individuals to develop new agricultural technology, authorizes research on new uses for farm and forest products, and directs the secretary to conduct demonstration projects on the development or commercialization of crops that would supply strategic industrial products.

8. Food Assistance. The food program section of the bill extends the food stamp program for five years, improves some program benefits, and requires all states to set up special employment and training programs to help move jobless stamp recipients onto payrolls. The bill also extends through fiscal 1987, the Temporary Emergency Food Assistance Program authorizing distribution of government surplus foods to the needy.

9. Other Provisions. An animal welfare title revises standards for the humane handling of animals by research facilities, dealers, and exhibitors. It directs the secretary to develop standards with minimum

requirements in areas including housing, feeding, shelter, veterinary care, and experimental procedures that minimize pain and distress.

The policy directives of the 1985 farm bill will not provide an overnight turnaround in the agricultural economy. But it does begin the process of making farmers less dependent on the government for income and more dependent on the market. Congress must allow the pricing mechanisms in the bill to work to increase agricultural exports. If we are willing to give these mechanisms time to work, we will succeed in the attempt to move agriculture closer to the market, and the 1985 farm bill will eventually be judged as the turning point in restoring a healthy agricultural sector.

TOM HARKIN

The Save the Family Farm Act

TWO CENTURIES AGO, Thomas Jefferson said: "I trust the good sense of our country will see that its greatest prosperity depends on a due balance between agriculture, manufacturing and commerce." Today, if he were alive, he would say we have taken leave of our senses. Our nation is out of balance. Agriculture is mired in recession and far more than our "greatest prosperity" is threatened. Who we are as a people—and what we stand for as a nation—are at stake.

Is there a moral obligation to save the family farm? We might as well ask, do we, as American citizens, have a shared responsibility for the well-being and happiness of our fellow citizens. If America is every man for himself, every woman for herself, then why should we care what happens to farm families? That's their problem. But I do not believe that that is what America is about. Throughout our history, our highest aspirations have been realized by uniting as one nation, not by dividing and pitting one region or one sector of the economy against the other. Since we share this common bond, we owe it to ourselves to ask some very basic questions about the direction we are going, and if that is not the direction we want to be going, then yes, we have a moral obligation to change our course.

Are we preserving or abandoning those cultural and social institutions that make us a stronger nation? Are we adding to or decreasing real economic opportunity for all our citizens? Are we strengthening or weakening our democratic process and democratic foundations? Keeping these questions in mind, let's take a look at the farm policies being pursued today—the so-called market-oriented approach.

This year, 1986, between 90 thousand and 110 thousand farm families will be forced off their land. If current trends continue, between one-

---

TOM HARKIN is a U.S. senator from Iowa.

fifth and one-fourth of the nation's farmland will end up on the auction block in the next few years. The February 12 issue of *American Banker* reported that due to accelerated farm foreclosures, Iowa lenders — banks, insurance companies, the Farm Credit System, and the Farmers Home Administration — now own 250 thousand acres of farmland, the equivalent of one Iowa county. Forty percent of the farmers in the Midwest are either insolvent or rapidly sliding toward bankruptcy. Bank closings and small business bankruptcies are occurring at rates unheard of since the Great Depression. Our small communities are being wiped off the map. Schools and churches are struggling to make ends meet as local tax bases decrease and people move away to find jobs in the cities. But the job market is no better in the cities. Entire plants have closed down and thousands of workers in agriculture-related industries have been thrown out of work.

What is happening in rural America is nothing less than the greatest forced migration in our nation's history. It's a modern-day "Grapes of Wrath." There's no reason we have to let this happen. The marketplace should not be the sole determinant of agriculture's future. We need to decide what kind of society we want to have, whether we value the family farm system of agriculture, whether we want to keep alive our rural communities. This is not softheaded idealism. It's hardheaded realism. Because if we don't decide in a democratic fashion and then act on our decisions, those decisions will be made for us in an undemocratic fashion.

The Japanese have half our population, yet they have twice as many farmers. Now the Japanese are well known to be highly efficient, hardheaded businesspersons. Do they have a soft spot in their hearts for farms? I doubt it. More likely, they have decided it is worth it to them as a nation to keep landownership as widespread as possible and to have their farms owned and operated by families who live on the land.

The exact opposite happened in Ethiopia. When the Marxists came into power, their strength was in the cities, so they purposely created a cheap food policy. The result was the destruction of their system of agriculture.

We face that same choice. We can support and nourish the family farm system of agriculture, or we can turn our backs on it and give a green light to the consolidation of farmland ownership in the hands of a few. The key to the family farmers' plight is low farm prices. The reason we have low farm prices is chronic overproduction.

How are these policies justified? By a blind adherence to a supposed free-market philosophy. Let the marketplace work, the free-marketers say, so we can reward the most efficient farms. But this argument has very little to do with the reality of the farm crisis. I would argue that many of our most efficient units of production in agriculture are the medium-sized farms here in the Midwest, which rely to a limited extent on labor outside of the immediate family. However, it is many of these farms that are failing today. Often it is farmers who have farmed successfully for twenty or thirty years and who a few years ago expanded their operations to incorporate a family member into the operation. Every institution at that time, whether it be the Federal Land Bank, land grant schools, or USDA, told the farmers that borrowing money was the prudent business decision.

As we know now, it was exactly the wrong thing to do. Should government pursue a policy of benign neglect or should it provide targeted assistance to help these medium-sized farming operations get back on their feet again? I believe the latter. Chronic overproduction doesn't reward efficiency. But it does affect large and small farms differently. When farm prices fall, the big corporate farms don't go under. They shift their resources elsewhere, close down parts of their operation, lay off workers, or just write it off in their tax returns.

Small- and medium-sized farmers don't have that luxury. When prices plummet and interest rates skyrocket year after year, they have only two choices: in former Agriculture Secretary Ezra Taft Benson's words, "Get big or get out." Under this kind of free-market approach, the freest are the richest. That will continue to be the case as long as we have chronic overproduction. The large agribusiness concerns, the fertilizer and chemical companies, the food processors, have a vested interest in keeping it that way. Since I've been in Congress, I have seen the influence of the agribusiness interests grow by leaps and bounds, while the farmers' influence has decreased.

The other justification we always hear for overproduction is the spectre of world hunger. Millions starve in famines every year. So American farmers must produce all out in order to feed the world. But the facts indicate otherwise: decades of careless overproduction in America have not ended world hunger. Today, there is less hunger in the world, not because we are producing more and doing a better job at feeding the world, but because the green revolution we started and the appropriate technology we helped develop is enabling the Third World to feed itself—and then some. China has increased its food production by 40 percent over the past five years. Even with bad weather, China could export two hundred million bushels of corn this year. India, once the world's basket case, is now exporting rice. Bangladesh, the same country whose famine inspired the first rock-music relief concert in 1971, is now self-sufficient in food grains.

So far, most of this increase has come from higher yields. Countries like Brazil and Argentina are already major exporters in wheat and soybeans, and they have not even begun to use all their available land. Argentina has enough fertile land not yet under cultivation to equal five Nebraskas.

I am not advocating we abandon our export markets, but let's not believe that driving down world commodity prices will help one iota in fighting world hunger. It will only drive our farmers off the land, and eventually, it will drive under Third World farmers as well. Take away these countries' ability to produce food, and you increase the likelihood of famine, eliminate their primary source of foreign currency, make it harder for them to pay back their debts to Western banks, destabilize their economies, and set the stage for violent unrest and revolution.

The fact is that unfettered production is a surefire formula for only one thing: the concentration of landownership and agricultural wealth. And that in turn threatens the very basis of our democracy. Justice Louis Brandeis once said: "We can have democracy in this country or we can have great wealth concentrated in the hands of a few, but we can't have both."

Ironically, the USDA conducted a study that backs up Justice Brandeis's assertion. A USDA researcher, Walter Goldschmidt, compared two rural communities in California — Arvin and Denuba (Goldschmidt 1975, 70–73). Both communities had roughly equivalent levels of agricultural production in terms of dollars. There was one crucial difference: Denuba's economy was based on a large number of small family farms. Arvin's economy was based on a few giant producers.

Goldschmidt compared these two communities back in the 1940s and then he came back again in the 1970s to see how they had changed. Here's what he found.

In the 1940s, Denuba, the community with many small family-sized farms, had 20 percent more people than Arvin. Denuba's population had a higher level of income, and more people in Denuba were self-employed. Denuba had far more civic and voluntary organizations, schools, and churches, and a much higher level of democratic participation in these organizations by local residents. Denuba had twice as many small businesses and 60 percent more retail businesses.

Thirty years later, when the study was resumed, the gap between Denuba and Arvin had widened even further. In the 1940s, Denuba residents had a 12 percent higher level of income than the people in

Arvin. By the 1970s, that gap had increased to 28 percent. In every other way, Denuba was a much more vibrant and healthy community. And its citizens were much more active in their local democratic institutions.

Realizing that this study contradicted all their arguments for bigger and bigger agriculture, the USDA quietly closed the book on that research project. But there you have a glimpse of the future of Iowa's rural communities if the family farm continues to decline. If we want to avert that future, we'd better start doing something now.

Those who believe only exports will save the farmers are chasing after a pot of gold at the end of the rainbow. What possible good does it do us to regain the export markets of the seventies if that means forcing tens of thousands of family farmers off their land? At the same time, we cannot afford to ask the taxpayers to bear a greater and greater share of the burden. Over the next four years, four out of every five dollars in net farm income will come from the government treasury. We're faced with a choice of either bankrupting the farmer or bankrupting the treasury—or both.

Clearly, we need a new direction in farm policy. And first on the agenda is ending the chronic overproduction that is keeping farm prices low. But to do that, we've got to give farmers a voice in their own destiny. As long as farm policy is decided in Washington, the agribusiness interests that benefit from overproduction will control the agenda. So our only hope is a nationwide farmer referendum—one farmer, one vote—to allow farmers to choose what kind of farm program they want. This is the basis of the Save the Family Farm Act that I introduced in 1985.

Here's how it would work. One referendum would be held to include the producers of all the major commodities. If the referendum passed, every farmer would be required to set aside 15 percent of his or her tillable crop acres. If a greater set-aside was needed for certain commodities, additional acreage would be retired by producers whose gross income exceeded two hundred thousand dollars. This additional set-aside would require larger producers to set aside a greater percentage of their land through a progressive formula. The bill also provides for a thirty-million-acre conservation reserve and a strong sodbuster provision.

In return, the loan rate would be substantially increased ($3.60 for corn, $8.98 for soybeans). Because of restricted supply, market prices would approach or exceed loan levels. Target-price payments would be eliminated, thus drastically reducing the cost to the government. Farmers would get a substantial income boost through the market rather than a government paycheck.

This bill is a true "market-oriented approach" (matching supply with anticipated demand) rather than the administration's "production only"-oriented program (lower prices that force farmers to produce more per given unit, thus lowering prices further).

Without a doubt, higher commodity prices would mean somewhat higher food prices – about 4 percent across the board. But isn't food too cheap now? It is time consumers began to pay the real cost of food, rather than continuing to enjoy low government-subsidized prices brought about by our cheap-food policy.

And what about exports? Will we have to forfeit the export market or erect barriers against imports? I believe this argument is a bugaboo. After all, the United States supplies 60 percent of the feed grains and 40 percent of the wheat in the world market. Other countries already price their grain through government actions at just below our cost. They will continue to do so whether our corn is two dollars a bushel or four dollars a bushel.

We will be a price maker, which I believe the rest of the world will follow in short order. Why? Because it will be in their best interest economically to do so, especially Third World countries who rely upon the export earnings of their agricultural products to help pay their foreign debts. It would also go a long way toward making Third World countries more self-sufficient in food, which can only improve world peace and order. With an increased standard of living, these Third World countries will become better markets, not just for our raw goods, but for our "value-added" products.

There is no reward in flooding the world with cheap, surplus grain, either for us or for Third World countries who need to increase their standard of living. Our goal should be to maximize export earnings, not just volume.

To summarize the major features of the proposal:

1. Elimination of subsidy payments. Producers would receive a fair price for their crops in the marketplace, not from the government. Costly subsidy payments would be eliminated, and the price floor (the price-support loan rate) for each commodity would be set at a level approximating the full cost of production for that product.

As of November 1985, for reference, the program formula would have set the price floors for eight major storable commodities at these levels: corn – $3.71/bu.; wheat – $5.21/bu.; grain sorghum – $6.29/bu.; soybeans – $9.10/bu.; oats – $2.14/bu.; barley – $3.47/bu.; rice – $14.21/cwt.; upland cotton – $.868/lb.

In each subsequent year of the program, the price floors would be gradually raised until, in the eleventh year of the program, they reach the level at which farmers' returns on equity and labor are on a par with the rest of our economy.

2. Balancing production with need. Mandatory production controls (subject to a producer referendum) would limit U.S. production of storable commodities to actual demand, including domestic consumption, export demand, humanitarian need, and strategic reserve requirements.

3. Targeted benefits to family farmers. Each producer would have a single-acreage base comprised of any acres on which any of the designated commodities were grown in any of the last four years. Each producer would be required to set aside 15 percent of this acreage base, but in times of surplus, giant operators would be required to set aside a progressively higher percentage of their base (a disincentive to conglomerate and tax-loss ventures).

4. Promotion of sound conservation practices. Locally approved conservation practices would be required on all set-aside land. In addition, farmers could participate in a National Conservation Reserve, allowing for the voluntary, long-term retirement of fragile land. The act would also contain prohibitions against "sodbusting" and provisions to encourage better protection of scarce groundwater resources.

5. Elimination of current disaster payments and disaster loan programs. The current myriad of programs would be consolidated into one simplified approach that offers income protection to producers and protects both consumers and livestock producers from shortage-induced price increases. Each producer would annually contribute a portion (probably 3 to 4 percent) of production as an "insurance premium" into a national Farmers' Disaster Reserve (FDR). In the event of a disaster, a producer would receive commodities from the FDR to compensate for up to 90 percent of the loss.

The current Federal Crop Insurance Corporation should be expanded to cover perishable commodities, which would be insured for a percentage of the previous year's marketings.

6. Increased funding for humanitarian food aid. The USDA would be directed to enter into multilateral agreements with other food-exporting nations to fulfill food aid requirements to needy countries. These multilateral agreements should also mandate that additional emphasis be placed on helping needy nations develop food self-sufficiency to the degree possible, with each exporting nation allocating an amount of cash or other resources consistent with their level of food aid.

7. Increased promotion of export markets. Market development would include increases in export credits, negotiation of more multiyear export contracts and a monetary adjustment program allowing foreign

buyers to receive, when available, surplus commodities from government stocks to offset the negative impact of the overvalued dollar.

8. Strong support for domestic food assistance. The program would address the right of every American to a nutritiuous diet and, through the food stamp program, the women, infants and children program, and other elderly and child nutrition programs would provide adequate assistance to eligible needy families and individuals.

9. Farm credit and debt restructuring. Congress should immediately enact a temporary moratorium on farm foreclosures until the act takes effect, at which time any foreclosed borrowers would be offered first right of refusal to repurchase any of their land or equipment not yet disposed of. The act would contain provisions to allow deferral of principal payments for one to five years if the borrowers can project an adequate cash flow by the end of the deferral period to resume payments on the principal.

In addition, the higher price-support levels and the FDR crop insurance program in the act should greatly increase lenders' willingness to make operating loans to farmers and should halt the decline in the value of most farmland.

I was pleased with the support this bill received in 1985: thirty-six votes out of one hundred in the Senate. And I am pleased to note that John F. Kennedy advocated a similar approach in his 1960 debate with Richard Nixon: "In my judgment," Kennedy said, "the only policy that will work will be for effective supply and demand to be in balance, and that can only be done through governmental action. I, therefore, suggest that in those basic commodities which are supported, that the federal government, after endorsement by the farmers in that commodity, attempt to bring supply and demand into balance (and) attempt effective production controls."

Unless we are willing to overhaul our commodity programs, to bring supply in line with demand, and unless we are willing to make some fundamental changes in our tax code, then the question of whether we have a moral obligation to save the family farmer will be moot. Economic forces already in motion will leave us with no one left to save. The farmers of America do not want handouts. They only want the opportunity to live and work in dignity on the land they love, raise families in peace and security, support their churches, schools, and small communities and pass on to their children the culture — the agri-culture — of being stewards of God's creation. Is it our ethical duty to grant them that opportunity?

As long as we are a free society with free democratic institutions, as long as we remain one nation, sharing a common history of fighting against injustice and inequality, as long as America remains a beacon of hope and opportunity for the world, as long as we hold the American dream in our hearts, it *is* our moral obligation to save the family farm.

Reference

Goldschmidt, Walter. "A Tale of Two Towns." In *Food for People, Not for Profit,* edited by Catherine Lerza and Michael Jacobson, 70–73. New York: Ballantine, 1975.

SUGGESTED READINGS

Alternative Agricultural and Food Policies and the 1985 Farm Bill, edited by Gordon C. Rausser and Kenneth R. Farrell. Washington, D.C.: National Center for Food and Agricultural Policy, 1984.

Congressional Quarterly, Inc. *Farm Policy: The Politics of Soil, Surpluses, and Subsidies.* Washington, D.C.: CQ Press, 1984.

Hadwiger, Don, and William Browne, eds. *The New Politics of Food.* Lexington, Mass.: Lexington Books, 1978.

Lowi, Theodore. *The End of Liberalism: The Second Republic of the United States.* 2nd ed. New York: Norton, 1979.

McConnell, Grant. *Private Power and American Democracy.* New York: Random House/Vintage, 1966.

Rasmussen, Wayne, and Gladys Baker. *The Department of Agriculture.* New York: Praeger, 1972.

United States Agricultural Policy, 1985 and Beyond: A Series of Seminars, edited by Jimmye S. Hillman. Tucson: Department of Agricultural Economics, University of Arizona, 1985.

CONCLUSION: Moral Arguments for Family Farms

WHAT MORAL OBLIGATIONS, if any, do we have toward family farmers? The diversity of viewpoints expressed in this volume brings home the complexity of the question. The variety of attitudes about the history of medium-sized farms and about the roles that the university, government, and church should play in the present crisis is nothing less than bewildering. Our social scientists and theologians, agricultural economists and philosophers, our politicians and ethicists only seem to agree about one thing: the family farm is perishing.

In such a situation, it would be foolish to try to hammer out a single conclusion from the variety of testimonies found here; there are simply too many conflicting judgments. Nonetheless, some sort of summary should be hazarded. That is what I wish to do in this last chapter. For all of their many disagreements, it seems that the authors in this work do agree on two points. And, running through the extreme pluralism of their views, I find at least four ethical arguments for the preservation of the family farm. These two points of agreement, and the four moral arguments, are summarized and criticized below. The discussion ends with a statement of my own, an argument from responsibility for preserving family farms.

Two Points of Agreement

Is there a moral obligation to save the family farm? There seem to be as many different answers as there are authors. Yet, in these pages, there are at least two things on which everyone seems to agree: (1) the playing field on which farmers have been competing is not level, and (2) there can be no moral obligation to save the family farm if there is not a workable plan that could save it.

With respect to the playing field, most of these writers agree that government tax laws have, to one degree or another, favored large farms. Haw correctly points out that an advantage of corporate farms is their ready access to capital; Tweeten says that in the two areas of investment tax credit law and rapid depreciation allowances, bigger farms have clearly had an advantage. Tweeten, who does not believe — all things considered — that family farms have been treated unjustly, does believe that these two advantages are unfair. He recommends that the offending laws be removed.

It should be noted that in 1986 Congress passed a tax reform bill that promises to accomplish the things Tweeten recommends. Bipartisan supporters have said that the new law should close virtually all of the loopholes that have made farming a major tax shelter for wealthy, outside investors. The bill, for example, puts a cap of twenty-five thousand dollars on the amount of paper losses a "passive" investor can deduct from taxable income, ends the special treatment given to capital gains, and terminates the investment tax credit. It also eliminates special deductions for land clearing and development while adding a longer depreciation schedule for single purpose farm structures (e.g., hog confinement buildings). Ending special treatment for capital gains should help in a small way to relieve some of our oversupply problems, and eliminating special deductions for land developers should help to slow the increasing amount of environmentally fragile land that is being put into use.

It is safe to say that commentators from the left and the right have noticed injustices in the tax laws regarding family farmers. Where taxes are concerned, the smaller farmers have been forced to run uphill, while bigger farmers were coasting down. But with the new tax laws, there is hope that these inequities may change.

With respect to the second matter (regarding moral obligations), we must be realistic. Moralists teach that the first order of business in solving any practical dilemma is to ask whether it is humanly possible to do what is in question. If it is not, then one ought to give it up. We may have very extensive duties toward offended parties, but we cannot work miracles. As in the case of a person we have erroneously convicted of a crime and executed, we cannot have a duty to bring the wronged person back to life. Whatever is beyond the pale of human possibility is beyond the pale of moral obligation.

Is it possible to save the family farm? Kirkendall tells us that slightly more than 1 percent of Americans now live on farms of this description. Paul Lasley predicts that percentage will almost certainly continue to decline. With so few Americans presently in this category, and with demographic trends continuing to shrink the number in it, is there *any-thing* that can be done?

In these pages, no one goes so far as to claim that there is no remedy imaginable. But many suggest it would take a Herculean effort to stem the tide, an effort that would be unreasonable in terms of energy and cost. Kirkendall, Johnson, Tweeten, Oster, and Boehlje all seem to think that the family farm is beyond salvation, practically speaking. The forces of the modern market and international competition seem to dictate that it will soon be impossible for a farmer to support a family on a farm grossing under two hundred thousand dollars a year. Oster thinks many family farmers are already underemployed, particularly if they work less than two thousand acres.

If they are right, then the answer to the initial question must be no. There is no moral duty to save the family farm because — even though it may be logically possible to save it — it is not humanly possible to do so. If these skeptics are on solid empirical ground in their forecasts, they have provided us with the answer to our ethical question. We might as well close the book, counsel family farmers to take Earl Butz's advice to get big or get out, and ask Wendell Berry to compose a eulogy for the once proud, now dead, family farm.

But, there are still believers. Among others, Carol Hodne, Denise O'Brien, Marty Strange, Jim Hightower, and Tom Harkin have all expressed an abiding faith in the family farm. However embattled it might now be, the farm remains an object of hope for them. Each of them has at least a rudimentary rescue plan in mind, some of which converge on the Harkin bill. If these proponents are right in thinking that it is practically possible for America to save its farms, then the moral question is still alive.

For the purpose of discussion, let us assume that the believers are right. Proceeding on that assumption, the next question becomes one of moral weighting. What will it cost to save it? What will be the benefits? Who will be helped by it? Who will have to pay? Will poor black farmers in Mississippi profit? Will consumers living below the poverty line have to pay higher food prices? How much will our collective soul gain?

Here again there is a variety of opinion. Some think that the family farm is essential to our spiritual health and cultural well-being. Harkin, the Catholic bishops, David Ostendorf, and Berry, for example, all seem willing to pay any price to save it, believing that giving it up would be equivalent to selling the heart of our nation. Detailed strategies for saving it have been suggested including various forms of governmental intervention. Harkin's plan would eliminate subsidy payments, institute a farm referendum to control production, target benefits to family farmers, and promote sound conservation practices. Professor Harl would form an agricultural credit corporation to deal with the immediate problem of debt, and give first crack at buying repossessed land to those presently working it.

Skeptics think that such plans might destroy the free exchange of commodities. Oster, Tweeten, Johnson, and Boehlje all appear to think Harkin's price is too high. They argue that our moral duties fall in the area of helping to retrain displaced farmers, while admitting that family farms are no longer the best solution to the problem of feeding a hungry world. Instead of asking for schemes to save farms, they ask, how can we assist healthy farmers as they seek to increase their economic base while reducing their vulnerability? How can we smooth the transition out of farming for those so financially burdened that they cannot survive?

The argument between the skeptics and the believers is not settled in these pages. One cannot call the game and declare a winner. But this situation should not be lamented; it shows that the farm crisis, though unresolved, still weighs heavily on our minds. Believers can take heart that public debate of the issue continues.

Is there a moral obligation to save the family farm? The only conclusion we can draw from these essays is, unfortunately, an ambiguous one; some say there is, some say there isn't. In short, the experts disagree. That, of course, is not a very satisfactory statement on which to end. So, in conclusion, I wish to offer my own assessment of the various ethical arguments offered by believers.

Five Moral Arguments on Behalf of Family Farms

I find authors making at least four different cases for the family farm. Each one defends the institution on ethical or religious grounds. Emotion, efficiency, stewardship, and cultural identity are the key concepts on which these arguments turn.

THE ARGUMENT FROM EMOTION. An argument from emotion is laced through many of these essays, but is felt most strongly in the voices of Carol Hodne, Denise O'Brien, Wendell Berry, and Tom Harkin. The gist of it is that while few Americans now live on family farms, a large percentage have parents or grandparents who came from them. Just as our forebears loved their farms, so ought we to love the farms on which some of their offspring now live. We have a duty to preserve things we love, and to do what we can to minimize suffering. Since family farms are "ours," since they are objects of love, and since they are now sources of considerable anguish, we ought to rescue them.

The argument makes sense if we think about other things that are especially meaningful to us. Symbols and institutions like the flag, the Bible, the Statue of Liberty, baseball, and libraries tie us to the past in extraordinary ways, not the least of which are the positive emotions they evoke—happiness, pride, a sense of belonging, hope for the future, devotion to one another. The things that provoke these emotions are worth preserving because we value the emotions. The family farm, according to this first argument, is worth preserving for the same reason— it stirs significant feelings in us.

But such an argument lacks persuasive force when closely examined. It is true that we preserve many things from our past, and we ought to value many of them. But we should not save them just because they have emotional value for us, or just because they are ours. We should preserve them only because they have *unique* value—for example, if they are uniquely humanizing for us, or if they symbolize principles larger than us. The flag may remind us of those who died for our freedom, the biblical story may remind us of a covenant God made with God's people. Many things in our past do not have this kind of universal significance, and we rightly let them fade from view: a flag with forty-eight stars, an outdated chemistry text, the game of marbles. We have no moral duty to save every old thing to which we have emotional attachment. By the same token, society does not have a universal moral duty to save any particular farm just because it is loved by a family.

Now I hardly think that Hodne, O'Brien, Harkin, and the rest mean to say what I have just attributed to them. They mean something more, something much richer. I will get into this in more detail in the argument from cultural identity. But some of it is relevant here. They seem to be telling us that the family farm represents a unique "way of life," a sort of calling in the Calvinist sense or a type of profession in the modern business sense. As a way of life, it does have symbolic importance, extraordinary value, and the emotions and practices it nurtures may well be lost forever if it dies.

There is much to be said for this argument, but it still seems unconvincing. There are many unique ways of life that we have permitted to die out. And, while we may miss the emotions that accompanied them, those ways of life did not leave us dehumanized when they disappeared. As Lasley observes, the days of the mom-and-pop grocery stores, the blacksmith shops, and cobbler benches are gone. Few of us felt moral obligations to keep those ways of life artificially in existence even though some lamented their passing.

Arguments based solely on appeal to emotion will not do. And it is not enough to establish a moral argument on the slim ground that family farms are a national heritage (Tweeten) or a unique way of life (Hodne). There must be more to the claim, and we must try to spell out exactly what the *more* is, if we are to construct a convincing ethical case.

Before moving on to the next argument, I wish to be very clear about one thing. I have called the present argument an argument "from emotion" because it appeals, at least on its surface, to *feelings* (of attachment, devotion, pride) and little else. I think it is not a good argument; but this is not because it is based on emotion, but rather because the particular emotions to which it appeals cannot support its general conclusion. I mention this to avoid misunderstanding; many ethicists used to claim that moral philosophy must be "rational" and "objective" and not subjective or emotional. In this old view, emotions were to be overcome in ethical inquiry, displaced by dispassionate thought. I disagree. Emotions are very important to moral judgment, and we should not think that ethics tries to get rid of them. Rather, we need to learn how to bring our feelings into moral deliberation *in the right way.*

This is a crucial matter. The academic separation of reason and emotion has often served to support sexist practices in the real world. Driving a wedge between the head and the heart, and then privileging the head, was an intellectual maneuver that went hand in hand with patriarchy—men being the head (intellect, reason) and women the heart (emotion, passion). When the realm of objectivity is defined in terms of reason and the realm of subjectivity is defined in terms of passion, the control of political and economic institutions will fall into the hands of men while women are granted control only over the home. According to the old scheme, objective ethical reflection—men's domain—was expected to discipline and govern subjective feelings and emotions—women's domain.

We are learning slowly. As we do, we realize that the problems attending this dichotomy extend far beyond philosophical dualisms of reason/emotion and male/female. They reach into agriculture. Notice how natural it is to refer to farmers in the masculine gender. Use of the locution "farmer's wife" to refer to married women farmers is not disappearing rapidly from our speech. The ease with which it falls from our tongues should alert us to the extent of the problem. What is equally harmful is that we usually construe our relationship to farmland the way we used to construe the relationship of men to women. As women are for men's use, so land is for "man's" domination.

The Christian stewardship ethic, which one might expect to help overcome this problem, may actually be part of the difficulty. It instructs Christians to give special care to the land, but not because the land has any intrinsic value of its own. They are to give it special care because they depend on it for their existence. That is, the land is cared for not for its own sake, but for our sake, for what it can do for us. This deep-seated theological view, we might note, underwrites our attempts to find technological answers to all agricultural problems. We do not worry about groundwater pollution now because it does not directly threaten our species. When it does, we will probably address the problem (as we have in past years) only insofar as it affects us. Chances are the frogs, fish, and worms of the future will have to fend for themselves. But this is not how we should proceed. When we separate head and heart, men and women, spirit and earth, we wind up privileging the interests of certain beings (whether men or humans) over those of others.

To fight what James Gustafson (1981) calls "anthropocentrism," contemporary theologians try to rehabilitate emotion, body, and land. They try to restore to consciousness the fact that we are material and maternal creatures, dust to dust, with passions and pasts. This is useful for agricultural policy insofar as we realize that farmland is a gift. It is valuable not because it can be used for our ends, but because it is God's, and God has given it to us as a trust. It is a gift, and we are rooted in it. We are not, strictly put, entitled to it; and yet we cannot live without it.

What does this have to do with the argument from emotion? It is significant that the argument in favor of family farms is heard forcefully in the voices of the writers in touch with emotions, people, land, and a sense of place and time. Hodne and O'Brien, Hendrickson and Berry talk about disappointment, love, depression, pride, joy, and other human emotions. They remind us that our moral reflections should be broad and lively, not thin and dead. If this particular part of their argument does not suffice to carry the day, the reason is not because emotion fails to be morally significant.

Gregg Easterbrook introduces the idea that family farms are important for emotional reasons, but then dismisses this as irrelevant. I think that Tim Carter is right to call him on this point. The fact that a farmer loves the land of his great-grandmother may very well be a sound ethical reason for society to help to preserve that family on that land. Wendell Berry emphasizes the point. Emotions are important in moral reflection. The problem is, *how* important? I suggest that our emotional investment in the family farm, strong as it may be, must not be expected to carry the whole burden of the argument. That weight is simply too great for pride or devotion or even love to bear.

THE ARGUMENT FROM EFFICIENCY. Some arguments try to defend the family farm by claiming that we must have it if we are to attain certain ends. Here the farm is defended for its economic benefits. The argument from efficiency is the clearest example and is used by Jim Hightower in this volume. It runs like this. The American economy prizes efficiency. The more cheaply you produce quality goods the more comparative advantage you have, and the more successful you will be. The family farm produces vast amounts of food at very cheap prices; it captures all or most of the economies of scale of agricultural producing units. Thus, it ought to be successful. If it is not, it is the fault of government policy, not farmers. Because it is failing today, the reason must be that politicians have tampered with the market system. But that means that family farmers *deserve* remedial political action.

This argument rests on the USDA study cited by Hightower, Harkin, and others, a study that concluded that the technology of tractors, pesticides, and hybrid seeds now available allows one person to plant and harvest enough grain and beef to support a family on a medium-sized plot. Because the family farm is the most efficient unit, we ought to see that it endures. Now, this argument is strong as long as its empirical claims remain valid. But there's the rub.

The USDA study has its critics; Boehlje, for one, says it is dated. Oster simply does not believe it and cites an Iowa State University study that claims a farmer and son who grow corn and beans now need two thousand acres to be fully employed. The fact that farming may outgrow the original USDA study points to the weakness of the argument from efficiency; it is contingent on variable conditions like the state of agricultural science. When technology advances — and perhaps it already has — the family farm may no longer make the best use of the relevant technologies. In that case, the argument withers. But suppose we agree that the study is still applicable; farms grossing between forty thousand dollars and two hundred thousand dollars a year continue to be the most efficient size. We still should not think of this as a conclusive moral reason to preserve those farms because it rests too heavily on contingent considerations. What we have here is an economic argument, not an ethical one.

It is worth noting that this sort of argument, to the extent that it is valid, is precisely the sort that would be convincing to utilitarians like Tweeten and Johnson. It defends family farms according to all of the values dear to them: competition, free enterprise, autonomy of producers and consumers, individual responsibility, and risk in the marketplace. It plays by the rules of the capitalist game, just as Novak and Oster would have it. Whatever is most efficient in the economic sphere wins in that sphere.

But not everyone is convinced that we should play by these rules, and the reasons are worth stating in some detail. American individualistic utilitarianism fails to distinguish between intrinsic and extrinsic values. It can only factor into its calculus things with measurable value, things that can serve as means to some end. Tweeten variously describes such ends as "well-being, satisfaction, quality of life, welfare, avoidance of pain, pursuit of happiness, or the greatest good for the greatest number." (See Chap. 21.) But not everything we value is valuable because it *leads to* happiness or well-being. We treasure some things not because they lead to satisfaction or the greatest good but because they are good, in and of themselves. Of what further use is my daughter Krista's smile? None. It is good in and of itself. What other end does ten more years of continued good health serve? None. It is valuable just for what it is. What beneficial consequences does a Mozart symphony or Van Gogh painting have? They may soothe us, or delight us, but that is not what makes them valuable; they are valuable in and of themselves.

Tweeten is right to say that utilitarianism is the "dominant moral philosophy in America." It is probably the ethical theory of choice among the silent majority of philosophers; it appears to be the only ethical theory available to agricultural economists. But even if most Americans agree, that still does not make it right to say that utilitarianism is the appropriate moral norm for public policy decisions. When we adopt a rigid utilitarian view, we blind ourselves to the intrinsic and inestimable value of things like smiles, health, life, music, art, land, religion, and family. For, try as we might, we cannot accurately translate these things into economic values.

Another reason to shy away from the utilitarian argument from efficiency is that efficiency is only one criterion among many that ought to come into play as we assess farm policy. There are also need, effort, fairness, and stewardship. Suppose we were to decide the question of who should own farmland by the single utilitarian concern of efficiency. What would we be overlooking in the process but the history of what has happened to disadvantaged groups? Consider American Hispanics, who live in disproportionate numbers under the poverty level. They may need farmland more than wealthy investors do. But the investors might be more efficient farmers and thus, by the narrow rules of utilitarian economics, be entitled to the land. Decisions based on efficiency alone blind us to the criterion of *need*. That seems wrong.

There is also the matter of initiative; some people may be willing to work harder at farming than even the most efficient of today's farmers. Should we not give them the opportunity to try their hand at it? Decisions based only on efficiency ignore the criterion of *effort*. That seems shortsighted. Then there is the matter of fairness; some groups may have the right to own land even if they are not as skilled at the trade. Loss of farmland ownership by blacks in the South has accelerated at an alarming rate. The Farm Security Administration of the New Deal period was designed to stop this trend and to help minority tenants become owners. But this program was scrapped over forty years ago. Is justice being done here? Decisions based only on efficiency do not allow us to consider questions of *fairness*. That seems unjust.

Finally, economic analyses judge farmers by the amount of their output. Wendell Berry suggests that we should judge them by their inputs; the Amish do not produce massive amounts when compared to intensive chemical farmers, but they do produce astonishing yields when compared to standard organic farmers. In the long run, they take better care of the land, paying it back the way Job did; their air and water, in the meantime, remain unpolluted. Should we not give these farmers special consideration? But since *stewardship* cannot be given an economic value, those of us with efficiency on the brain cannot factor it into our decision-making charts. That seems just plain dumb.

Utilitarianism focuses all attention on one value while ignoring others (Wunderlich 1984). For this reason, it is not the only appropriate moral norm for public policy, and we should not consent to play by its rules alone. We must be able to bring into our discussion questions such as these: Is a certain policy fair? Is it a prudent use of resources? Does it respond to the injustices of past history? What will be its long-term environmental consequences?

To see the shortcomings of decisions based only on the values factored into what economists call "Pareto-optimal" analyses, think about what things we could justify in the name of the greatest good for the greatest number. We could justify kidnapping a healthy child, killing her, and using her blood to save the life of another child. Her heart and kidneys could be used to save the lives of three others; and her corneas could go to restore the sight of two poor blind kids. We would dramatically improve six lives while only sacrificing one; the total balance of well-being or happiness in the world would increase sixfold. Yet, as Gilbert Harman (1977) points out, anyone who thought this proposal morally permissible would surely be a beast. I do not think that any utilitarian would find it anything but abominable. Yet that is precisely what utilitarianism, strictly taken, would allow. We could steal one wealthy farmer's property, parcel it out to three bankrupt farmers, and save their homesteads. We could give the farmer's machinery to two others, and sign the deed to his or her house over to a migrant California grape picker. This would give society six healthy farmers where once there was only one, maximizing well-being in the world and increasing the total balance of pleasure over pain six times. But this scheme too is deeply offensive morally.

Any theory that would justify either of the above cases must be deficient. Utilitarianism does justify these actions, and this is one more reason that we should not continue to base public policy solely on it. By the same token, losing the argument from efficiency may not be much of a loss. For even if the family farm was, is, and ever will be the most efficient farm, this way of defending it is not an ethical way. It is an economic way, plain and simple.

THE ARGUMENT FROM STEWARDSHIP. Like the last one, the argument from stewardship defends family farms because of the ends those farms serve. Unlike the argument from efficiency, however, the argument from stewardship claims that this sort of farm serves the best spiritual, rather than economic, ends. Clearly a religious line of attack, it rests on a view that may or may not be shared by most Americans. We are told that some 98 percent of Americans believe in God, but the specific creed of "American civil religion" (Bellah 1967) is notoriously vague. So it is not at all certain that the following argument, resting as it does on specific theological views, would be persuasive to a majority of those in this country. Nonetheless, it is often heard and deserves attention.

The actual claims are two. First, farmland has rights of its own. (See Austin 1977.) Second, animals have rights. Family farms respect these rights better than any other sort of farm. Hence, they ought to be preserved. The argument rests, as I have said, on religious warrants.

It is extremely difficult to generate philosophical arguments that would convince most persons that land and animals have rights. Some are beginning to argue that animals do have rights (Regan 1983), but I believe that the case is extremely difficult to make. Kant could make no sense of such a claim, and utilitarians like Mill and Bentham—while vastly more sympathetic to nonhumans than Kant—would also stop short of ascribing rights to them. There are even fewer philosophers who would ascribe rights to dirt. But there is a growing number who think that the natural environment does have intrinsic value of its own (Scherer and Attig 1983).

The view that cows and pigs and soil have value in and of themselves depends, I believe, on theological foundations. Once those are granted, the argument gets going. If we believe that there is a God, that God is good, created the world, and established a covenant with us, then it is easy to believe that God gave us the land we farm. The claim that we have an obligation to respect the land receives support from the assertion that land is God's gift. Whatever we may think, however, anthropologists tell us that gifts are not really free; they inevitably carry implicit expectations and obligations. God gives us the land not for us to use as we please, but for us to care for, preserve, and cherish. We are to pass it on to our children because the gift is given not just to our generation but to those who came before and those who will come after.

Suppose that the Catholic bishops and David Ostendorf are right about this. Land is a divine gift, has intrinsic value, and ought to be respected. How can we factor this religious claim into secular policy discussions? It is a difficult question, because as pluralists we do what we can to minimize our reliance on theology in public debate. Rationality, not revelation, must be the guide in a nation that separates church and state. I do not know how to solve this problem quickly. I believe as a Christian that it is a religious fact that farmland is a gift from God. Yet I believe as an American that we should make decisions based on claims that are reasonably, not divinely, disclosed. All I can suggest is that the Jewish and Christian traditions, which inform our national identity, may have a message to send to Washington; farmland is much more valuable than any economic calculus can make out.

The intrinsic value of animals, on the other hand, is another matter. While the vast majority of philosophers continue to think that the idea of animal rights is a futile one, Tom Regan—a well-respected philosopher—has made a strong case otherwise. I am not persuaded that he has shown that animals have rights, but I think he has given good philosophical reasons for believing something I believe on theological grounds; some animals are valuable in and of themselves. They may not have rights, but they do have intrinsic worth. We should relate to these beings in the way that God relates to them: God cares for them. As God's stewards, we should do what we can to prevent suffering and promote well-being in the animal kingdom.

What we find with regard to animals (but not land) is that we can get the theologian's conclusion without invoking religious arguments. If Regan is right, we can conclude on rational grounds that we have specific moral duties toward certain nonhumans. Some philosophers have given eloquent arguments maintaining that one of these obligations is not to kill animals for food (Regan 1975). If they are right, the future family farm will have to look very different indeed! But stopping short of this radical conclusion, suppose that we have no duty to preserve animal life at all costs. Suppose that there is nothing wrong with killing animals to feed ourselves so long as we do not cause them to suffer in the meantime. Could there still be something wrong with raising them in conditions that are inhumane?

On many contemporary farms, egg-laying chickens are held in such small cages that they turn to feather pecking and cannibalism. On modern dairy farms, calves are often separated from their mothers within twenty minutes of birth. Veal calves are kept in pens less than two feet wide, prevented from turning around and from chewing roughage. Without ascribing rights to these animals, we can say that they have intrinsic value and ought for that reason to be treated better. Let us not say that we have no right to eat them, to breed them for food, to confine them, herd them, milk them. Those are very unpopular—and highly contentious—conclusions.

Let us, however, consider their animal natures. Let us do say that we ought not to keep chickens in conditions that lead them to eat themselves or to deprive cows prematurely of their offspring. Let us raise animals in such a way that they appear to us happy and healthy. And healthiness is not just freedom from physical disease; we should also attend to their psychological well-being. This calls for careful study of animal behavior. We need to know about the habits and desires of the animals we raise so we can raise them in appropriate conditions.

We can take two things from the argument from stewardship. First, land may be a gift from a higher power, a gift that belongs not to us but to God. Therefore, secular policies should be developed that will encourage land uses that respect this gift and fulfill obligations to future recipients. Second, nonhumans high on the evolutionary scale may have intrinsic value. Consequently, policies should be developed that encourage farming methods that allow animals a maximum amount of movement, health care, family and communal life, and freedom from suffering. If family farms are the best arrangements for achieving these ends, then we surely ought to save them.

Is this a convincing argument? It is, as long as it is stated in the contingent form: *if* family farms are the best way to do thus and so, *then* they are worthy of preservation. Notice, this does not establish a moral obligation to save family farms. For even if one believes that land is a gift from a beneficent power and that animals have intrinsic value, it still does not follow that family farms must be saved. Even huge, corporate farms could be legislatively required to treat land and animals along the most rigorous lines of good stewardship.

As far as the family farm is concerned, nothing logically follows from the theological claim that God created the environment and gave us farmland. Even if we grant that all of God's creatures have immense worth, we still cannot get the conclusion that privately owned, family-worked farms are more sacred than corporate or state farms. For American society at the present time, the case can be made, I believe, that family farmers treat their animals best and are the best stewards of the land. But if other agricultural arrangements could be invented that would lead to even better stewardship practices, then those new forms would be preferable.

THE ARGUMENT FROM CULTURAL IDENTITY. Having examined two utilitarian arguments for saving the farm, we return to that which was left unpacked in the argument from emotion. In the first argument, we sensed more than met the eye. What was implicit there is as old as Thomas Jefferson. He claimed that democracy will not work unless America has a majority of its inhabitants living and working on small family farms. Those farms give us our national character and supply us with virtuous, honest, hardworking, patriotic citizens.

The strength of this argument, I believe, is not in its claim that farmers are morally superior to urbanites, but in its claim that widespread land resources are necessary if political and economic power is to be democratically distributed. Otherwise, fewer and fewer will become stronger and stronger, and the populace at large will be rendered voiceless. Kirkendall traces the history of Jeffersonian agrarianism and also raises the objection that comes readily to mind. If Jefferson is right, how is it that we still have a democracy when less than 2 percent of us live on family farms?

Kirkendall's skepticism is understandable. Hasn't history proven Jefferson wrong? There seems to be no causal relationship between the demise of family farms and the demise of democracy. Is democracy any weaker today when there are two million farms than it was in 1935 when there were six million? Either Jefferson is wrong—family farms are not necessary for democracy—or else we no longer have a democracy. Most of us would find the latter proposition ludicrous; if any country on earth is democratic, ours is. But before we allow Kirkendall's rhetorical question to bury the argument from cultural identity, it is worth asking whether we really do have a democracy.

There are those who think otherwise. Some of our most sensitive critics have alleged that contemporary America is ruled by a handful of powerful people. Writers such as Christopher Lasch, Alasdair MacIntyre, and Robert Bellah have argued that things are not, democratically speaking, in good working order. Political decisions are made by special interest groups and Washington-based power brokers. Narrowly focused lobbies determine policy direction, while political action committees elect candidates. If these claims are right, then American democracy may not be so healthy after all.

The difference between those who think democracy is eroding and those who think it is flourishing can be seen by comparing the views of our last two presidents. Jimmy Carter cautioned the nation to tighten its collective belt, moderate its expectations, and recognize its limits. In a speech of July, 1979, he claimed that the nation "had dangerously strayed from the ideal vision of the founding generation." If we are to survive, he preached, we need to "again come to represent a cohesive spiritual commonwealth rather than a fragmented society of individuals and groups selfishly pursuing their own narrow interests" (Shi 1984, 271).

If Carter chastised us in spiritual metaphors for being self-indulgent, his successor praised us in no less religious terms for being aggressive. Ronald Reagan did not shy away from using Carter's Christian rhetoric, but he took exactly the opposite view of America. Preaching economic abundance and "bigger is better," Reagan "bullishly reaffirmed the expansive ethic of liberal capitalism" (Shi 1984, 271). As David Shi puts it, while Reagan may have been "Jeffersonian in his sentimental political vision of a decentralized, self-governing republic, he was Hamiltonian in his refusal to emphasize any moral limits on the individual pursuit or consumption of wealth or on the size and influence of corporations" (p. 274).

There is no question about which mirror Americans prefer to gaze into. In landslide elections, Reagan defeated Carter in 1980 and Mondale in 1984. A majority of us prefer to think with Reagan that monopoly capitalism and unrestrained economic growth work hand in hand with democracy. But at a deeper level, many of us are uneasily aware that the two may not be compatible. More and more, we sense the middle class sliding into the lower class. More and more, we hear reports about millions of people feeling disenfranchised, powerless. Ironically, from the heart of Reagan's political support — right wing Christian fundamentalism — come sermons decrying the spiritual vacuity of America. As evidence, preachers cite the growing rates of alcohol and drug abuse, divorce, and suicide.

Worries about the nation's health did not evaporate with Jimmy Carter's quiet retreat to Georgia. The conflicted state of our national soul is difficult to interpret, but it does not seem outrageous to suggest the following conclusion. Adherents of the old Jeffersonian agrarianism may not be so far off when they claim that family farms are essential to our cultural identity. They are simply wrong to push the idea that peasants in the countryside are inherently more virtuous than suburbanites and city folk. And they should not suggest that doubling the number of family farms in America will magically cut in half the amount of drug abuse. But when they make the following point, they appear to be ex-

actly right; as family farms have disappeared, the control of the resources generated by agriculture has slowly slipped out of the grasp of
the majority of Americans. Earl Butz is right; land is power. As more
Americans have found themselves with less land, they have found themselves with less political might as well.

If critics are right, then our democracy is in ill health, and the
demise of family farming may be a contributing cause.

The Jeffersonian argument from equitable distribution of power is
the strongest argument we have encountered. But it is not universally
persuasive; one needs to be sympathetic with critics like Lasch, Bellah,
and Tony Smith (who see democratic capitalism failing rather than saving us) if one is to be convinced. Others, like Novak and Helms who see
America through Ronald Reagan's eyes, will not agree. The argument
depends heavily on a specific reading of recent American history.

If not all would agree with the agrarianizers' view of contemporary
America, we might, however, agree on their weaker claim: that the widespread distribution of farmland ownership is a good thing. This would
not prove that family farms ought to be saved, but that as many people
as possible ought to own farmland. Carried to its extreme, of course,
this might not necessarily support private ownership of farms. It might
argue for state ownership of farmland; here, everyone would own a
piece.

While I do not believe that the argument from cultural identity is
unassailable, I do believe it holds two important lessons. (1) In order to
chart a course for the future, we must not only know where we are, but
we must also remember where we have been. Hodne, Ostendorf, and
Berry remind us of the importance of the past and urge us to preserve
prior generations in our memory, rooting ourselves in the lives and
places of our ancestors. That seems like good advice. (2) We must be
diligent in attempting to distribute land and power fairly. We ought not
to be satisfied with the fact that since 1940, most funds supplied by
Washington have benefited large farmers (Chap. 9). And just because
William Jennings Bryan made the same criticism during the Populist
Revolt some ninety years ago is no reason for us to stop complaining
about the increasingly undemocratic concentration of economic and political power in this country (Chaps. 14 and 19). We ought to guard
against the formation of the monopolies and oligopolies that gnaw at
democracy. If, as my grandpa used to say, "the middle man gets the
dollar," why should that be so? Why not restore the money that was
taken from consumers by oligopolistic food manufacturing industries?
Why not pass some of it along to young black farmers in Mississippi

who wish to raise vegetables and market them locally? If antitrust laws are shelved, and union-busting schemes encouraged, then who can fault us when, without being able to give academically sound arguments explaining our emotions, we nonetheless feel we have been deprived of what is ours? It is the task of a democratic nation to fight resentment of this sort. One way is to make policies that help oppressed people.

If the family farm is a structure that keeps us in touch with our emotions and our past, and thereby with ourselves, it is well worth saving. If it is a symbol for a vast network of small rural businesses in which economic power is widely distributed, it is well worth saving. If, on the other hand, the modern family farm is owned by members of an increasingly restricted class in which ownership depends primarily on the accident of birth, then we ought to change this. If there are other, better, ways of benefiting *all* members of society, we should find them.

THE ARGUMENT FROM RESPONSIBILITY. I have reviewed four arguments claiming to show we have a moral obligation to save the family farm. None of them firmly establishes the case. Each one either rests on questionable premises or leaps to conclusions that do not follow. Is there a strong argument, based on religious or ethical considerations, for preserving farms loved, worked, and owned by families? I think that the best one can do is to make the case a contingent one; *if* family farms help us to meet certain obligations, then we have good reasons to save them. This final argument "from responsibility" draws upon elements of the arguments from emotion, stewardship, and cultural identity.

As human beings who live for only a short time on this good earth, we have duties not only to take care of one another, but also to remember our ancestors and provide for our offspring. Edmund Burke wrote that a nation is a partnership of the living, the dead, and the unborn (Shakeshaft 1986). Our obligations include a pact with the dead—the covenant to remember their achievements and misdoings. Our obligations also include duties to unconceived generations; our responsibility to pass this world on to them in at least as good a shape as we received it.

We have obligations to each other. One of these is to treat each other equally, meeting one another's needs, being fair in the distribution of the earth's resources.

And we have obligations to ourselves. We are responsible for figuring out who we are, where we are going, and how our paths will affect others. This is not something we can do entirely on our own. If we do not receive immense help from our parents and society when we are young, it seems that no amount of education will teach us the moral virtues of love and justice. Thus, it is our duty to discern the lessons of the past, to teach them to our children, and to form a community that can accept the American story as its own.

Being a family farmer means caring for one's land. Such love cannot be taught in agricultural colleges; it is a practice that one learns at the feet of a master. It is knowledge of the heart, not the head, and it is best passed from generation to generation, not from agribusiness expert to agricultural student. This does not mean that newcomers cannot love the land; only that their doing so requires that they learn right emotions and intentions, not just right equations and ratios. This sort of care comes from lived experience and tradition—from memories, from the past. This provides a clear moral justification for giving preferential treatment to those farms that have long histories of having been family undertakings.

Because of our duties to future generations, we have a moral obligation to preserve the fertility of soil and the purity of air and water. This requires anticipation, imagination, and vision; it requires looking to the future. But there are two possible objections to this claim. First, we do not know how to represent the interests of future generations. This philosophical objection should not detain us; just because we do not now know how to represent the interests of unborn people does not mean that we should not try. Second, future generations may develop hydroponic—soilless—techniques of raising food. This objection rests on a technological possibility that is remote. Even if university scientists were to develop the relevant technologies, the financial burden of constructing that sort of agriculture to feed the world's billions would be prohibitive. For reasons of this sort, I think it safe to say that we have the obligation to look ahead and to do everything we can to preserve clean air, water, and soil.

When it comes to agriculture, the family farm seems to be a reasonable arrangement for meeting these obligations. It does not seem improbable that a grandparent envisioning great-grandsons and daughters will have a stronger motive in adopting conservation tillage practices than a corporate manager whose only concern is to turn a profit this year. To the extent that family farms are trying to return to certain organic farming methods, we have good reasons for keeping them in existence. They represent a forward-looking, sustainable agriculture.

Because of our duties to each other, we should try to adopt procedures by which power and land are distributed democratically. No one should hold proportionately too much land, no matter how efficient at using it one might be. Efficiency is not our only value.

I believe that the owner of a farm should also be the farmer of that land. This does not follow strictly from the principle of fair distribution, but it is consistent with ideas of responsibility. The right to administer and work land should preferably be given to those who are closest to it. They are the ones most dependent on it, and they are the ones most sensitive to the limits and potentials of the new corner section, the old river bottom.

Finally, because of our Socratic duty to know ourselves, we must not forget our past nor restrict our vision of the future. We need to learn from our ancestors the practices that embody moral wisdom, and we need to learn from our children the unlimited horizons of what might be. To the extent that family farms are the places where the stories of our ancestors lie, it is in our own interests not to be too hasty in tearing down old homesites. To the extent that new agricultural arrangements might help more families own and care for farmland, we ought not to be too hesitant to adopt new farming methods and rural structures.

I do not believe there is a moral obligation to save the family farm. Not everything about it is worth saving. It is not worth saving a patriarchal institution that exploits the labor of women and children and men, nor is it worthwhile saving an institution whose ownership ranks are almost exclusively white males. It is not worth saving a way of farming increasingly dependent on petroleum-based chemicals. It is not worth saving "the family farm" when that phrase represents nothing more than political rhetoric used to advance the self-interests of large commodity organizations.

Yet the alternative seems clear—fifty thousand superfarms, producing 75 percent of agricultural output. That is just more of the same: oligopolistic capitalism, vertically integrated corporations, multinational food conglomerates. That is no way to plan a democratic future.

I believe that we have moral obligations to each other, to our children, and to the natural environment. In order to live justly and fairly, to respect the material world that sustains us, and to preserve its resources for those who come after, we must discover economic structures and communal arrangements that allow us to meet these obligations. If the alternatives are fifty thousand superfarms or half a million family farms, our responsibility seems clear. To the extent that family farms help us fulfill our duties to one another, to unborn generations, and to God's created world, it is our duty to help them survive.

References

Austin, Rich L. "Three Axioms for Land Use." *Christian Century,* 12 October 1977, 910–15.

Bellah, Robert. "Civil Religion in America." *Daedalus* 96 (Winter 1967): 1–21.

Gustafson, James M. *Ethics from a Theocentric Perspective,* vol. 1. Chicago: University of Chicago Press, 1981.

Harman, Gilbert. *The Nature of Morality.* New York: Oxford University Press, 1977.

Regan, Tom. *The Case for Animal Rights.* Berkeley: University of California Press, 1983.

_____. "The Moral Basis of Vegetarianism." *Canadian Journal of Philosophy* 5 (October 1975): 181–214.

Scherer, D., and T. Attig, eds. *Ethics and the Environment.* Engelwood Cliffs, N.J.: Prentice-Hall, 1983.

Shakeshaft, Jerry. "The Politics and Ethics of Power." Lecture, 29 May 1986, Iowa State University, Ames.

Shi, David. *The Simple Life: Plain Living and High Thinking in American Culture.* New York: Oxford University Press, 1984.

Singer, Peter. *Animal Liberation: A New Ethics for Our Treatment of Animals.* New York: New York Review/Random House, 1975.

Wunderlich, Gene. "Fairness in Landownership." *American Journal of Agricultural Economics* 66 (December 1984): 802–6.

SUGGESTED READINGS

Bellah, Robert. *Habits of the Heart: Individualism and Commitment in American Life.* Berkeley: University of California Press, 1985.

Lasch, Christopher. *The Culture of Narcissism: American Life in an Age of Diminishing Expectations.* New York: Norton, 1979.

MacIntyre, Alasdair. *After Virtue: A Study in Moral Theory.* Notre Dame, Ind.: University of Notre Dame Press, 1981.

INDEX